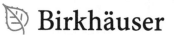

Monographs in Mathematics

Volume 109

The foundations of this outstanding book series were laid in 1944. Until the end of the 1970s, a total of 77 volumes appeared, including works of such distinguished mathematicians as Carathéodory, Nevanlinna and Shafarevich, to name a few. The series came to its name and present appearance in the 1980s. In keeping its well-established tradition, only monographs of excellent quality are published in this collection. Comprehensive, in-depth treatments of areas of current interest are presented to a readership ranging from graduate students to professional mathematicians. Concrete examples and applications both within and beyond the immediate domain of mathematics illustrate the import and consequences of the theory under discussion.

More information about this series at http://www.springer.com/series/4843

Masao Nagasawa

Markov Processes
and Quantum Theory

 Birkhäuser

Masao Nagasawa
Kawasaki, Japan

ISSN 1017-0480   ISSN 2296-4886   (electronic)
Monographs in Mathematics
ISBN 978-3-030-62690-7   ISBN 978-3-030-62688-4   (eBook)
https://doi.org/10.1007/978-3-030-62688-4

Mathematics Subject Classification: 46–02, 46E35, 42C40, 42B35

This book is published under the imprint Birkhäuser, www.birkhauser-science.com by the registered
company Springer Nature Switzerland AG
The registered company address is: Gewerbestrasse 11, 6330 Cham, Switzerland

Preface

Quantum Theory is more than one hundred years old, but its mathematical foundations have been gradually clarified only in the last half-century. We will tell the story in this book. The whole contents of quantum theory will be unified under the name of *theory of stochastic processes*. As a matter of fact, in our view, Quantum Theory is an application of the theory of Markov (stochastic) processes to the analysis of the motion of small particles in physics (such as electrons and photons).

It is correctly considered that quantum theory is radically different from the classical theory (i.e., Newton's mechanics). In fact, quantum theory is obtained by means of the quantization of the classical theory. Quantization means that we replace classical physical quantities with operators and subject them to the so-called commutation relations. This procedure itself is the core of the Born-Heisenberg quantum mechanics and is responsible for the successes of quantum theory. But why we must proceed in this way was never clearly explained. It has been accepted that success itself justifies the method. However, this incomprehensible so-called quantization is a technical matter, and quantum theory should be understood in the context of the theory of stochastic processes, as will argue in the sequel.

In this book we adopt the theory of Markov processes as a mathematical foundation of quantum theory instead of quantization.

In the conventional quantum theory, it has been *implicitly* assumed that the paths of the motion of a particle are smooth, like in the classical Newtonian mechanics. In this book we assume that, on the contrary, the paths of the motion of a small particle are not smooth; rather, the particle performs a random zigzag motion. We will explain and prove that *the theory of random motion of small particles covers all of what has been called quantum theory*. In the conventional quantum theory the notion of random motion of small particles is somehow neglected and has never been considered seriously. We will attempt to make as clear as possible the fact that quantum theory is precisely the theory of random motion. To develop the new quantum theory, we will extend slightly the conventional theory of Markov processes, and we will do it step by step in each chapter, as required.

To make things clearer, let us quickly look back at the historical development of *quantum theory* . Here we call the whole discussion on the motion of electrons and photons (or other small particles) *quantum theory*. The origin of this theory is Planck's *quantum hypothesis*, formulated in 1900. In deriving his heat radiation formula, Planck postulated that light energy can only be emitted and absorbed in discrete bundles, called quanta. However, what this actually means was not clear until Einstein's particle theory of light appeared (Einstein (1905, a)). Particles of light are then called photons. The physical objects that quantum theory treats are small particles such as electrons and photons.

If so, we must then determine how such small particles move under the action of external forces. In the conventional quantum theory, discussions on the trajectories of small (quantum) particles have been neglected. In fact, if you look for discussions on the trajectories of small (quantum) particles in the available quantum theory books, you will certainly be disappointed, because you will find nothing. You will find there mainly methods for computing energy values. In fact, the existing so-called theory of quantum mechanics offers no possibility of computing trajectories of the motion of electrons (small particles). One might argue that the Schrödinger equation tells us about the motion of electrons under external forces (potentials). But how? As a matter of fact, the Schrödinger equation is an equation for an "amplitude of probability" and does not give information on trajectories of electrons. For this we must resort to our imagination. This is the only way one proceeds to understand the so-called quantum mechanics.

In this book, we will develop **a theory that enables us to compute and describe the motion, namely, trajectories of electrons (of small particles), and not only to compute energy.** In fact, small particles such as electrons perform a Brownian motion, which is an unavoidable, intrinsic motion of small particles, because it is caused by vacuum noise. Hence, the paths of small particles are not differentiable everywhere. We will accept this fact and call it **random hypothesis**.

As a consequence, Newton's classical equation of motion is not applicable to the random motion, and we need a **new equation of motion** which can be applied to particles performing random motions.

We call the theory of random motion of particles that is based on these new equations of motion **mechanics of random motion**, i.e., a **new quantum theory**. It is an elaboration of quantum theory in terms of the theory of Markov processes. This is the main theme of the present book. However, to formulate the new equations of motion we must extend the conventional theory of Markov processes.

If one applies the *mechanics of random motion* to the physical phenomena that have been called *mysteries of quantum mechanics*, the mystery is removed. This kind of quantum mechanical phenomena will be explained clearly in the theory of random motion. Moreover, problems that could not be handled in quantum mechanics will be brought under the clear and bright light of the theory of random motion and solved unequivocally, because we can now compute accurate trajectories (paths) of the motion of electrons and see how electrons move under the influence of external forces.

Now, if one looks at the history of quantum theory and the history of the theory of stochastic processes, one finds that they both started at the beginning of the twentieth century, and developed in parallel independently, except for Schrödinger's analysis in 1931. Nevertheless, the two theories were intimately related, although most of people did not recognize this clearly.

In quantum theory, the Schrödinger equation was discovered in 1926. Schrödinger investigated the Brownian motion which is symmetric with respect to time reversal (Schrödinger (1931)) in order to deepen his understanding of the Schrödinger equation. On the other hand, in the theory of stochastic processes, Kolmogorov laid the foundations of probability theory and discovered the equation of diffusion processes that today bears his name (Kolmogoroff (1931, 1933)). The two equations are remarkably similarity.

In fact, although this is not well known, Kolmogorov's elaboration of probability theory and Kolmogorov's equation of diffusion processes were strongly influenced by Schrödinger's investigation of the Brownian motion and the Schrödinger equation. Moreover, motivated by Schrödinger's investigation of time reversal of Brownian motion (Schrödinger (1931, 1932)), Kolmogorov discussed the time reversal and the duality of the diffusion processes with respect to the invariant measure (Kolmogoroff (1936, 1937)).

There is another remarkable similarity. The Schrödinger equation contains the first-order time derivative multiplied by the imaginary unit i. On the other hand, Kolmogorov's equation also contains the first-order time derivative, but without the imaginary unit i. Hence, the Schrödinger equation is an equation for complex-valued functions, whereas the Kolmogorov equation is one for real-valued functions. Nevertheless, the two equations describe the same physical phenomena, namely, the random motion of small particles under external forces, although this is not self-evident. In fact, Schrödinger recognized this fact and tried to clarify it in his work (Schrödinger 1931), but that attempt was not so successful. This is a difficult problem. We will analyze Schrödinger's problem and solve it in this book.

In short, we will clarify the intimate relationship between Schrödinger's work and that of Kolmogorov, and unify the two theories into a single theory. This unified theory is

the new quantum theory i.e., **mechanics of random motion.**

In other words, we will argue that **quantum theory is nothing but mechanics of random motion under the influence of external forces.**

Now, as we have explained, in the theory of random motion the **random hypothesis** is an indispensable basic notion. In the quantum mechanics of Born and Heisenberg, however, this recognition of the importance of the random hypothesis is missing. This will be explain further in Chapter 3. (Cf. also Luis de la Peña et al. (2015).)

We will first present in Chapter 1 some aspects of the theory of stochastic processes, in particular, Kolmogorov's equation, and Itô's stochastic differential equations and Itô's formula. This part is, in a sense, a fast introduction to the theory of Markov processes. Then, we will introduce the **equation of motion for stochastic processes**, which is a new notion, and provide mathematical tools to handle it.

We will then discuss the superposition of complex evolution functions of stochastic processes. Superposition induces the so-called entangled random motion. As an application of superposition (the entangled motion), we solve the problem of the double slit experiment, which was one of unsettled problems of quantum mechanics. In quantum mechanics people speak of *interference* to explain the double-slits problem (cf., e.g., Merli P. G., Missiroli G. F., and Pozzi G. (1976), Rosa (2012)). However, it is clear that interference is not possible if we consider an experiment in which we send electrons one by one, with a sufficiently long time separation.

There is one more point deserving attention. As is well known, in the conventional theory of diffusion processes one specifies an initial value for an evolution equation. In our theory of mechanics of random motion, instead, we specify both an initial value *and* a terminal value for an evolution function. We will see that this is one of the most important facts and advantages of our theory.

Chapter 2 is devoted to applications of the theory. We will show that we can follow exact trajectories of electrons moving even in inside atoms. In particular, we will see that Bohr's transition between energy levels is not a jump, but a continuous change of motion described by the Schrödinger equation.

In Chapter 3, we argue that Heisenberg's uncertainty principle is erroneous, and that Einstein's locality holds, contrary to Bell's claim. An example satisfying the locality property is given.

In Chapter 4 we provide the Feynman-Kac and Maruyama-Girsanov formulas, and also explain the time reversal in stochastic processes and its applications.

Chapter 5 introduces the concept of the relative entropy of stochastic processes, and discusses the so-called "propagation of chaos" of Kac as an application in this context.

The creation and annihilation of random particles will be discussed in the framework of the theory of Markov processes in Chapter 6.

The contents of this volume overlaps to some extent with the contents of two other monographs Nagasawa (1993 and 2000). In Nagasawa (1993) emphasis is placed on clarifying the relationships between the Schrödinger equation and diffusion processes. Nagasawa (2000), on the other hand, provides an exposition of the theory of Markov processes and also of time reversal in Markov processes, which plays an essential role in quantum theory of random motion. In fact, the system of equations of motion of quantum particles consists of a pair of equations, one in which time runs forward, and a second one in which time runs backward, so time reversal plays a key role. In the present volume the entire material is systematically presented as a theory of Markov processes in quantum theory.

Contents

Chapter 1

Mechanics of Random Motion

Small particles, such as electrons, perform a random zigzag motion (i.e., Brownian motion), which is unavoidable and intrinsic for small particles. We introduce this as a hypothesis (**random hypothesis**) and adopt it in this book. Then, classical mechanics no longer describes the motion of such particles, because their paths are not smooth, i.e., do not admit a tangent. Therefore, a new equation of motion is needed for such small particles. We will develop here a **new mechanics of random motion**, based on the theory of Markov processes.

1.1 Smooth Motion and Random Motion

The notion of a *quantum*, introduced by Max Planck at the beginning of twentieth century, was the first step on the way to leaving the setting of classical physics, and subsequently physical phenomena were classified into two categories:

First category: classical.

Second category: quantum.

However, since the *quantum* category is not well defined, the principles guiding the classification of phenomena into these two categories remain ambiguous. Because of this we will, instead, classify physical phenomena as:

First category: deterministic,
in other words, the paths $X(t, x)$ of the motion are smooth (i.e., at least twice differentiable everywhere);

Second category: random,
in other words, the paths $X(t, x)$ of the motion are nowhere differentiable.

There is no ambiguity in this last classification. We now give the definition of the quantum setting considered in this book:

© The Author(s), under exclusive license to Springer Nature Switzerland AG 2021
M. Nagasawa, *Markov Processes and Quantum Theory*,
Monographs in Mathematics 109, https://doi.org/10.1007/978-3-030-62688-4_1

The New Quantum Theory is a theory of motion of small particles under random noise, in other words, **a theory of random motion with paths** $X(t, x)$ **that are nowhere differentiable.**

Then we will see that:

"Newtonian mechanics" is applicable to the analysis of smooth paths of the first category; this is the well-know, classical theory.

Correspondingly,

"the mechanics of random motion" is applicable to non-smooth paths of the second category.

We are interested in the second category, *motions that are nowhere differentiable*, that is, *random motions*. To treat them, we resort to the theory of stochastic processes, in particular, the theory of Markov processes. The idea of using Markov processes is relatively new; as a matter of fact, reaching it took a long time. Although this is a long and complicated story, we will nevertheless tell it, because it will help considerably the understanding of the importance of the concept of *random motion* in quantum theory.

Thus, let us look back at the history of quantum theory, and describe its development and meaning. We begin with the wave mechanics of Schrödinger, and then continue with the quantum mechanics of Born and Heisenberg. Both theories are well known.

On Schrödinger's wave mechanics:

Following the idea of de Broglie that electrons can be regarded as waves, in 1926 Schrödinger proposed to describes electrons by means of the equation

$$i\hbar\frac{\partial\psi(t, x)}{\partial t} + \frac{\hbar^2}{2m}\triangle\psi(t, x) - V(x)\psi(t, x) = 0, \tag{1.1.1}$$

where m is the mass of the electron, $\hbar = h/2\pi$, and h is the Planck constant. Equation (1.1.1) is called the **Schrödinger equation**. In the equation, x stands for the three coordinates (x, y, z), and $\triangle$ denotes the Laplacian

$$\triangle\psi = \frac{\partial^2\psi}{\partial x^2} + \frac{\partial^2\psi}{\partial y^2} + \frac{\partial^2\psi}{\partial z^2}.$$

Equation (1.1.1) was the starting point of a revolution in physics.

Schrödinger regarded the solution $\psi(t, x)$ of the equation (1.1.1) as a complex-valued wave and identified it with an electron, with the intensity given by $|\psi(t, x)|^2$. Importantly, the intensity can be observed, since it is a real number. This is the so-called Schrödinger's **wave theory of electrons** (Schrödinger (1926)).

There are some ambiguities concerning the Schrödinger equation (1.1.1). We mention here some of them.

(a) First of all, the wave equation (1.1.1) contains the imaginary number i. Can a physical law contain the imaginary number i? This is a serious issue, since before Schrödinger there was no example of physical lows that involved i.

(b) The solution $\psi(t, x)$ of the Schrödinger equation (1.1.1), i.e., the *wave*, is a complex-valued function. Before Schrödinger's work there were not many examples of physically relevant quantities that are complex-valued functions.

(c) Moreover, equation (1.1.1) is not a conventional classical wave equation. The classical wave equation is of course a real-valued equation and a second-order partial differential equation with respect to the time variable t:

$$\frac{\partial^2 u}{\partial t^2} - c^2 \left(\frac{\partial^2 u}{\partial x^2} + \frac{\partial^2 u}{\partial y^2} + \frac{\partial^2 u}{\partial z^2} \right) = 0.$$

By contrast, the time derivative in equation (1.1.1) is of the first order. Therefore, there remains the serious question of whether equation (1.1.1) is really a wave equation.

Schrödinger's wave theory of electrons was highly successful in computing the light spectrum of hydrogen atoms and in other. problems. Nevertheless, it run into a serious contradiction. An electron has an electric charge e and is always observed at a point, never spreading out in space. This is not compatible with the idea that the electron is spreading in space as a wave.

If so, the claim that **the Schrödinger equation is a wave equation and an electron is a wave** was wrong.

After some efforts to ssettle this contradiction, Schrödinger abandoned the wave theory and turned his attention to the theory of Brownian motion (a particle-type theory).

He found motivation in an original idea of the famous astronomer Arthur Eddington:

"The complex conjugate $\overline{\psi}(t, x)$ of the wave function satisfies the equation

$$-i\hbar \frac{\partial \overline{\psi}(t, x)}{\partial t} + \frac{\hbar^2}{2m} \triangle \overline{\psi}(t, x) - V(x)\overline{\psi}(t, x) = 0. \tag{1.1.2}$$

The wave equation (1.1.1) and its complex conjugate (1.1.2) are time-reversed forms of one another. Therefore, $|\psi(t, x)|^2 = \psi(t, x)\overline{\psi}(t, x)$ predicts the physical state at the present time t as a product of the information from the past $\psi(t, x)$ and the information from the future $\overline{\psi}(t, x)$."

Inspired by Eddington's idea, Schrödinger considered the Brownian motion and introduced the twin (paired) equations

$$\begin{cases} \dfrac{\partial \phi}{\partial t} + \dfrac{1}{2}\sigma^2 \triangle \phi = 0, \\[2mm] -\dfrac{\partial \widehat{\phi}}{\partial t} + \dfrac{1}{2}\sigma^2 \triangle \widehat{\phi} = 0, \end{cases} \tag{1.1.3}$$

where in the first (resp., second) equation time runs from the past to the future (resp., from the future to the past). In fact, if in the first equation we replace the time variable t by $-t$, we get the second equation. In this setting, the distribution density in a time interval $a \leq t \leq b$ is given by the product $\phi(t,x)\widehat{\phi}(t,x)$ (Schrödinger (1931)).

Schrödinger's attempt can be described as follows: The wave theory is defective in some sense, but there is no doubt that the Schrödinger equation (1.1.1) is effective in computing energy values in quantum theory. Therefore, if we can show that the twin equations (1.1.3) are effective in computing energy values, or that the Schrödinger equation can be derived from the twin equations (1.1.3), then we obtain a quantum theory based on Brownian motion (Schrödinger (1931)). However, he could not establish this in a satisfactory manner.

In fact, Schrödinger was aware of the similarity existing between the solution pair $\{\phi(t,x),\widehat{\phi}(t,x)\}$ of equations (1.1.3) and the pair $\{\psi(t,x),\overline{\psi}(t,x)\}$ formed by the wave function and its complex conjugate, and he tried to find the relation between equations (1.1.3) and the wave equation (1.1.1), but was not successful. We will solve this problem in Section 1.9.

In conclusion:

Schrödinger introduced complex-valued physical quantities in physics via the Schrödinger equation and the complex-valued wave functions. This a major contribution to both physics and the theory of stochastic processes.

Let us reiterate that the Schrödinger equation is actually not a wave equation, but a complex-valued evolution equation. This point will be discussed in the sequel.

On the quantum mechanics of Born and Heisenberg:

Based on the original idea of Heisenberg (Heisenberg (1925)), Born developed the matrix mechanics (Born and Jordan (1925), Born, Heisenberg and Jordan (1926)). In matrix mechanics all physical quantities were represented by matrices. This was called *quantization*, but its physical meaning was not clear at all. In this theory the energy of electrons was computed by means of an eigenvalue problem. Since the theory of matrices was not well known at that time, Born, Heisenberg and their collaborators faced difficulties.

Soon after Born and Heisenberg's works, Schrödinger published his papers on wave mechanics. Born immediately recognized the importance of the wave equation introduced by Schrödinger, because computations based on differential equations were much more familiar to him and easier to handle compared to computations with matrices.

Born then applied Schrödinger's wave mechanics to the dispersion problem and claimed that $|\psi(t,x)|^2$ should be interpreted not as the intensity of a wave, but as the distribution density of electrons (small particles). This was called "Born's statistical interpretation." In this way Born brought the wave equation and the wave function $\psi(t, x)$ into the matrix mechanics (Born (1926)). The amalgamation of matrix mechanics and the Schrödinger wave equation with its wave function $\psi(t, x)$ and the associated statistical interpretation of Born is called the Born-Heisenberg "quantum mechanics".

While Born introduced the Schrödinger function $\psi(t, x)$ in quantum mechanics and claimed that $|\psi(t, x)|^2$ is the distribution density of small particles, no claim was made concerning the function $\psi(t, x)$ itself. It seems that Born did not clearly understand what it was and what it meant in the Born-Heisenberg theory.

Born asserted that an electron is a particle and not a wave, and denied Schrödinger's wave theory. However, if an electron is not a wave, what does Schrödinger's equation describe? Without a clear answer to this question, the notion of "wave" nevertheless made its way in quantum mechanics implicitly, via the Schrödinger equation. As a result, it became unclear whether in quantum mechanics an electron is a particle or a wave.

Bohr claimed that an electron is both a particle and a wave, and called this "complementarity". However, how to interpret this is unclear, because the notions of "particle" and "wave" belong to different categories and cannot be identified.

If, according to Born, $|\psi(t, x)|^2$ is to be regarded not as the intensity of a wave, but as the distribution density of electrons (small particles), then is there a way to see the paths (trajectories) of electrons which induce such a distribution?

Quantum mechanics cannot provide an answer to this question. Its setting does not allow use to discuss and analyze the paths of electrons, because it provides no suitable mathematical tools.

Heisenberg claimed that the location and the momentum of an electron are always blurred and uncertain, and hence there is no path of motion. (In Section 3.6 we will prove that his claim is erroneous.)

There still remains the question: Is there no path of the motion for an electron, as Heisenberg claimed? In fact, we cannot discuss or compute paths of the motion of an electron in quantum mechanics: a **new theory** is needed.

The new theory must allow computations as in quantum mechanics, and also endow the Schrödinger equation (1.1.1) and Schrödinger function $\psi(t, x)$ with a clear physical meaning. The theory that can achieve these objectives is the **"mechanics of random motion"**.

Furthermore, "mechanics of random motion" will provide a clear answer, in the framework of theory of stochastic processes, to an unanswered question in quantum mechanics: Is moving matter a particle or a wave?

We now close our introductory considerations and return to the main exposition.

(i) The notion of the smooth motion in the first category requires no further explanations.

(ii) Concerning the non-smooth zigzag random motions, i.e., the theory of stochastic processes, explanations are needed. A typical example of stochastic
processes is Brownian motion.

Brownian motion was observed as paths of small dust particles in the air
under bright light. Let us denote by B_t the position of a small dust particle. Consider two moments of time $s < t$. When the particle occupies a point x at time s,
we cannot predict its exact position at time t, but we can compute the probability that the particle lies in a region Γ, which is called the transition probability
$\mathbf{P}\left[B_s = x, \ B_t \in \Gamma\right]$ of the Brownian motion.

Figure 1.1 depicts a path of two-dimensional Brownian motion.

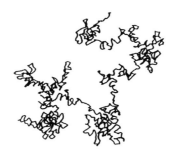

Figure 1.1

For simplicity, we consider first the Brownian motion B_t in dimension one.
If the transition probability density $q(s, x, t, y)$ is known, namely,

$$q(s, x, t, y) = \frac{1}{\sqrt{2\pi(t - s)}} \exp\left(-\frac{(y - x)^2}{2(t - s)}\right), \qquad (1.1.4)$$

then the transition probability is given by

$$\mathbf{P}\left[B_s = x, \ B_t \in \Gamma\right] = \int_\Gamma q\left(s, x, t, y\right) dy. \qquad (1.1.5)$$

In general, we set

$$u(t, x) = \int_{\mathbf{R}} q\left(t, x, b, y\right) f\left(y\right) dy. \qquad (1.1.6)$$

If we take the function $f(x) = \mathbf{1}_\Gamma(x)$ (the indicator function of the set Γ), equal
to 1 on Γ and 0 otherwise, the right-hand side of formula (1.1.6) coincides with
the right-hand side of formula (1.1.5).

Then the function $u(t, x)$ of the variables t, x satisfies the parabolic partial differential equation

$$\frac{\partial u}{\partial t} + \frac{1}{2}\frac{\partial^2 u}{\partial x^2} = 0. \tag{1.1.7}$$

We will prove this in a general form as Theorem 1.2.2 of Kolmogorov.

Einstein (1905, b) considered the transition probability density in (1.1.4) and showed that it satisfies the partial differential equation (1.1.7). However, he did not show that a path B_t is a random zigzag motion — the mathematics known to physicists in 1905 did not allow it, although Lebesgue's measure theory emerged already in 1900.

On the other hand, zigzag functions as arise in the description of the paths of Brownian motion were investigated as nowhere differentiable functions from the end of nineteenth century on. However, it was not known that Lebesgue's measure theory and nowhere differentiable functions relate to random motion. This was clarified by N. Wiener and P. Lévy during the years 1920–1930, but their work was not well known among physicists at that time.

What we have described to this point is the "kinematics" of Brownian motion. The "mechanics" of the Brownian motion was first considered by Schrödinger in 1931.

In the following two sections we will treat random motions (stochastic processes) and their path analysis. The equations governing random motions will be discussed in Section 1.4.

1.2 On Stochastic Processes

Based on the mathematical research by N. Wiener and P. Lévy on Brownian motion, the theory of stochastic processes was established as rigorous mathematical theory by Kolmogorov and K. Itô during the period 1930–1950. This provides the mathematical apparatus that lies at the foundations of the mechanics of random motion.

Let us examine this theory. We denote by $\omega(t)$ or ω a continuous, but nowhere differentiable zigzag function. It will represent a sample path of random motion. To discuss random motion, we must consider all such sample paths. We denote by Ω the set of all such samples, and call it *sample space*.

We denote by $X_t(\omega)$ the motion of an electron following a sample path $\omega(t)$, which is a function of time t: $X_t(\omega) = \omega(t)$. If we allow ω to change, then $X_t(\omega)$ becomes a function of ω which is defined on the sample space Ω and depends on the parameter t. A mathematical object that describes the random motion of an electron is thus defined.

If we collect all sample paths of an electron that is located at the point x at time s and at a later time t is found in a region Γ, we get a subset

$$\{\omega : X_s(\omega) = x,\ X_t(\omega) \in \Gamma\}$$

of the sample space Ω. Such a subset is called an "event" in probability theory. The the probability that this event occurs is then

$$\mathbf{P}[\{\omega : X_s(\omega) = x,\ X_t(\omega) \in \Gamma\}].$$

Thus, to discuss the random motion X_t we need the following mathematical tools:

(i) **a sample space** Ω,

(ii) **a family** $\mathcal{F}$ **of subsets (events) of the sample space** Ω,

(iii) **a probability measure P, which measures the probability of events**.

These three objects are standard mathematical ingredients of probability theory. We denote the triple as $\{\Omega, \mathcal{F}, \mathbf{P}\}$, and call it a **probability space**; the word "space" has no special meaning, it is just a tradition.

Then $X_t(\omega)$ is a function with time parameter t defined on a probability space $\{\Omega, \mathcal{F}, \mathbf{P}\}$. The aggregate $\{X_t(\omega)\,;\ t \in [a, b]\,, \mathbf{P}\}$ is called a **stochastic process**. This is the mathematical substance of the theory of stochastic processes we treat in this book.

The most typical stochastic process is Brownian motion. We next give the definition of the one-dimensional Brownian motion $B_t(\omega)$.

Definition 1.2.1. Let $\{\Omega, \mathcal{F}, \mathbf{P}\}$ be a probability space. A function $B_t(\omega)$ that depends on the time parameter t and is defined on the probability space $\{\Omega, \mathcal{F}, \mathbf{P}\}$ is called **Brownian motion**, if the following conditions are satisfied:

(i) For each fixed $\omega \in \Omega$, $B_t(\omega)$ is a continuous function of t and $B_0(\omega) = 0$.

(ii) For any $0 < t$, the distribution of $B_t(\omega)$ is the normal (Gaussian) distribution, i.e.,

$$\mathbf{P}\left[\{\omega : B_t(\omega) \in dx\}\right] = \frac{1}{\sqrt{2\pi t}} \exp\left(-\frac{x^2}{2t}\right) dx. \qquad (1.2.1)$$

(iii) For any values of time $0 \le r < s \le s' < t$, the random variables $B_t(\omega) - B_{s'}(\omega)$ and $B_s(\omega) - B_r(\omega)$ are independent.

It is not self-evident that these mathematical objects exist. Specifically, the problem is whether such a probability measure $\mathbf{P}$ exists. The existence was established by N. Wiener and P. Lévy in the 1920th. The measure N. Wiener constructed is now called the "Winer measure" (Wiener (1923), Paley-Wiener (1934), Lévy (1940)). This was the dawn of the theory of stochastic processes. Continuing this work, Kolmogorov laid the foundations of probability theory and gave an analytic method for studying stochastic processes. This subsequently led to the "path analysis of random motion" by K. Itô. These contributions will be presented below.

Remark. We next give some formulas connected with (1.2.1). Let $a > 0$. Differentiating both sides of the equality

$$\int_{-\infty}^{\infty} \exp\left(-ax^2\right) dx = \sqrt{\frac{\pi}{a}}$$

with respect to a and taking $a = 1/(2t)$, we get

$$\int_{-\infty}^{\infty} x^2 \frac{1}{\sqrt{2\pi t}} \exp\left(-\frac{x^2}{2t}\right) dx = t. \tag{1.2.2}$$

Next, differentiating twice both sides of (1.2.2) with respect to a and taking $a = 1/(2t)$, we get

$$\int_{-\infty}^{\infty} x^4 \frac{1}{\sqrt{2\pi t}} \exp\left(-\frac{x^2}{2t}\right) dx = 3t^2. \tag{1.2.3}$$

We claim that the expectation $\mathbf{P}\left[|B_t|^2\right]$ of $|B_t|^2$ is

$$\mathbf{P}\left[|B_t|^2\right] = t. \tag{1.2.4}$$

This is one of the important properties of Brownian motion. Indeed, the distribution of the Brownian motion is

$$\mathbf{P}[B_t \in dx] = \frac{1}{\sqrt{2\pi t}} \exp\left(-\frac{x^2}{2t}\right) dx.$$

Since the expectation $\mathbf{P}\left[|B_t|^2\right]$ is the integral of $|x|^2$ with respect to the distribution $\mathbf{P}[B_t \in dx]$, we have

$$\mathbf{P}\left[|B_t|^2\right] = \int |x|^2 \, \mathbf{P}\left[B_t \in dx\right] = \int_{-\infty}^{\infty} x^2 \frac{1}{\sqrt{2\pi t}} \exp\left(-\frac{x^2}{2t}\right) dx,$$

and by formula (1.2.2), the right-hand side is equal to t, as claimed.

As an application of formula (1.2.4), we show that the paths of Brownian motion B_t are nowhere differentiable.

Since the distribution of the random variable $B_t - B_s$ coincides with that of B_{t-s}, formula (1.2.4) implies that

$$\mathbf{P}\left[|B_t - B_s|^2\right] = |t - s|.$$

Therefore,

$$\lim_{t-s\downarrow 0} \mathbf{P}\left[\left|\frac{B_t - B_s}{t - s}\right|^2\right] = \lim_{t-s\downarrow 0} \frac{1}{t - s} = \infty.$$

Thus, the first derivative of B_t with respect to the time variable t diverges. Strictly speaking, the time derivative of the Brownian motion does not exist in the function space $L^2(\Omega)$. More generally,

$$\mathbf{P}[\{\omega : B_t(\omega) \text{ is nowhere differentiable w.r.t. the time variable } t\}] = 1.$$

This result is due to N. Wiener.

Next we will explain the notion of the **"transition probability of a stochastic process"**.

Kolmogorov defined the transition probability $Q(s, x, t, \Gamma)$ as follows:

Definition 1.2.2. The function $Q(s, x, t, \Gamma)$, where $s, t \in [a, b]$, $s \leq t$, are times, x is a point in $\mathbf{R}^d$, and Γ is a region in $\mathbf{R}^d$, is called a **transition probability** if it satisfies the following two conditions:

(i) the Chapman-Kolmogorov equation:

$$Q(s, x, t, \Gamma) = \int_{\mathbf{R}^d} Q(s, x, r, dy) Q(r, y, t, \Gamma), \quad s \leq r \leq t; \qquad (1.2.5)$$

(ii) the normalization condition:

$$Q(s, x, t, \mathbf{R}^d) = \int_{\mathbf{R}^d} Q(s, x, t, dy) = 1. \qquad (1.2.6)$$

We will prove that, **given a transition probability $Q(s, x, t, \Gamma)$, there exists a stochastic process $X_t(\omega)$ with the transition probability $Q(s, x, t, \Gamma)$.**

We first observe the motion of a particle over a closed time interval $[a, b]$. The probability that a particle starting at the point x is found in a small region dy at time t is $Q(a, x, t, dy)$.

The probability that a particle starting at the point y at time t is found in a region Γ_b at time b is $Q(t, y, b, \Gamma_b)$.

Therefore, the probability that a particle goes through a region Γ at time t and is found in a region Γ_b at time b is

$$\int_{\Gamma} Q(a, x, t, dy)Q(t, y, b, \Gamma_b), \quad a < t < b.$$

Figure 1.2 depicts a sample path $\omega(t)$ that starts at a point x at time a, passes through a region Γ at time t and arrives in a region Γ_b at time b. There are many such samples, forming the set

$$\{\omega : \omega(a) = x, \ \omega(t) \in \Gamma, \ \omega(b) \in \Gamma_b\}.$$

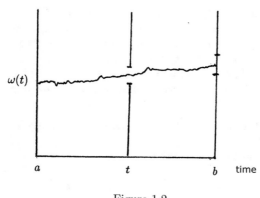

Figure 1.2

The probability of this set of samples is therefore

$$\mathbf{P}\left[\{\omega : \omega(a) = x,\ \omega(t) \in \Gamma,\ \omega(b) \in \Gamma_b\}\right] = \int_{\Gamma} Q(a, x, t, dy)Q(t, y, b, \Gamma_b). \quad (1.2.7)$$

Now let $a < s < t < b$. If a particle starts at a point x at time a, goes through a region Γ_1 at time s, with $a < s < b$, and through a region Γ_2 at time t, with $s < t < b$, then we get a sample path like the one shown in Figure 1.3.

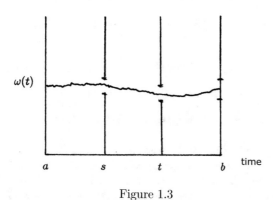

Figure 1.3

Let us calculate the probability of the collection of all such sample paths,

$$\{\omega : \omega(a) = x,\ \omega(s) \in \Gamma_1,\ \omega(t) \in \Gamma_2,\ \omega(b) \in \Gamma_b\}.$$

We know that the probability of the set of sample paths $\omega(t)$ that start a the point x at time a and pass through a small region dx_1 at time s is $Q(a, x, s, dx_1)$.

Similarly, the probability of the set of sample paths $\omega(t)$ that start at a point x_1 at time s and pass through a small region dx_2 at time t is $Q(s, x_1, t, dx_2)$; and the probability of the set of sample paths $\omega(t)$ that start at a point x_2 at time t and arrive at a region Γ_b at time b is $Q(t, x_2, b, \Gamma_b)$.

Then the probability of the set of sample paths $\omega(t)$ that start at a point x at time t and pass through a region Γ_1 at time s and through another region Γ_2 at time t is

$$\int Q(a, x, s, dx_1)\mathbf{1}_{\Gamma_1}(x_1)Q(s, x_1, t, dx_2)\mathbf{1}_{\Gamma_2}(x_2).$$

Therefore, the sought-for probability is

$$\mathbf{P}\left[\{\omega : \omega(a) = x, \ \omega(s) \in \Gamma_1, \ \omega(t) \in \Gamma_2, \ \omega(b) \in \Gamma_b\}\right]$$

$$= \int Q(a, x, s, dx_1)\mathbf{1}_{\Gamma_1}(x_1)Q(s, x_1, t, dx_2)\mathbf{1}_{\Gamma_2}(x_2)Q(t, x_2, b, \Gamma_b). \qquad (1.2.8)$$

Generally, we consider times

$$a < t_1 \leq t_2 \leq \cdots \leq t_n < b$$

and regions

$$\Gamma_1, \Gamma_2, \ldots, \Gamma_n,$$

and define the probability of the set of sample paths

$$\{\omega : \omega(a) = x, \omega(t_1) \in \Gamma_1, \omega(t_2) \in \Gamma_2, \ldots, \omega(t_n) \in \Gamma_n, \omega(b) \in \Gamma_b\}$$

$$\mathbf{P}[\{\omega : \omega(a) = x, \omega(t_1) \in \Gamma_1, \omega(t_2) \in \Gamma_2, \ldots, \omega(t_n) \in \Gamma_n, \omega(b) \in \Gamma_b\}]$$

$$= \int Q(a, x, t_1, dx_1)\mathbf{1}_{\Gamma_1}(x_1)Q(t_1, x_1, t_2, dx_2)\mathbf{1}_{\Gamma_2}(x_2)$$

$$\times \ Q(t_{n-1}, x_{n-1}, t_n, dx_n)\mathbf{1}_{\Gamma_n}(x_n)Q(t_n, x_n, b, \Gamma_b), \qquad (1.2.9)$$

where if $\Gamma_b = \mathbf{R}^d$, the last factor does not appear, as $Q(t_n, x_n, b, \mathbf{R}^d) = 1$.

The subset of the sample space Ω considered above and appearing on the left-hand side in formula (1.2.9) is called a cylinder. Therefore, formula (1.9.2) defines a set function $\mathbf{P}$ on cylinder subsets of the sample space Ω, which can be extended to a **probability measure** on the smallest σ-algebra $\mathcal{F}$ generated by such cylinders. This statement is called the Kolmogorov extension theorem.

For each t, we set $X_t(\omega) = \omega(t)$, which defines a function $X_t(\omega)$ of ω. This yields a stochastic process $\{X_t(\omega); \ t \in [a, b], \mathbf{P}\}$.

Formula (1.2.9) defines what is known as the "finite-dimensional distribution" of the stochastic process $\{X_t(\omega); \ t \in [a, b], \mathbf{P}\}$.

When the starting time is s and the starting point is x, we use the explicit notation $\mathbf{P}_{(s,x)}$, i.e., we write

$$\mathbf{P}_{(s,x)}[\{\omega : \omega\,(t_1) \in \Gamma_1,\ \omega\,(t_2) \in \Gamma_2, \ldots, \omega\,(t_n) \in \Gamma_n\}]$$

$$= \int Q(s, x, t_1, dx_1)\mathbf{1}_{\Gamma_1}(x_1)Q(t_1, x_1, t_2, dx_2)\mathbf{1}_{\Gamma_2}(x_2)$$

$$\times\, Q(t_{n-1}, x_{n-1}, t_n, dx_n)\mathbf{1}_{\Gamma_n}(x_n), \tag{1.2.10}$$

where we are not concerned with the motion in the time interval $t_n < t \leq b$.

When an initial distribution $\mu(dx)$ is given, we define the probability measure $\mathbf{P}_\mu$ by

$$\mathbf{P}_\mu[\{\omega : \omega\,(t_1) \in \Gamma_1,\ \omega\,(t_2) \in \Gamma_2, \ldots, \omega\,(t_n) \in \Gamma_n\}]$$

$$= \int \mu(dx)Q(a, x, t_1, dx_1)\mathbf{1}_{\Gamma_1}(x_1)Q(t_1, x_1, t_2, dx_2)\mathbf{1}_{\Gamma_2}(x_2)$$

$$\times\, Q(t_{n-1}, x_{n-1}, t_n, dx_n)\mathbf{1}_{\Gamma_n}(x_n). \tag{1.2.11}$$

When there is no need to emphasize the initial distribution $\mu(dx)$, we simply write $\mathbf{P}$ instead of $\mathbf{P}_\mu$.

The stochastic process $\{X_t\,(\omega)\,; t \in [a, b], \mathbf{P}_\mu\}$ provides the mathematical description of the random motion with an initial distribution $\mu(dx)$ and a transition probability $Q(s, x, t, \Gamma)$.

Moreover, the stochastic process $\{X_t\,(\omega)\,; t \in [a, b], \mathbf{P}_\mu\}$ defined above is a **Markov process** with the transition probability $Q(s, x, t, \Gamma)$.

Let us explain **what the Markov property means.**

Consider a random motion which starts at a point $x \in \mathbf{R}^d$ at time r, and define the probability measure $\mathbf{P}_{(r,x)}$ by (1.2.10). Then, for times $r < t_1 \leq t_2 \leq \cdots \leq t_n < b$ starting with r and functions $f(x_1, x_2, \ldots, x_n)$ on $\left(\mathbf{R}^d\right)^n$, it holds that

$$\mathbf{P}_{(r,x)}[f(X_{t_1}, X_{t_2} \ldots, X_{t_n})$$

$$= \int Q(r, x, t_1, dx_1)Q(t_1, x_1, t_2, dx_2)\,Q(t_2, x_2, t_3, dx_3) \times \cdots \tag{1.2.12}$$

$$\times\, Q(t_{n-1}, x_{n-1}, t_n, dx_n)f(x_1, \ldots, x_n).$$

In the simple case of a function $f(x)$ on $\mathbf{R}^d$ we have

$$\mathbf{P}_{(s,y)}[f(X_t)] = \int Q(s, y, t, dz)f(z). \tag{1.2.13}$$

Using this, we can rewrite the equation (1.2.12) as follows.

Take $r < t_1 \leq t_2 \leq \cdots \leq t_n = s < t$. Then

$$\mathbf{P}_{(r,x)}[g(X_{t_1}, X_{t_2}, \ldots, X_s)f(X_t)]$$

$$= \int Q(r, x, t_1, dx_1)Q(t_1, x_1, t_2, dx_2)Q(t_2, x_2, t_3, dx_3) \times \cdots$$

$$\times Q(t_{n-1}, x_{n-1}, s, dy)g(x_1, \ldots, x_{n-1}, y) \int Q(s, y, t, dz) f(z)$$

$$= \mathbf{P}_{(r,x)} \left[g(X_{t_1}, X_{t_2}, \ldots, X_s) \mathbf{P}_{(s, X_s)}[f(X_t)] \right],$$

where we used (1.2.13). The equality

$$\mathbf{P}_{(r,x)}[g(X_{t_1}, X_{t_2}, \ldots, X_s)f(X_t)] = \mathbf{P}_{(r,x)} \left[g(X_{t_1}, X_{t_2}, \ldots, X_s) \mathbf{P}_{(s, X_s)}[f(X_t)] \right]$$
(1.2.14)

is called the **"Markov property"** of the stochastic process $\{X_t(\omega); t \in [a, b], \mathbf{P}\}$.

You will understand the Markov property gradually, once you use it often enough. Now let us interpret the equality (1.2.14).

First of all, the term $g(X_{t_1}, X_{t_2}, \ldots, X_s)$ that appears on both sides represents information about the history of the motion of a particle.

On the left-hand side we observe $f(X_t)$ at time t, which depends on the history of the motion of the particle. Differently from the term $f(X_t)$ on the left-hand side, in the right-hand side, $\mathbf{P}_{(s, X_s)}[f(X_t)]$ is the expectation of $f(X_t)$ under the condition that a particle starts at the point X_s at time s.

Here we see the difference: the term $\mathbf{P}_{(s, X_s)}[f(X_t)]$ on the right-hand side is determined exclusively by the position X_s at time s and is independent of the history of the motion from time r to time s.

Thus, equality (1.2.14), i.e., the **Markov property** of the stochastic process $\{X_t(\omega); t \in [a, b], \mathbf{P}\}$ asserts that

"the behavior of the motion from time s to time t is determined by the position X_s of the particle and is independent of the history before time s."

We formulate our conclusions as a theorem.

Theorem 1.2.1 (Analytic construction of Markov processes). *Let $Q(s, x, t, \Gamma)$ be a transition probability (see Definition 1.2.2). Then there exists a stochastic process $\{X_t(\omega); t \in [a, b], \mathbf{P}\}$ such that the equation (1.2.9) of finite-dimensional distributions is satisfied. In particular,*

$$Q(s, x, t, \Gamma) = \mathbf{P}[\{\omega : X_s(\omega) = x, \ X_t(\omega) \in \Gamma\}].$$

Therefore, the stochastic process $\{X_t(\omega); t \in [a, b], \mathbf{P}\}$ is a Markov process with the transition probability $Q(s, x, t, \Gamma)$.

This theorem describes Kolmogorov's analytic method for constructing Markov processes. It is the **fundamental theorem that guarantees the existence of Markov processes**.

Remark. Theorem 1.2.1 has a gap: it does not state that the probability measure **P** lives on the sample space Ω, i.e., $\mathbf{P}[\Omega] = 1$. To ensure that this will hold, the transition probability $Q(s, x, t, \Gamma)$ must satisfy a continuity condition (see the next Theorem 1.2.2). In fact, these two theorems should be combined.

As already mentioned, N. Wiener proved the existence of a probability measure **P** for the Brownian motion, which means that the Brownian motion is a well-defined mathematical object. Kolmogorov generalized the existence theorem for a wider class of stochastic processes.

If we apply Kolmogorov's analytic method to the transition probability

$$q(s, x, t, y) = \frac{1}{\sqrt{2\pi(t - s)}} \exp\left(-\frac{(y - x)^2}{2(t - s)}\right),$$

then the Brownian motion $\{B_t(\omega) \, ; \, t \in [0, \infty), \mathbf{P}\}$ is realized as a stochastic process.

Next, we explain some terminology that is frequently used in probability theory.

We can imagine arbitrary subsets of the sample space Ω, but in fact we do not operate with all of them. What we treat is a so-called σ-additive family (σ-algebra) $\mathcal{F}$ of subsets, where σ indicates that $\mathcal{F}$ is closed under countable set operations.

If **P** is a probability measure, i.e., a set-function defined on a σ-algebra $\mathcal{F}$, then the probability of an event $A \in \mathcal{F}$ is denoted by $\mathbf{P}[A]$. The triple $\{\Omega, \mathcal{F}, \mathbf{P}\}$ will be called a probability space.

A function $X(\omega)$ defined on the sample space Ω is said to be measurable, or $\mathcal{F}$-measurable, if $\{\omega : X(\omega) \le a\} \in \mathcal{F}$ for arbitrary a. In probability theory measurable functions are referred to as random variables.

The expectation $\mathbf{P}[X]$ of a random variable $X(\omega)$ is defined as

$$\mathbf{P}[X] = \int_\Omega X(\omega) \, d\mathbf{P},$$

i.e., as the Lebesgue integral of $X(\omega)$ with respect to the measure **P**.

Remark. On the Lebesgue integral: The classical Riemann integral is defined for continuous functions on the finite-dimensional Euclidean space. On the other hand, the Lebesgue integral can be defined for functions on infinite-dimensional spaces, such as the sample space Ω, a typical infinite-dimensional space in the theory of stochastic processes. The Lebesgue integral with respect to a probability measure **P** is defined as follows: For a non-negative random variable $0 \le X(\omega)$, we consider first the step functions

$$X_n = \sum_{k=0}^{\infty} \frac{k + 1}{2^n} \mathbf{1}_{\Gamma_{n,k}},$$

where $\Gamma_{n,k} = \left\{ \omega : \frac{k}{2^n} < X(\omega) \le \frac{k+1}{2^n} \right\}$. The integral of the step function X_n with respect to the probability measure $\mathbf{P}$ is defined as

$$\mathbf{P}[X_n] = \sum_{k=0}^{\infty} \frac{k+1}{2^n} \, \mathbf{P}[\Gamma_{n,k}],$$

where the right-hand side is well defined, because $X(\omega)$ is measurable and hence $\Gamma_{n,k} \in \mathcal{F}$. The integral of $X(\omega)$ is then defined as the limit

$$\mathbf{P}[X] = \lim_{n \to \infty} \mathbf{P}[X_n].$$

Now if X is an arbitrary random variable, we decompose it as $X = X^+ - X^-$ and set

$$\mathbf{P}[X] = \mathbf{P}[X^+] - \mathbf{P}[X^-].$$

We next define the notion of random motion, or, in other words, stochastic process.

Definition 1.2.3. A family $\{X_t(\omega) ; t \in [a,b]\}$ of random variables defined on a probability space $\{\Omega, \mathcal{F}, \mathbf{P}\}$ is called a **stochastic process with time parameter** $t \in [a,b]$. In this book we will always assume that $X_t(\omega)$ is continuous in $t \in [a,b]$, $-\infty < a < b < \infty$.

On transition probabilities and parabolic partial differential equations:

We have seen that a transition probability determines a Markov process. We next show that a Markov process is characterized by a parabolic partial differential equation. This is one of the most important theorems in the theory of Markov processes.

Assume that the transition probability $Q(s,x,t,\Gamma)$ has a density function $q(s,x,t,y)$:

$$Q(s,x,t,\Gamma) = \int_{\Gamma} q(s,x,t,y)\,dy, \quad t > s.$$

The Chapman-Kolmogorov equation (1.2.5) can be written as

$$q(s,x,t,y) = \int_{\mathbf{R}^d} q(s,x,r,z)\,dz\,q(r,z,t,y), \quad s < r < t. \tag{1.2.15}$$

The normalization condition takes on the form

$$\int_{\mathbf{R}^d} q(s,x,t,y)\,dy = 1. \tag{1.2.16}$$

Define

$$u(t,x) = \int_{\mathbf{R}^d} q(t,x,b,y)f(y)dy, \tag{1.2.17}$$

and let us look for an equation satisfied by the function $u(t, x)$.

By the Chapman-Kolmogorov equation (1.2.15),

$$u(t - h, x) = \int_{\mathbf{R}^d} q(t - h, x, t, y)\, dy \int_{\mathbf{R}^d} q(t, y, b, z)\, f(z)\, dz$$
$$= \int_{\mathbf{R}^d} q(t - h, x, t, y) u(t, y)\, dy.$$

Therefore, if we take a neighborhood $U_\varepsilon(x)$ of the point x, then

$$\frac{u(t - h, x) - u(t, x)}{h} = \frac{1}{h} \int_{\mathbf{R}^d} q(t - h, x, t, y)\, (u(t, y) - u(t, x))\, dy$$
$$= \frac{1}{h} \int_{U_\varepsilon(x)^c} q(t - h, x, t, y)\, (u(t, y) - u(t, x))\, dy \quad (1.2.18)$$
$$+ \frac{1}{h} \int_{U_\varepsilon(x)} q(t - h, x, t, y)\, (u(t, y) - u(t, x))\, dy.$$

We assume that the probability that our particle leaves immediately the neighborhood $U_\varepsilon(x)$ is small, i.e.,

$$\lim_{h \downarrow 0} \frac{1}{h} \int_{U_\varepsilon(x)^c} q(t - h, x, t, y)\, dy = 0.$$

This assumption guarantees that our stochastic process X_t is continuous in t.

Then the limit of the first integral on the right-hand side of formula (1.2.18) is

$$\lim_{h \downarrow 0} \frac{1}{h} \int_{U_\varepsilon(x)^c} q(t - h, x, t, y)\, (u(t, y) - u(t, x))\, dy = 0.$$

Moreover, using the Taylor expansion, for $y \in U_\varepsilon(x)$,

$$u(t, y) - u(t, x) = \sum_{i=1}^{d} \left(y^i - x^i\right) \frac{\partial u}{\partial x^i} + \frac{1}{2} \sum_{i,j=1}^{d} \left(y^i - x^i\right)\left(y^j - x^j\right) \frac{\partial^2 u}{\partial x^i \partial x^j} + o(\varepsilon),$$

which we substitute in the second integral in (1.2.18).

Now, let $x \in \mathbf{R}^d$, $x = (x^1, x^2, \ldots, x^d)$. The expectation of the drift $y^i - x^i$ from x^i to y^i in a short time interval h is

$$\int_{U_\varepsilon(x)} q(t - h, x, t, y)\left(y^i - x^i\right) dy,$$

and vanishes as $h \downarrow 0$. We then set

$$\lim_{h \downarrow 0} \frac{1}{h} \int_{U_\varepsilon(x)} q(t - h, x, t, y)\left(y^i - x^i\right) dy = b(t, x)^i. \quad (1.2.19)$$

The vector with the coordinates $b(t, x)^i$ represents the **instantaneous drift**.

On the other hand,

$$\lim_{h \downarrow 0} \frac{1}{h} \int_{U_\varepsilon(x)} q(t - h, x, t, y)(y^i - x^i)^2 dy = \left(\sigma^2(t, x)\right)^{ii}$$

represents the **randomness intensity**, which is the instantaneous variance of the i-th coordinate.

Moreover, for $i, j = 1, 2, \ldots, d$, we set

$$\lim_{h \downarrow 0} \frac{1}{h} \int_{U_\varepsilon(x)} q\left(t - h, x, t, y\right)\left(y^i - x^i\right)\left(y^j - x^j\right) dy = \left(\sigma^2(t, x)\right)^{ij}. \qquad (1.2.20)$$

Since $o(\varepsilon)$ can be made arbitrarily small, we have

$$-\lim_{h \downarrow 0} \frac{u(t, x) - u(t - h, x)}{h} = \sum_{i=1}^{d} (b(t, x))^i \frac{\partial u}{\partial x^i} + \frac{1}{2} \sum_{i,j=1}^{d} \left(\sigma^2(t, x)\right)^{ij} \frac{\partial^2 u}{\partial x^i \partial x^j},$$

where the left-hand side is $-\dfrac{\partial u}{\partial t}$. Hence

$$\frac{\partial u}{\partial t} + \frac{1}{2} \sum_{i,j=1}^{d} \sigma^2(t, x)^{ij} \frac{\partial^2 u}{\partial x^i \partial x^j} + \sum_{i=1}^{d} b(t, x)^i \frac{\partial u}{\partial x^i} = 0.$$

Thus, we found a partial differential equation for the function (1.2.17).

Theorem 1.2.2 (Kolmogorov's Theorem). *Assume that the transition probability density $q(s, x, t, y)$ satisfies the continuity condition*

$$\lim_{h \downarrow 0} \frac{1}{h} \int_{U_\varepsilon(x)^c} q(t - h, x, t, y) dy = 0.$$

Define the **drift coefficient** *by*

$$b(t, x)^i = \lim_{h \downarrow 0} \frac{1}{h} \int_{U_\varepsilon(x)} q\left(t - h, x, t, y\right)\left(y^i - x^i\right) dy,$$

and the **randomness intensity** *by*

$$\sigma^2(t, x)^{ij} = \lim_{h \downarrow 0} \frac{1}{h} \int_{U_\varepsilon(x)} q\left(t - h, x, t, y\right)\left(y^i - x^i\right)\left(y^j - x^j\right) dy.$$

Then the function $u(t, x)$ defined by formula (1.2.7) satisfies the parabolic partial differential equation

$$\frac{\partial u}{\partial t} + \frac{1}{2} \sum_{i,j=1}^{d} \sigma^2(t, x)^{ij} \frac{\partial^2 u}{\partial x^i \partial x^j} + \sum_{i=1}^{d} b(t, x)^i \frac{\partial u}{\partial x^i} = 0, \qquad (1.2.21)$$

called the **Kolmogorov equation**.

Thus, according to Theorem 1.2.2, the transition probability density $q(s, x, t, y)$ satisfies the parabolic equation (1.2.21). This is one of the most important results in the theory of Markov processes. Conversely, if $q(s, x, t, y)$ is a solution of equation (1.2.21), then applying Theorem 1.2.1 we can construct a Markov process with the transition probability density $q(s, x, t, y)$.

As an example, we apply Theorem 1.2.2 to the transition probability density

$$q(s, x, t, y) = \frac{1}{\sqrt{2\pi(t-s)}} \exp\left(-\frac{(y-x)^2}{2(t-s)}\right),$$

which was considered by Einstein.

We first compute the drift $b(t, x)$ by formula (1.2.19). Since

$$\int_{-\infty}^{\infty} x \frac{1}{\sqrt{2\pi t}} \exp\left(-\frac{x^2}{2t}\right) dx = 0,$$

we have

$$b(t, x) = \lim_{h \downarrow 0} \frac{1}{h} \int_{U_\varepsilon(x)} q(t-h, x, t, y)(y-x) dy$$

$$= \lim_{h \downarrow 0} \frac{1}{h} \int_{U_\varepsilon(x)} \frac{1}{\sqrt{2\pi h}} \exp\left(-\frac{(y-x)^2}{2h}\right)(y-x) dy = 0,$$

where we used the fact that the distribution is almost concentrated in $U_\varepsilon(x)$ when we let $h \downarrow 0$. Therefore, $b(t, x) = 0$.

Next, we compute $\sigma^2(t, x)$. In view of formula (1.2.2), equation (1.2.20) takes on the form

$$\sigma^2(t, x) = \lim_{h \downarrow 0} \frac{1}{h} \int_{U_\varepsilon(x)} q(t-h, x, t, y)(y-x)^2 dy$$

$$= \lim_{h \downarrow 0} \frac{1}{h} \int_{U_\varepsilon(x)} \frac{1}{\sqrt{2\pi h}} \exp\left(-\frac{(y-x)^2}{2h}\right)(y-x)^2 dy = 1,$$

where as above we used the fact that the distribution is almost concentrated in $U_\varepsilon(x)$ when $h \downarrow 0$. Hence, $\sigma^2(t, x) = 1$.

By Theorem 1.2.2, in the present case Kolmogorov's equation (1.2.21) is

$$\frac{\partial u}{\partial t} + \frac{1}{2}\frac{\partial^2 u}{\partial x^2} = 0. \tag{1.2.22}$$

This coincides with a result obtained by Einstein in 1905. Thus, Kolmogorov's Theorem 1.2.2 contains Einstein's result on the Brownian motion as a special case.

In dimension one, equation (1.2.21) reads

$$\frac{\partial u}{\partial t} + \frac{1}{2}\sigma^2(t,x)\frac{\partial^2 u}{\partial x^2} + b(t,x)\frac{\partial u}{\partial x} = 0, \tag{1.2.23}$$

where the coefficient $b(t,x)$ indicates the presence of a drift.

The random motion described by equation (1.2.22) is the Brownian motion. A question arises: How do the sample paths of the random motion described by equation (1.2.23) differ from those of the Brownian motion? Answering this question was K. Itô's motivation for initiating the path analysis of random motions. This will be explained in the next section.

Remark. Our discussion in this section did not touch upon the problem of external forces. Random motion under external forces will be discussed in Section 1.4.

1.3   Itô's Path Analysis

In the preceding section we constructed a stochastic process $X_t(\omega)$ with a prescribed transition probability. However, we do not know much about how to actually compute $X_t(\omega)$.

Equation (1.2.23) suggests that the random motion $X_t(\omega)$ might be given by a differential equation of the from

$$dX_t = \sigma(t,X_t)dB_t + b(t,X_t)dt, \tag{1.3.1}$$

where one notes that the coefficient of dB_t is not $\sigma^2(t,x)$, but $\sigma(t,x)$. The reason for this will be soon clear.

It was in 1930–1940 that K. Itô developed this idea. Equation (1.3.1) is not self-evident, it is Itô's original contribution, and it is now called the "Itô stochastic differential equation."

Integrating the differential form of Itô's equation, we obtain

$$X_t = X_a + \int_a^t \sigma(s,X_s)dB_{s-a} + \int_a^t b(s,X_s)ds, \tag{1.3.2}$$

where $B_t \left(= B_t(\omega) = B(t,\omega)\right)$ is a given d-dimensional Brownian motion defined on a probability space $\{\Omega, \mathcal{F}, \mathbf{P}\}$. The starting point X_a is a random variable.

Itô's path analysis deals directly with the **sample paths of a stochastic process**. Namely, we construct the stochastic process $X_t(\omega)$ directly as a solution of the stochastic differential equation (1.3.1). This was not possible in Kolmogorov's theory.

In the d-dimensional Euclidean space, equation (1.3.1) becomes

$$dX_t = \boldsymbol{\sigma}(t,X_t)\,dB_t + \mathbf{b}(t,X_t)\,dt, \tag{1.3.3}$$

where

$$\boldsymbol{\sigma}(t,x) = \{\sigma(t,x)^i_j\} : [a,b] \times \mathbf{R}^d \to \mathbf{R}^{d^2},$$

is a matrix-function, $\mathbf{b}(t,x)$ is a drift vector and B_t is a d-dimensional Brownian motion. In coordinates,

$$dX^i_t = \sum_{j=1}^{d} \sigma(t, X_t)^i_j dB^j_t + b(t, X_t)^i dt, \quad i = 1, 2, \dots, d.$$

We will often write equation (1.3.3) in the integral form

$$X_t = X_a + \int_a^t \boldsymbol{\sigma}(s, X_s)\, dB_{s-a} + \int_a^t \mathbf{b}(s, X_s)\, ds, \tag{1.3.4}$$

where the second term on the right-hand side is Itô's stochastic integral of $\boldsymbol{\sigma}(s, X_s)$ with respect to the Brownian motion B_s.

The stochastic differential equation (1.3.4) can be solved by successive approximations. For $n \geq 1$, set

$$X^{(n)}_t = X_a + \int_a^t \boldsymbol{\sigma}\left(s, X^{(n-1)}_s\right) dB_{s-a} + \int_a^t \mathbf{b}\left(s, X^{(n-1)}_s\right) ds.$$

We then get a solution of (1.3.4) as $X_t = \lim\limits_{n \to \infty} X^{(n)}_t$, where $X^{(0)}_t = X_a$.

The transition probability density of the stochastic process $X_t(\omega)$ obtained as a solution of equation (1.3.3) or (1.3.4) satisfies the Kolmogorov equation (1.2.21). To show this, we need some preparation, so the proof will be provided later on.

Remark. On Itô's stochastic integral: The second integral on the right-hand side of (1.3.4) is not the conventional Riemann-Stieltjes or Lebesgue integral. It was first introduced by Itô (see K. Itô (1942)); it is now called the Itô integral or stochastic integral.

Let us explain Itô's stochastic integral in the one-dimensional case. The Brownian motion B_t is defined on a probability space $\{\Omega, \mathcal{F}, \mathbf{P}\}$.

Let $a = s_0 < \cdots < s_k < \cdots < s_n = t$ and consider the sum

$$\sum_{k=0}^{n} \sigma(s_k, X_{s_k}(\omega))\left(B_{s_{k+1}}(\omega) - B_{s_k}(\omega)\right), \quad \omega \in \Omega.$$

This sum does not converge as $n \to \infty$, $|s_{k+1} - s_k| \downarrow 0$, because for almost every $\omega \in \Omega$,

$$\lim_{n \to \infty} \sum_{k=0}^{n} \left|B_{s_{k+1}}(\omega) - B_{s_k}(\omega)\right| = \infty.$$

Thus, the classical Riemann-Stieltjes integral cannot be defined with respect to the Brownian motion B_t.

Nevertheless, in view of (1.2.2),

$$\sum_{k=0}^{n} \mathbf{P}\left[\left|B_{s_{k+1}} - B_{s_k}\right|^2\right] = \sum_{k=0}^{n}(s_{k+1} - s_k) = t - a < \infty.$$

Therefore, the sequence

$$\sum_{k=0}^{n} \sigma\left(s_k, X_{s_k}\right)\left(B_{s_{k+1}} - B_{s_k}\right)$$

converges in the space $L^2\left(\Omega, \mathbf{P}\right)$ with the norm $\|f\| = \left(\mathbf{P}\left[|f\left(\omega\right)|^2\right]\right)^{\frac{1}{2}}$: there exists an element $Y_t \in L^2\left(\Omega, \mathbf{P}\right)$ such that

$$\left\|\sum_{k=0}^{n} \sigma\left(s_k, X_{s_k}\right)\left(B_{s_{k+1}} - B_{s_k}\right) - Y_t\right\| \to 0.$$

Then we set

$$\int_a^t \sigma\left(s, X_s\right) dB_{s-a} := Y_t,$$

which is continuous function of t. (For details, see, e.g., Chapter XVI in Nagasawa (2000).)

We next state one of the important properties of the stochastic integral as a lemma.

Lemma 1.3.1. *The expectation of the stochastic integral vanishes, that is,*

$$\mathbf{P}\left[\int_a^t \sigma\left(s, X_s\right) dB_s\right] = 0.$$

Proof. Let Γ_s be a bounded $\mathcal{F}_s$-measurable function. Then, by the Markov property

$$\mathbf{P}\left[\Gamma_s B_t\right] = \mathbf{P}\left[\Gamma_s \mathbf{P}_{(s,B_s)}\left[B_t\right]\right],$$

which in terms of the transition probability density $q(s, x, t, y)$ equals

$$\mathbf{P}\left[\Gamma_s \int q(s, B_s, t, y) y\, dy\right] = \mathbf{P}\left[\Gamma_s B_s\right];$$

here we used the fact that $\int q(s, x, t, y) y\, dy = x$, the proof of which is left to the reader. Therefore, for a bounded $\mathcal{F}_{s_k}$-measurable function Γ_{s_k} we have

$$\mathbf{P}\left[\Gamma_{s_k} B_{s_{k+1}}\right] = \mathbf{P}\left[\Gamma_{s_k} B_{s_k}\right],$$

so

$$\sum_{k=0}^{n} \mathbf{P}\left[\Gamma_{s_k}\left(B_{s_{k+1}} - B_{s_k}\right)\right] = 0.$$

This implies that the assertion of the lemma holds for any positive measurable function $\sigma\left(s, x\right)$. $\square$

Writing equation (1.3.3) in coordinates:

$$dX_t^i = \sum_{j=1}^{d} \sigma(t, X_t)_j^i dB_t^j + b(t, X_t)^i dt, \quad i = 1, 2, \ldots, d, \tag{1.3.5}$$

we denote its solution by X_t^i, $i = 1, 2, \ldots, d$.

To discuss solutions of the stochastic differential equation (1.3.5), we must first of all compute the differential $df(t, X_t)$ of $f(t, X_t)$ for twice differentiable functions $f \in C^2\left([a, b] \times \mathbf{R}^d\right)$.

In classical differential calculus, we neglect the second-order differential $dX_t^i dX_t^j$, because it is of small order. Therefore, the classical differential formula reads

$$df(t, X_t) = \frac{\partial f}{\partial t} dt + \sum_{i=1}^{d} \frac{\partial f}{\partial x^i} dX_t^i. \tag{1.3.6}$$

On the other hand, if X_t is a random motion, the formula for the differential of a function is not self-evident, since we cannot neglect the second-order differential $dX_t^i dX_t^j$. Therefore, we need use Itô's formula

$$df(t, X_t) = \frac{\partial f}{\partial t} dt + \sum_{i=1}^{d} \frac{\partial f}{\partial x^i} dX_t^i + \frac{1}{2} \sum_{i,j=1}^{d} \frac{\partial^2 f}{\partial x^i \partial x^j} dX_t^i dX_t^j. \tag{1.3.7}$$

As a matter of fact, the classical formula (1.3.6) does not hold, since the assumptions of the classical differential calculus are not satisfied. Let us consider the Brownian motion B_t as an important fundamental case. Then we have

"P. Lévy's formulas"

$$dB_t^i dB_t^i = dt, \quad dB_t^i dB_t^j = 0, \quad i \neq j, \tag{1.3.8}$$

$$dB_t^i dt = 0, \quad (dt)^2 = 0. \tag{1.3.9}$$

P. Lévy's formulas are symbolic formulas of practical use in the **Path Analysis** of random motion. The frequently used one is $dB_t^i dB_t^i = dt$.

We first prove that $dB_t^i dB_t^i = dt$, which rigorously speaking means that, for any $f(s, \omega)$,

$$\int_0^t f(s) \, dB_s^i dB_s^i = \int_0^t f(s) \, ds,$$

where the equality is understood in $L^2(\Omega, \mathbf{P})$.

Proof. For simplicity, we write B_s instead of B_s^i. Then, assuming that $|f| \leq 1$, we have

$$\mathbf{P}\left[\left(\sum_{k=0}^{n} f(s)(B_{s_{k+1}} - B_{s_k})^2 - \sum_{k=0}^{n} f(s)\,(s_{k+1} - s_k)\right)^2\right]$$

$$\leq \sum_{k=0}^{n} \mathbf{P}\left[\left(|B_{s_{k+1}} - B_{s_k}|^2 - (s_{k+1} - s_k)\right)^2\right]$$

$$= \sum_{k=0}^{n} \left(\mathbf{P}\left[|B_{s_{k+1}} - B_{s_k}|^4\right] - (s_{k+1} - s_k)^2\right)$$

$$= 2\sum_{k=0}^{n} (s_{k+1} - s_k)^2 \leq 2t\,\Delta_n,$$

where we used formula (1.2.3). Since $\Delta_n \downarrow 0$,

$$\sum_{\kappa=0}^{n} f(s_k)\,(B_{s_{k+1}} - B_{s_k})^2 \to \int_a^t f(s)\,ds$$

in $L^2(\Omega, \mathbf{P})$, which proves the first of P. Lévy's formula (1.3.8). $\square$

We emphasize that the differential on the left-hand side of P. Lévy's formula $dB_t^i dB_t^i = dt$ is equal to the first-order differential dt on the right-hand side and hence cannot be neglected.

The second formula in (1.3.8) asserts that

$$\int_0^t f(s)\,dB_s^i dB_s^j = 0, \quad i \neq j.$$

Proof. Assuming again that $|f| \leq 1$, we have

$$\mathbf{P}\left[\left(\sum_{k=0}^{n} f(s)(B_{s_{k+1}}^i - B_{s_k}^i)(B_{s_{k+1}}^j - B_{s_k}^j)\right)^2\right]$$

$$\leq \sum_{k=0}^{n} \mathbf{P}\left[(B_{s_{k+1}}^i - B_{s_k}^i)^2(B_{s_{k+1}}^j - B_{s_k}^j)^2\right],$$

where we used the fact that the cross term vanishes in view of the definition of the Brownian motion. Moreover, since the random variables $(B_{s_{k+1}}^i - B_{s_k}^i)^2$ and $(B_{s_{k+1}}^j - B_{s_k}^j)^2$ are independent, the last expression is equal to

$$\sum_{k=0}^{n} \mathbf{P}\left[(B_{s_{k+1}}^i - B_{s_k}^i)^2\right] \mathbf{P}\left[(B_{s_{k+1}}^j - B_{s_k}^j)^2\right] = \sum_{k=0}^{n} (s_{k+1} - s_k)^2 \leq t\,\Delta_n \to 0$$

because $\Delta_n = \sup |s_{k+1} - s_k| \downarrow 0$. Therefore, $\int_0^t f(s)\, dB_s^i dB_s^j = 0$ in $L^2(\Omega, \mathbf{P})$. The remaining formulas in (1.3.9) are clear. $\qquad\qquad\qquad\square$

Now, if we substitute the solution $dX_t^1, dX_t^2, \ldots, dX_t^d$ of the stochastic differential equation (1.3.5) and use P. Lévy's formulas (1.3.8) and (1.3.9), we obtain Itô's formula, the most fundamental formula in path analysis.

Itô's Theorem. *Let X_t be a solution of the stochastic differential equation* (1.3.3) (*or* (1.3.5)):

$$dX_t^i = \sum_{j=1}^d \sigma(t, X_t)_j^i dB_t^j + b(t, X_t)^i dt.$$

Then **"Itô's formula"** *holds*:

$$df(t, X_t) = Lf(t, X_t)\, dt + \sum_{i,j=1}^d \sigma(t, X_t)_j^i \frac{\partial f}{\partial x^i}(t, X_t)\, dB_t^j. \tag{1.3.10}$$

Here L denotes the partial differential operator

$$L = \frac{\partial}{\partial t} + \frac{1}{2} \sum_{i,j=1}^d \sigma^2(t, x)^{ij} \frac{\partial^2}{\partial x^i \partial x^j} + \sum_{i=1}^d b(t, x)^i \frac{\partial}{\partial x^i}, \tag{1.3.11}$$

where $\sigma^2(t, x)^{ij} = \sum\limits_{k=1}^d \sigma(t, x)_k^i \sigma(t, x)_k^j$.

Proof. Consider the differential expression (1.3.7) and substitute it in (1.3.5). Then we have

$$\sum_{i=1}^d \frac{\partial f}{\partial x^i} dX_t^i = \sum_{i=1}^d \frac{\partial f}{\partial x^i} \left(\sum_{j=1}^d \sigma(t, x)_j^i\, dB_t^j + b(t, X_t)^i dt \right)$$

$$= \sum_{i,j=1}^d \sigma(t, X_t)_j^i \frac{\partial f}{\partial x^i}(t, X_t)\, dB_t^j + \sum_{i=1}^d b(t, X_t)^i \frac{\partial f}{\partial x^i} dt.$$

Next, expanding $dX_t^i dX_t^j$ as

$$\left(\sum_{k=1}^d \sigma(t, x)_k^i\, dB_t^k + b(t, X_t)^i dt \right) \left(\sum_{\ell=1}^d \sigma(t, x)_\ell^j\, dB_t^\ell + b(t, X_t)^j dt \right)$$

and applying P. Lévy's formulas (1.3.8) and (1.3.9), we find that

$$dX_t^i dX_t^j = \sum_{k=1}^d \sigma(t, x)_k^i\, dB_t^k \sum_{\ell=1}^d \sigma(t, x)_\ell^j\, dB_t^\ell = \sum_{k=1}^d \sigma(t, x)_k^i\, \sigma(t, x)_k^j dt,$$

because the other terms vanish. Hence, upon denoting

$$\sigma^2(t,x)^{ij} = \sum_{k=1}^{d} \sigma(t,x)_k^i \, \sigma(t,x)_k^j,$$

we have that

$$\frac{1}{2} \sum_{i,j=1}^{d} \frac{\partial^2 f}{\partial x^i \partial x^j} \, dX_t^i dX_t^j = \frac{1}{2} \sum_{i,j=1}^{d} \sigma^2(t,x)^{ij} \frac{\partial^2 f}{\partial x^i \partial x^j} \, dt.$$

Summarizing,

$$df(t,X_t) = \left(\frac{\partial f}{\partial t} + \frac{1}{2} \sum_{i,j=1}^{d} \sigma^2(t,x)^{ij} \frac{\partial^2 f}{\partial x^i \partial x^j} + \sum_{i=1}^{d} b(t,X_t)^i \frac{\partial f}{\partial x^i} \right) dt$$

$$+ \sum_{i,j=1}^{d} \sigma(t,X_t)_j^i \frac{\partial f}{\partial x^i}(t,X_t) \, dB_t^j,$$

i.e., Itô's formula (1.3.10) holds.   □

Integrating both sides of (1.3.10), we obtain the

"Integral form of Itô's formula"

$$f(t,X_t) - f(s,X_s) - \int_s^t Lf(r,X_r) \, dr$$

$$= \sum_{i,j=1}^{d} \int_s^t \sigma(r,X_r)_j^i \frac{\partial f}{\partial x^i}(r,X_r) \, dB_r^j, \tag{1.3.12}$$

where L is the operator given by (1.3.11).

Itô's formulas (1.3.10) and (1.3.12) are **"fundamental formulas of path analysis"** and play a powerful role in differential and integral calculus of path analysis.

Remark. To justify the claim that Itô's formula is the "formula of differential and integral calculus" in path analysis, we consider the simple integral

$$\int_0^t (B_s)^n dB_s.$$

According to classical calculus, we should have

$$\int_0^t (B_s)^n dB_s = \frac{1}{n+1}(B_t)^{n+1},$$

which is incorrect. That is, we cannot use classical calculus when we deal with dB_t.

To get the correct expression, we must apply Itô's formula (1.3.12) to $f(x) = x^{n+1}$ and $X_t = B_t$. In this case $\sigma = 1$ and $b \equiv 0$. Therefore,

$$(B_t)^{n+1} - \frac{1}{2} \int_0^t n(n+1)(B_s)^{n-1} ds = \int_0^t (n+1)(B_s)^n dB_s.$$

That is, the correct Itô integral formula is

$$\int_0^t (B_s)^n dB_s = \frac{1}{n+1}(B_t)^{n+1} - \frac{n}{2} \int_0^t (B_s)^{n-1} ds,$$

where the second term on the right-hand side is the typical correction term of Itô's integral. If $n = 1$, then

$$\int_0^t B_s \, dB_s = \frac{1}{2}(B_t)^2 - \frac{1}{2}t,$$

where now the correction term is $-\frac{1}{2}t$.

Closing the story about correction terms, we will prove that the solution X_t of Itô's stochastic differential equation (1.3.4) is a Markov process.

Proof. If we divide the integration interval in the right-hand side of equation (1.3.4) into $a \leq r$ and $r < t$, we have

$$X_t = X_a + \int_a^r \boldsymbol{\sigma}(s, X_s) \, dB_{s-a} + \int_a^r \mathbf{b}(s, X_s) \, dt$$
$$+ \int_r^t \boldsymbol{\sigma}(s, X_s) \, dB_{s-a} + \int_r^t \mathbf{b}(s, X_s) \, dt.$$

Since the sum of the first three terms is just X_r, we can rewrite equation (1.3.4) as

$$X_t = X_r + \int_r^t \boldsymbol{\sigma}(s, X_s) \, dB_{s-r} + \int_r^t \mathbf{b}(s, X_s) \, dt, \quad a \leq r < t \leq b.$$

In other words, X_t depends on the position X_r at time r, but does not depend on the history before r. This is the Markov property of X_t. $\qquad\square$

Equation (1.2.14) presents the Markov property in terms of expectation, which is a powerful tool in computations. We have explained this in the previous section. The expression of the Markov property of X_t in terms of stochastic differential equations provided above is intuitive and easy to understand.

We have done enough preparation. We are now ready prove that the solution of the stochastic differential equation (1.3.3) coincides with the stochastic process $X_t(\omega)$ discussed in Kolmogorov's Theorem 1.2.1 and Theorem 1.2.2.

Thus, let us start from a point x at time s in the stochastic differential equation (1.3.4). Then

$$X_t = x + \int_s^t \boldsymbol{\sigma}\left(r, X_r\right) dB_{r-s} + \int_s^t \mathbf{b}\left(r, X_r\right) dt. \tag{1.3.13}$$

Further, let $\left\{X_t\left(\omega\right); t \in [s,b], \mathbf{P}_{(s,x)}\right\}$ be a Markov process that starts from a point x at time s, and set

$$u(s,x) = \mathbf{P}_{(s,x)}[f(X_b)] = \int q(s,x,b,z)f(z)dz, \tag{1.3.14}$$

where $q\left(s,x,t,z\right)$ is the transition probability density of the Markov process. We claim that the function $u(s,x)$ thus defined satisfies the Kolmogorov equation

$$\frac{\partial u}{\partial s} + \frac{1}{2}\sum_{i,j=1}^d \sigma^2(s,x)^{ij}\frac{\partial^2 u}{\partial x^i \partial x^j} + \sum_{i=1}^d b(s,x)^i \frac{\partial u}{\partial x^i} = 0.$$

First of all, by the Markov property,

$$\begin{aligned}
u(s,x) = \mathbf{P}_{(s,x)}[f(X_b)] &= \mathbf{P}_{(s,x)}\left[\mathbf{P}_{(s+h,X_{s+h})}[f(X_b)]\right] \\
&= \mathbf{P}_{(s,x)}[u(s+h, X_{s+h})],
\end{aligned} \tag{1.3.15}$$

where we used the defining equation (1.3.14) for $s + h$. Let us apply Itô's formula (1.3.12) to $u(s,x)$ and X_t and compute the expectation. Since the right-hand side of (1.3.12) is a stochastic integral, its expectation vanishes by Lemma 1.3.1. Consequently,

$$\mathbf{P}_{(s,x)}[u(t,X_t)] - u(s,x) = \mathbf{P}_{(s,x)}\left[\int_s^t Lu(r,X_r)dr\right], \tag{1.3.16}$$

where L is given in (1.3.11). Setting here $t = s + h$, we have

$$\mathbf{P}_{(s,x)}[u(s+h, X_{s+h})] - u(s,x) = \mathbf{P}_{(s,x)}\left[\int_s^{s+h} Lu(r,X_r)dr\right].$$

By (1.3.15), the first term on the left-hand side is $u(s,x)$, and hence

$$u(s,x) - u(s,x) = \mathbf{P}_{(s,x)}\left[\int_s^{s+h} Lu(r,X_r)dr\right].$$

Hence,

$$\mathbf{P}_{(s,x)}\left[\int_s^{s+h} Lu(r,X_r)dr\right] = 0. \tag{1.3.17}$$

Dividing the left-hand side by h and letting $h \downarrow 0$, we see that

$$\lim_{h \downarrow 0} \frac{1}{h} \mathbf{P}_{(s,x)} \left[\int_s^{s+h} Lu(r, X_r) dr \right] = Lu(s, x),$$

Therefore, $u(s, x)$ satisfies the equation $Lu(s, x) = 0$. Thus, we have proved

Theorem 1.3.1 (Stochastic Differential Equation and Parabolic Partial Differential Equation).

(i) *Let $\boldsymbol{\sigma}(t, x) = \{\sigma(t, x)_j^i\}$, $i, j = 1, 2, \ldots, d$, be a positive matrix-function, and X_t be a solution of the stochastic differential equation*

$$X_t = X_a + \int_a^t \boldsymbol{\sigma}\left(r, X_r\right) dB_{r-a} + \int_a^t \mathbf{b}\left(r, X_r\right) dr, \qquad (1.3.18)$$

or, in coordinates,

$$X_t^i = X_a^i + \int_a^t \sum_{j=1}^d \sigma\left(r, X_r\right)_j^i dB_{r-a}^j + \int_a^t \mathbf{b}(r, X_r)^i dr, \quad i = 1, 2, \ldots, d,$$

$$(1.3.19)$$

where the space variable is $x = (x^1, x^2, \ldots, x^d)$, $d = 1, 2, \ldots$.

Then the function

$$u(s, x) = \mathbf{P}_{(s,x)}[f(X_b)] = \int q(s, x, b, z) f(z) dz,$$

expressed in terms of the transition probability density $q\left(s, x, t, z\right)$, satisfies the parabolic partial differential equation

$$\frac{\partial u}{\partial s} + \frac{1}{2} \sum_{i,j=1}^d \sigma^2(s, x)^{ij} \frac{\partial^2 u}{\partial x^i \partial x^j} + \sum_{i=1}^d \mathbf{b}(s, x)^i \frac{\partial u}{\partial x^i} = 0, \qquad (1.3.20)$$

where $\sigma^2(t, x)^{ij} = \sum_{k=1}^d \sigma(t, x)_k^i \sigma(t, x)_k^j$. That is, $q\left(s, x, t, z\right)$ is the fundamental solution of equation (1.3.20).

Conversely, the Markov process $\{X_t\left(\omega\right); \ t \in [a, b], \mathbf{P}\}$ constructed by means of the fundamental solution $q\left(s, x, t, z\right)$ of the parabolic partial differential equation (1.3.20) is uniquely determined and X_t satisfies the stochastic differential equation (1.3.18). In this sense, the stochastic differential equation (1.3.18) and the parabolic partial differential equation (1.3.20) are equivalent.

(ii) *If σ is a constant and $\mathbf{b}\left(s, x\right)$ is a d-dimensional drift vector, then the Markov process $\{X_t\left(\omega\right); \ t \in [a, b], \mathbf{Q}\}$ determined by the fundamental solution $q(s, y, t, x)$ of the equation*

$$\frac{\partial u}{\partial t} + \frac{1}{2} \sigma^2 \triangle u + \mathbf{b}\left(t, x\right) \cdot \nabla u = 0 \qquad (1.3.21)$$

is given by the solution X_t of the stochastic differential equation

$$X_t = X_a + \sigma B_{t-a} + \int_a^t \mathbf{b}\,(s, X_s)\,ds. \qquad (1.3.22)$$

Conversely, the solution X_t of the above stochastic differential equation is a Markov process and its transition probability density coincides with the fundamental solution $q(s, y, t, x)$.

Theorem 1.3.1 is the main theorem in the theory of random motion treated in this book and plays an important role in the applications discussed in Chapter 2.

Example. For simplicity, assume that the space dimension is one, $\sigma^2(t, x) = 1$ and $b(t, x) = b$ is a constant. Then equation (1.3.21) becomes

$$\frac{\partial u}{\partial t} + \frac{1}{2}\frac{\partial^2 u}{\partial x^2} + b\frac{\partial u}{\partial x} = 0.$$

According to Theorem 1.3.1, the paths X_t of the random motion are given by the formula

$$X_t = X_a + B_{t-a} + b \times (t - a),$$

i.e., X_t is the Brownian motion with drift $b \times (t - a)$, or, in other words, the uniform motion with Brownian noise. In the preceding section we presumed this fact based on Kolmogorov's Theorem 1.2.2.

1.4   Equation of Motion for a Stochastic Process

Let us now consider the equation of motion of an electron (or small particle) subject to external forces.

We first look back at classical mechanics. Newton's equation of motion for a particle of mass m subject to an external force $F(x) = -\operatorname{grad} V(x)$ is

$$m\frac{dv(t)}{dt} = -\operatorname{grad} V(x),$$

where $V(x)$ is the potential function of the force $F(x)$. The potential function governs the evolution of the velocity $v(t)$ of the particle through the equation of motion, i.e., the velocity $v(t)$ is obtained by solving the equation of motion. This is "mechanics".

Once the velocity $v(t)$ is known, the path $x(t)$ of the motion of the particle can be computed by integrating the velocity:

$$x(t) = x_0 + \int_0^t v(t)\,dt.$$

This is a problem of "kinematics".

Historically, Newton invented the equation of motion for "mechanics", which belongs to physics, and invented the differential calculus for "kinematics", which belongs to mathematics.

In this context, Hamilton (Irish mathematician, in the 19th century) undertook the development of Newton's mechanics into two steps, "mechanics" and "kinematics". This formulation is called Hamilton's equation of motion.

Since the paths $X_t(\omega)$ of an electron admit no time derivative at almost of all the moments of time, Newton's mechanics is not applicable to such paths. Hence, we must find another way.

Random motion was discussed in Sections 1.2 and 1.3, but what we used there were Markov processes, which belong to the conventional theory of stochastic processes. In this context we did not consider "external forces".

In this section we will discuss the random motion of electrons (small particles) under external forces. However, this topic does not fit into the formalism of the conventional theory of Markov processes (stochastic processes). This point touches the main theme of Chapter 1.

In fact, according to Theorem 1.2.2, Markov processes are characterized by Kolmogorov's equation (1.2.21), which does not contain terms corresponding to external forces. In the case of the Brownian motion, we have

$$\frac{\partial u}{\partial t} + \frac{1}{2}\sigma^2 \triangle u = 0.$$

Let us add here the potential term $c(x)$, i.e., consider the equation

$$\frac{\partial u}{\partial t} + \frac{1}{2}\sigma^2 \triangle u + c(x)u = 0.$$

One would guess that this equation can be taken as the equation governing the motion of small particles under an external force. However, this is not the case: its fundamental solution cannot be used to construct a stochastic process, because it does not satisfy the normalization condition (1.2.16).

In fact, the motion of an electron with a potential term $c(x)$ is governed by the pair of equations

$$\begin{cases} \dfrac{\partial \phi}{\partial t} + \dfrac{1}{2}\sigma^2 \triangle \phi + c(x)\phi = 0, \\[2mm] -\dfrac{\partial \widehat{\phi}}{\partial t} + \dfrac{1}{2}\sigma^2 \triangle \widehat{\phi} + c(x)\widehat{\phi} = 0, \end{cases}$$

where $\sigma^2 = \frac{\hbar}{m}$, m is the mass of the electron, $\hbar = \frac{1}{2\pi}h$, and h the Planck constant.

More generally, the **"equations of motion of small particles"** are

$$\begin{cases} \dfrac{\partial \phi}{\partial t} + \dfrac{1}{2}\sigma^2(\nabla + \mathbf{b}\,(t,x))^2\phi + c\,(t,x)\,\phi = 0, \\[2mm] -\dfrac{\partial \widehat{\phi}}{\partial t} + \dfrac{1}{2}\sigma^2(\nabla - \mathbf{b}\,(t,x))^2\widehat{\phi} + c\,(t,x)\,\widehat{\phi} = 0; \end{cases} \tag{1.4.1}$$

we call them the **"Schrödinger-Nagasawa equations of motion"**. They were derived by Nagasawa (1993, 2000). In equations (1.4.1) the time parameter t runs in the interval $a \leq t \leq b$, $c(t,x)$ is a scalar function and $\mathbf{b}(t,x)$ is a vector function which appears in the presence of an electromagnetic potential.

The first equation in (1.4.1) is an evolution equation which describes physical phenomena (random motion under external forces) from the starting time a to the terminal time b, while the second equation in (1.4.1) does the same backwards, from the terminal time b to the starting time a. As we will explain, we are now able to construct stochastic processes by using the solution of the equations of motion because we are dealing with a pair, and not a single equation; this is not possible only one of the elements of the pair. The equations of motion (1.4.1) can also be referred to as **"twin (or paired) equations of motion"**.

When we solve the equations (1.4.1), we prescribe an initial function $\widehat{\phi}_a\,(z)$ at the initial time a, as well as a terminal function $\phi_b\,(y)$ at the terminal time b. We will call $\widehat{\phi}_a\,(z)$ **"entrance function"** and call $\phi_b\,(y)$ **"exit function"**.

When a solution of the equations of motion (1.4.1) is represented in terms of a function $p\,(s,x,t,y)$ in the form

$$\begin{cases} \phi\,(t,x) = \displaystyle\int p\,(t,x,b,y)\,\phi_b\,(y)\,dy, \\[2mm] \widehat{\phi}\,(t,x) = \displaystyle\int dz\,\widehat{\phi}_a\,(z)\,p\,(a,z,t,x), \end{cases}$$

$p\,(s,x,t,y)$ is called the **fundamental solution** of the equations of motion (1.4.1).

The solution $\phi\,(t,x)$ of the first equation in (1.4.1) is called **"evolution function"** and the solution $\widehat{\phi}\,(t,x)$ of the second equation is called **"backward evolution function"**.

In classical mechanics the motion is governed by a single equation, Newton's equation. By contrast, to describe the random motion we need a pair of equations. To explain why this is the case, we resort to a modification of Eddington's idea.

Each of the Schrödinger-Nagasawa equations (1.4.1) is the time-reversed form of the other. The distribution function $\mu(t,x)$ at the present time t is given as the product of the information $\widehat{\phi}(t,x)$ obtained by prescribing the entrance function $\widehat{\phi}_a\,(z)$ and the information $\phi(t,x)$ obtained by prescribing the exit function $\phi_b\,(y)$, i.e., $\mu(t,x) = \phi(t,x)\widehat{\phi}(t,x)$.

At the starting time a we begin an experiment by giving data in the form of an entrance function $\widehat{\phi}_a$ and at the terminal time b we specify an exit function ϕ_b. The theory predicts the motion of a particle in the time interval $a < t < b$.

Therefore, in our approach the motion is governed by a pair of equations. *The fact that we can choose the exit function ϕ_b besides the entrance function is the most fundamental advantage of our theory.* This was not known in the conventional theory of stochastic processes or in quantum mechanics.

As a matter of fact, it took a long time to reach a pair of equations with natural time flow and with the backward time flow. The history of this development was outlined in Section 1.1. By the way, if we look at one of the equations of motion (1.4.1) we recognize that is simply the Kolmogorov equation with a potential term added. However, a single equation is not enough to construct a stochastic process: we absolutely need the pair.

Let us explain the reason for this. To see the influence of the external forces on the random motion of an electron, we must first of all show that we can construct a stochastic process $\{X_t(\omega) ; t \in [a, b], \mathbf{Q}\}$ with the solution of the equations of motion (1.4.1). We will do it, but we cannot follows Kolmogorov's recipe that was presented in Section 1.2, because we cannot use equation (1.2.9) of finite-dimensional distributions.

More precisely, the fundamental solution $p(s, x, t, y)$ of the equations of motion (1.4.1) does not satisfy the normalization condition (1.2.16), that is,

$$\int p(t, x, b, y) dy \neq 1.$$

Therefore, we cannot use equation (1.2.9). This means that we cannot apply the theory of Markov processes discussed in the preceding section to the random motion determined by our equations of motion (1.4.1). This is a decisive point. A new way was suggested by Schrödinger (1931).

We will build "**a way to obtain stochastic processes from the equations of motion (1.4.1)**", which does not rely on the theory of Markov processes.

To this end, we need a **new normalization condition** for the fundamental solution $p(s, x, t, y)$ of the Schrödinger-Nagasawa equations of motion (1.4.1).

The triplet normalization condition:

If the fundamental solution $p(s, x, t, y)$ and a pair consisting of an entrance function $\widehat{\phi}_a(x)$ and exit function $\phi_b(y)$ satisfies

$$\int dx\, \widehat{\phi}_a(x)\, p(a, x, b, y) \phi_b(y)\, dy = 1, \tag{1.4.2}$$

we call $\{p(s, x, t, y), \widehat{\phi}_a(x), \phi_b(y)\}$ a **triplet**, and call equation (1.4.2) **the triplet normalization condition**.

Condition (1.4.2) says that "the probability that a particle starts at time a with an entrance function $\widehat{\phi}_a(x)$ and ends up at time b with an exit function $\phi_b(y)$ is equal to one".

Based on this observation, we extend (1.4.2) to cylinders

$$\{\omega : \omega(t_1) \in \Gamma_1, \omega(t_2) \in \Gamma_2, \ldots, \omega(t_n) \in \Gamma_n\},$$

where $a < t_1 < t_2 < \cdots < t_n < b$. The probability of this sample set is defined in terms of the triplet $\{p(s, x, t, y), \widehat{\phi}_a(x), \phi_b(y)\}$ by

$$\mathbf{Q}[\{\omega : \omega(t_1) \in \Gamma_1, \omega(t_2) \in \Gamma_2, \ldots, \omega(t_n) \in \Gamma_n\}]$$
$$= \int dx_0\, \widehat{\phi}_a(x_0) p(a, x_0, t_1, x_1) \mathbf{1}_{\Gamma_1}(x_1) dx_1\, p(t_1, x_1, t_2, x_2) \mathbf{1}_{\Gamma_2}(x_2) dx_2 \times \cdots$$
$$\times p(t_{n-1}, x_{n-1}, t_n, x_n) \mathbf{1}_{\Gamma_n}(x_n) dx_n p(t_n, x_n, b, y) \phi_b(y) dy; \qquad (1.4.3)$$

thus, it depends on both an exit function $\phi_b(y)$ and an entrance function $\widehat{\phi}_a(x_0)$. In our theory, equation (1.4.3) replaces the classical Kolmogorov equation (1.2.9), which does not contain the exit function $\phi_b(y)$. We note that equation (1.4.3) is invariant under time reversal.

Then the distribution of $\omega(t)$ is

$$\mathbf{Q}[\{\omega : \omega(t) \in \Gamma\}] = \int dx_0\, \widehat{\phi}_a(x_0) p(a, x_0, t, x_1) \mathbf{1}_{\Gamma}(x_1) dx_1\, p(t, x_1, b, y) \phi_b(y) dy,$$

which depends on the exit function $\phi_b(y)$ at the terminal time b in addition to the entrance function $\widehat{\phi}_a(x)$ at the entrance time a.

Moreover, the set function $\mathbf{Q}$ defined by (1.4.3) for cylinders

$$\{\omega : \omega(t_1) \in \Gamma_1, \omega(t_2) \in \Gamma_2, \ldots, \omega(t_n) \in \Gamma_n\} \subset \Omega$$

can be extended to events in the smallest σ-algebra $\mathcal{F}$ containing such subsets, and if we set $X_t(\omega) = \omega(t)$, we obtain a stochastic process $\{X_t(\omega) ; t \in [a, b], \mathbf{Q}\}$.

If we compare formula (1.4.3) with formula (1.2.9) of Kolmogorov's approach, we see that equation (1.4.3) has the remarkable feature that the random motion starts at time a with an entrance function $\widehat{\phi}_a(x)$ and *ends at the terminal time b with an exit function $\phi_b(y)$.* **Equation (1.4.3) represents a distinctive and powerful characteristic of our mechanics of random motion (new quantum theory).**

We established the equations of motion (1.4.1), which is physics, and we adopted the theory of stochastic process, which is mathematics. In other words, the theory of stochastic process serves as kinematics for the mechanics of random motion.

Thus, we constructed a stochastic process $\{X_t(\omega); t \in [a, b], \mathbf{Q}\}$ determined by the equations of motion (1.4.1), as promised. Presumably, if we investigate the process $X_t(\omega)$, we will be able to understand the random motion under external forces.

However, this is too optimistic. In classical mechanics, if we solve the equation of motion, we obtain the velocity. By contrast, in the case of our random motion, even if we solve the equations of motion, we do not get a useful physical quantity such as the "velocity". It seems that we have not yet extracted from the solution of the Schrödinger-Nagasawa equations of motion (1.4.1) all the necessary information. To extract more information from the solution pair $(\phi(t,x), \widehat{\phi}(t,x)$, we the following new notion.

Definition 1.4.1. With the evolution function $\phi(t,x)$ and the backward evolution function $\widehat{\phi}(t,x)$ we define

$$R(t,x) = \frac{1}{2} \log \phi(t,x)\, \widehat{\phi}(t,x), \tag{1.4.4}$$

and

$$S(t,x) = \frac{1}{2} \log \frac{\phi(t,x)}{\widehat{\phi}(t,x)}, \tag{1.4.5}$$

and call $R(t,x)$ the **generating function of distribution** and $S(t,x)$ the **drift potential**. Moreover, we call the pair $\{R(t,x), S(t,x)\}$ the **generating function** of our random motion.

The "drift potential" $S(t,x)$ **is a new notion in quantum theory, and also in the theory of stochastic processes.**

In fact, we did not see such a function $S(t,x)$ in the analysis of the Brownian motion carried out in the preceding section. The drift potential $S(t,x)$ is a physical quantity which is intrinsic to our stochastic process $\{X_t(\omega); t \in [a,b], \mathbf{Q}\}$. The random motion $X_t(\omega)$ "carries" the drift potential $S(t,x)$. This is a decisive feature, different from the conventional Brownian motion.

The **generating function of distribution** $R(t,x)$ and the **drift potential** $S(t,x)$ are fundamental physical quantities of random motion. We will dwell upon this in more detail.

First of all, if we use the generating function $\{R(t,x), S(t,x)\}$, we can represent the evolution function $\phi(t,x)$ and the backward evolution function $\widehat{\phi}(t,x)$ as exponential functions.

Definition 1.4.2. The evolution functions can be represented in terms of the generating function $\{R(t,x), S(t,x)\}$ as

$$\phi(t,x) = e^{R(t,x)+S(t,x)}, \quad \widehat{\phi}(t,x) = e^{R(t,x)-S(t,x)}. \tag{1.4.6}$$

These expressions will be called the **exponential form** of the evolution functions.

The "exponential form" is absolutely necessary for developing our theory, and plays an important role in what follows.

Next, we use the pair $\{R(t,x), S(t,x)\}$ to define a physical quantity which corresponds to the velocity in the classical mechanics.

Definition 1.4.3.

(i) The vector-function

$$\mathbf{a}(t,x) = \sigma^2 \nabla \log \phi(t,x) = \sigma^2(\nabla R(t,x) + \nabla S(t,x)) \qquad (1.4.7)$$

is called the **evolution drift**.

(ii) The vector-function

$$\widehat{\mathbf{a}}(t,x) = \sigma^2 \nabla \log \widehat{\phi}(t,x) = \sigma^2(\nabla R(t,x) - \nabla S(t,x))$$

is called the **backward evolution drift**.

The **evolution drift** $\mathbf{a}(t,x) = \sigma^2(\nabla R(t,x) + \nabla S(t,x))$ is one of the most important notions in the mechanics of random motion. It is a generalization of the concept of velocity in classical mechanics; it is a new concept, not known in the conventional theory of stochastic processes. The backward evolution drift will be needed together with the evolution drift to define the kinetic energy of random motion.

We must emphasize that the physical quantity which characterizes our stochastic process $X_t(\omega)$ is the drift potential $S(t,x)$ which figures in the evolution function $\phi(t,x) = e^{R(t,x)+S(t,x)}$. In the conventional theory of stochastic processes presented in the preceding section, the drift potential does not exist. This is an important point that we need to re-emphasize:

The stochastic process $X_t(\omega)$ is endowed with the drift potential $S(t,x)$.

The drift potential $S(t,x)$ generates the evolution drift $\mathbf{a}(t,x) = \sigma^2(\nabla R(t,x) + \nabla S(t,x))$. Summarizing, we state

Theorem 1.4.1. *Let $p(s,x,t,y)$ be the fundamental solution of the equations of motion (1.4.1). Then the triplet $\{p(s,x,t,y), \widehat{\phi}_a(x), \phi_b(y)\}$ which satisfies the triplet normalization condition (1.4.2) determines a stochastic process $\{X_t(\omega); t \in [a,b], \mathbf{Q}\}$. The stochastic process $X_t(\omega)$ evolves carrying the drift potential $S(t,x)$. Together with the generating function $R(t,x)$, the drift potential $S(t,x)$ defines the evolution drift vector*

$$\mathbf{a}(t,x) = \sigma^2(\nabla R(t,x) + \nabla S(t,x)).$$

The evolution drift vector $\mathbf{a}(t,x)$ determines the random motion of an electron under external forces.

In the mechanics of random motion, solving the equations of motion yields the evolution drift vector. In classical mechanics, we get the velocity. Then **"kinematics"** takes over the whole business. We will develop the kinematics of random motion in the following section. Moreover, we note that the random motion of a particle with mass has a momentum and a kinetic energy, which will be discussed in Chapter 3.

1.5 Kinematics of Random Motion

In the preceding section we saw that external forces determine the evolution drift vector $\mathbf{a}(t, x)$ through the equation of motion (1.4.1). Here we will discuss the kinematics of random motion. We will show that in our kinematics the random motion determined by the equation of motion is a Markov process with the evolution drift vector $\mathbf{a}(t, x)$. This is the main theme of this section.

We let us examine briefly a simple case. The Schrödinger-Nagasawa equations of motion

$$
\begin{cases}
\dfrac{\partial \phi}{\partial t} + \dfrac{1}{2}\sigma^2 \triangle \phi + c(x)\phi = 0, \\
-\dfrac{\partial \widehat{\phi}}{\partial t} + \dfrac{1}{2}\sigma^2 \triangle \widehat{\phi} + c(x)\widehat{\phi} = 0
\end{cases}
$$

determine a stochastic process $\{X_t(\omega); t \in [a, b], \mathbf{Q}\}$. We will show that

$$
\{X_t(\omega); t \in [a, b], \mathbf{Q}\}
$$

is a Markov process governed by the evolution equation

$$
\frac{\partial u}{\partial s} + \frac{1}{2}\sigma^2 \triangle u + \mathbf{a}(s, x) \cdot \nabla u = 0,
$$

where $\mathbf{a}(t, x) = \sigma^2(\nabla R(t, x) + \nabla S(t, x))$ is the evolution drift vector. Moreover, our random motion can be expressed as a solution X_t of the stochastic differential equation

$$
X_t = X_a + \sigma B_{t-a} + \int_a^t \mathbf{a}(s, X_s)\, ds.
$$

In short, the external forces $c(x)$ induce an evolution drift $\mathbf{a}(t, x)$ of the random motion X_t. This is what we would like to show; it represents our concept of **kinematics**.

We will proceed step by step and explain **the kinematic equation and the path equation in the setting of the new quantum theory** for the general Schrödinger-Nagasawa equations of motion (1.4.1).

According to the definition (1.4.3) of the probability measure $\mathbf{Q}$, the distribution of the random motion $X_t(\omega)$ is

$$
\mathbf{Q}[\{\omega : X_t(\omega) \in \Gamma\}] = \int dx_0\, \widehat{\phi}_a(x_0)p(a, x_0, t, x_1)\mathbf{1}_\Gamma(x_1)dx_1\, p(t, x_1, b, y)\phi_b(y)dy.
$$

It is not plausible that the random motion $\{X_t(\omega); t \in [a, b], \mathbf{Q}\}$ enjoys the Markov property, because it involves the exit function $\phi_b(y)$. The distribution of a Markov process depends on the initial distribution, but should not depend on any information from the future, so for the Markov property the presence of the exit function $\phi_b(y)$ is problematic.

Nevertheless, we will prove that the motion $\{X_t(\omega); t \in [a, b], \mathbf{Q}\}$ is a Markov process. To this end we first define

$$q(a, x_0, t, x_1) = \frac{1}{\phi(a, x_0)} p(a, x_0, t, x_1)\phi(t, x_1).$$

This function is the transition probability density of the stochastic process $X_t(\omega)$:

$$\mathbf{Q}[\{\omega : X_t(\omega) \in \Gamma\}]$$
$$= \int dx_0 \, \widehat{\phi}_a(x_0)\phi(a, x_0)\frac{1}{\phi(a, x_0)} p(a, x_0, t, x_1)\phi(t, x_1)dx_1 \, \mathbf{1}_\Gamma(x_1)$$
$$= \int dx_0 \, \widehat{\phi}_a(x_0)\phi(a, x_0)q(a, x_0, t, x_1)dx_1 \, \mathbf{1}_\Gamma(x_1).$$

Theorem 1.5.1 (Transition probability density). *Let the function $q(s, x, t, y)$ be defined in terms of the fundamental solution $p(s, x, t, y)$ of the equations of motion (1.4.1) and the evolution function $\phi(t, x)$ by the formula*

$$q(s, x, t, y) = \frac{1}{\phi(s, x)} p(s, x, t, y)\,\phi(t, y). \tag{1.5.1}$$

Then $q(s, x, t, y)$ satisfies the conditions (1.2.15) and (1.2.16), i.e., it is a **transition probability density**.

Proof. For $s < r < t$, we have

$$\int_{\mathbf{R}^d} q(s, x, r, z)dz \, q(r, z, t, y)$$
$$= \int_{\mathbf{R}^d} \frac{1}{\phi(s, x)} p(s, x, r, z)\phi(r, z)dz \frac{1}{\phi(r, z)} p(r, z, t, y)\phi(t, y)$$
$$= \int_{\mathbf{R}^d} \frac{1}{\phi(s, x)} p(s, x, r, z)dz \, p(r, z, t, y)\phi(t, y),$$

which due to the fact that $p(s, x, r, z)$ satisfies the Chapman-Kolmogorov equation is equal to

$$\frac{1}{\phi(s, x)} p(s, x, t, y)\phi(t, y) = q(s, x, t, y).$$

That is, the function $q(s, x, t, y)$ satisfies the Chapman-Kolmogorov equation. Moreover,

$$\int_{\mathbf{R}^d} q(s, x, t, y)dy = \frac{1}{\phi(s, x)} \int_{\mathbf{R}^d} p(s, x, t, y)\phi(t, y)dy,$$

where $\phi(s, x) = \int_{\mathbf{R}^d} p(s, x, t, y)\phi(t, y)dy$. Hence,

$$\int_{\mathbf{R}^d} q(s, x, t, y)dy = \frac{1}{\phi(s, x)}\phi(s, x) = 1,$$

i.e., $q(s, x, t, y)$ satisfies the normalization condition.   □

Let $p(s, x, t, y)$ be the fundamental solution of the equations of motion (1.4.1) and consider the stochastic process $\{X_t(\omega); t \in [a, b], \mathbf{Q}\}$ defined by formula (1.4.3) with the triplet $\{p(s, x, t, y), \widehat{\phi}_a(x), \phi_b(y)\}$. We will prove that $\{X_t(\omega); t \in [a, b], \mathbf{Q}\}$ is a Markov process with the transition probability $q(s, x, t, y)dy$ defined by (1.5.1) (Be careful! Not $p(s, x, t, y)dy$).

Theorem 1.5.2 (Markov property of $\{X_t(\omega); t \in [a, b], \mathbf{Q}\}$). *Let $p(s, x, t, y)$ be the fundamental solution of the equations of motion (1.4.1) and $\{X_t(\omega); t \in [a, b], \mathbf{Q}\}$ be the stochastic process defined by (1.4.3) with the triplet*

$$\{p(s, x, t, y), \widehat{\phi}_a(x), \phi_b(y)\},$$

which satisfies the triplet normalization condition (1.4.2).

Then $\{X_t(\omega); t \in [a, b], \mathbf{Q}\}$ is a Markov process with the initial distribution density

$$\mu_a(x) = \phi_a(x)\widehat{\phi}_a(x), \quad where \; \phi_a(x) = \int p(a, x, b, y)\phi_b(y)dy,$$

and with the transition probability density $q(s, x, t, y)$ defined by formula (1.5.1).

Proof. We already know that, for a given $\{p(s, x, t, y), \widehat{\phi}_a(x), \phi_b(y)\}$, the random motion $\{X_t(\omega); t \in [a, b], \mathbf{Q}\}$ exists and its finite-dimensional distribution (1.4.3) is given by

$$\begin{aligned}
\mathbf{Q}&\left[f(X_{t_1}, \ldots, X_{t_n})\right] \\
&= \int dx_0\, \widehat{\phi}_a(x_0)\, p(a, x_0, t_1, x_1)\, dx_1\, p(t_1, x_1, t_2, x_2)\, dx_2 \times \cdots \qquad (1.5.2) \\
&\quad \times p(t_n, x_n, b, y)\, \phi_b(y)dy\, f(x_1, \ldots, x_n).
\end{aligned}$$

for functions $f(x_1, \ldots, x_n)$ and $a < t_1 < t_2 < \cdots < t_n < b$.

To prove the Markov property of $\{X_t(\omega); t \in [a, b], \mathbf{Q}\}$, we apply a technique due to Doob. As we will see, this allows us to "hide" the dependence on the exit function $\phi_b(y)$ at the terminal time $b(> a)$.

Denote the evolution function $\phi(t, x)$ by

$$\phi_t(x) = \int p(t, x, b, y)\phi_b(y)dy.$$

Using $\phi_t(x)$, we rewrite the right-hand side of (1.5.2) as follows. At each time t_i, we multiply and divide by $\phi_{t_i}(x_i)$. Then

$$\mathbf{Q}\left[f(X_{t_1}, \cdots, X_{t_n})\right]$$
$$= \int dx_0\, \phi_a(x_0)\widehat{\phi}_a(x_0)\frac{1}{\phi_a(x_0)}p\,(a, x_0, t_1, x_1)\,\phi_{t_1}(x_1)dx_1\,\frac{1}{\phi_{t_1}(x_1)}$$
$$\times\, p\,(t_1, x_1, t_2, x_2)\,\phi_{t_2}(x_2)dx_2\,\frac{1}{\phi_{t_2}(x_2)}\,\times \cdots$$
$$\times\, \phi_{t_n}(x_n)dx_n\frac{1}{\phi_{t_n}(x_n)}\,p(t_n, x_n, b, y)\phi_b(y)dy\, f(x_1, \ldots, x_n).$$

This technique, first used by Doob (1953), is called Doob's super-harmonic transformation.

Rewriting the last equation with $q\,(s, x, t, y)$ defined in (1.5.1), we have

$$\mathbf{Q}\left[f(X_{t_1}, \cdots, X_{t_n})\right]$$
$$= \int dx_0\, \phi_a\,(x_0)\,\widehat{\phi}_a\,(x_0)\,q\,(a, x_0, t_1, x_1)\,dx_1\,q\,(t_1, x_1, t_2, x_2)\,dx_2\,\times \cdots$$
$$\times\, dx_{n-1}\,q(t_{n-1}, x_{n-1}, t_n, x_n)dx_n q\,(t_n, x_n, b, y)\,dy f\,(x_1, \ldots, x_n),$$

where $\int q\,(t_n, x_n, b, y)\,dy = 1$. Setting here $\mu_a(x) = \phi_a(x)\widehat{\phi}_a(x)$, we conclude that

$$\mathbf{Q}\left[f\,(X_{t_1}, \ldots, X_{t_n})\right]$$
$$= \int dx_0\, \mu_a\,(x_0)\,q\,(a, x_0, t_1, x_1)\,dx_1\,\times\,q\,(t_1, x_1, t_2, x_2)\,dx_2\,\times \cdots \qquad (1.5.3)$$
$$\times\, dx_{n-1}\,q(t_{n-1}, x_{n-1}, t_n, x_n)dx_n f(x_1, \ldots, x_n).$$

Thus, if we set $Q(s, x, t, dy) = q(s, x, t, y)dy$, this is nothing but the defining equation (1.2.12) of the Markov property. Therefore, the stochastic process $\{X_t(\omega),$ $t \in [a, b], \mathbf{Q}\}$ is a Markov process with the transition probability $Q(s, x, t, dy) = q(s, x, t, y)dy$ and the initial distribution density $\mu_a(x) = \phi_a(x)\widehat{\phi}_a(x)$, as claimed.
$\square$

Theorem 1.5.3 (Kinematics of random motion). *Let $p\,(s, x, t, y)$ be the fundamental solution of the equations of motion* (1.4.1). *Then:*

(i) *The transition probability density*

$$q(s, x, t, y) = \frac{1}{\phi\,(s, x)}p\,(s, x, t, y)\,\phi\,(t, y)$$

defined by formula (1.5.1) *is the fundamental solution of the parabolic partial differential equation*

$$\frac{\partial u}{\partial s} + \frac{1}{2}\sigma^2\triangle u + \left(\sigma^2\mathbf{b}(x, s) + \mathbf{a}\,(s, x)\right)\cdot\nabla u = 0, \qquad (1.5.4)$$

where

$$\mathbf{a}(t, x) = \sigma^2\nabla\log\phi(t, x) = \sigma^2(\nabla R(t, x) + \nabla S(t, x))$$

is the evolution drift vector determined by the evolution function

$$\phi(t,x) = e^{R(t,x)+S(t,x)}.$$

We call equation (1.5.4) the **"kinematics equation"**.

(ii) Moreover, $q(s,x,t,y)$ satisfies, as a function of the variables (t,y), the equation

$$-\frac{\partial \mu}{\partial t} + \frac{1}{2}\sigma^2 \triangle \mu - \nabla \cdot \left(\left(\sigma^2 \mathbf{b}(t,y) + \mathbf{a}(t,y)\right)\mu\right) = 0. \tag{1.5.4*}$$

Remark. We can compute the motion of small particles, for example, the paths (trajectories) of the motion of an electron in a hydrogen atom, by solving the kinematics equation (1.5.4). This was not possible in quantum mechanics. Details will be provided in the following chapters.

Proof of Theorem 1.5.3. We need a preliminary lemma.

Lemma 1.5.1 (Nagasawa (1989, a)). *Let $\phi(s,x)$ and $p(s,x)$ be differentiable functions, and let*

$$u(s,x) = \frac{p(s,x)}{\phi(s,x)}.$$

Then the function $u(s,x)$ satisfies the equation

$$Lu + \sigma^2 \frac{1}{\phi}\,\nabla\phi \cdot \nabla u = \frac{1}{\phi}\left(Lp + \overline{c}(s,x)\,p\right) - \frac{u}{\phi}\left(L\phi + \overline{c}(s,x)\,\phi\right), \tag{1.5.5}$$

where $\overline{c}(s,x)$ is an arbitrary function and L denotes the differential operator

$$L = \frac{\partial}{\partial s} + \frac{1}{2}\sigma^2 \triangle + \sigma^2 \mathbf{b}(s,x) \cdot \nabla.$$

Proof. For simplicity, we consider the one-dimensional case. Then we have

$$\left(\frac{p}{\phi}\right)' = \frac{p'}{\phi} - \frac{p\phi'}{\phi^2} \quad \text{and} \quad \left(\frac{p}{\phi}\right)'' = \frac{p''}{\phi} - \frac{p\phi''}{\phi^2} + 2\left(\frac{p(\phi')^2}{\phi^3} - \frac{p'\phi'}{\phi^2}\right).$$

Consequently,

$$Lu = L\left(\frac{p}{\phi}\right) = \frac{1}{\phi}\left(\frac{\partial p}{\partial s} + \frac{1}{2}\sigma^2 p'' + \sigma^2 bp'\right) - \frac{p}{\phi^2}\left(\frac{\partial \phi}{\partial s} + \frac{1}{2}\sigma^2 \phi'' + \sigma^2 b\phi'\right)$$
$$+ \sigma^2\left(\frac{p}{\phi^3}(\phi')^2 - \frac{1}{\phi^2}p'\phi'\right),$$

where

$$\sigma^2\left(\frac{p}{\phi^3}(\phi')^2 - \frac{1}{\phi^2}p'\phi'\right) = -\sigma^2 \frac{\phi'}{\phi}\left(\frac{p}{\phi}\right)' = -\sigma^2 \frac{\phi'}{\phi}(u)'.$$

Therefore,

$$Lu + \sigma^2 \frac{\phi'}{\phi}(u)' = \frac{1}{\phi}Lp - \frac{p}{\phi^2}L\phi.$$

Let us subtract and add $\frac{1}{\phi}\overline{c}p$ on the right-hand side. Then, since $u = \frac{p}{\phi}$, we have

$$Lu + \sigma^2 \frac{\phi'}{\phi}(u)' = \frac{1}{\phi}(Lp + \overline{c}p) - \frac{u}{\phi}(L\phi + \overline{c}\phi),$$

i.e., we obtained equation (1.5.5). □

Next, we rewrite the equations of motion (1.4.1) in expanded form as

$$\begin{cases} \dfrac{\partial \phi}{\partial t} + \dfrac{1}{2}\sigma^2 \triangle \phi + \sigma^2 \mathbf{b}(t,x) \cdot \nabla \phi + \overline{c}(t,x)\phi = 0, \\ -\dfrac{\partial \widehat{\phi}}{\partial t} + \dfrac{1}{2}\sigma^2 \triangle \widehat{\phi} - \sigma^2 \mathbf{b}(t,x) \cdot \nabla \widehat{\phi} + \widehat{c}(t,x)\widehat{\phi} = 0, \end{cases} \quad (1.5.6)$$

where

$$\overline{c}(t,x) = c(t,x) + \frac{1}{2}\sigma^2 \left(\nabla \cdot \mathbf{b}(t,x) + \mathbf{b}(t,x)^2\right),$$

and

$$\widehat{c}(t,x) = c(t,x) + \frac{1}{2}\sigma^2 \left(-\nabla \cdot \mathbf{b}(t,x) + \mathbf{b}(t,x)^2\right).$$

The terms containing $\mathbf{b}(t,x)$ on the right-hand side are correction terms.

Let $\phi(s,x)$ be the evolution function satisfying (1.5.6). Then

$$L\phi + \overline{c}(s,x)\phi = 0.$$

Therefore, the second term on the right-hand side of equation (1.5.5) vanishes. If we set

$$p(s,x) = p(s,x,t,y)\phi(t,y)$$

where $p(s,x,t,y)$ is the fundamental solution of equation (1.5.6), then

$$Lp + \overline{c}(s,x)p = 0.$$

Therefore, the first term on the right-hand side of equation (1.5.5) also vanishes. Moreover, by (1.5.1),

$$u(s,x) = \frac{p(s,x)}{\phi(s,x)} = q(s,x,t,y).$$

Thus, by (1.5.5), as a function of (s,x),

$$Lq + \sigma^2 \frac{1}{\phi}\nabla\phi \cdot \nabla q = 0.$$

We conclude that the function $q(s, x, t, y)$ defined by formula (1.5.1) satisfies the equation (1.5.4) with the evolution drift vector

$$\mathbf{a}(s, x) = \sigma^2 \frac{\nabla \phi(s, x)}{\phi(s, x)}.$$

This completes the proof of the first assertion of Theorem 1.5.3.

To prove equation (1.5.4*), we will use the following lemma.

Lemma 1.5.2. *Let $\phi(t, y)$ and $p(t, y)$ be differentiable functions and let*

$$\mu(t, y) = p(t, y) \phi(t, y).$$

Then $\mu(t, y)$ satisfies the equation

$$L^0 \mu - \nabla \cdot \left[\left(\sigma^2 \frac{1}{\phi} \nabla \phi \right) \mu \right] = -p \left(L\phi + \bar{c}(t, y) \phi \right) + \phi \left(L^0 p + \bar{c}(t, y) p \right), \quad (1.5.7)$$

where $\bar{c}(t, y)$ is given by the formula below equations (1.5.6) and the differential operator L is the same as in (1.5.5), but in the variables (t, y) instead of (s, x). The differential operator L^0 is given by

$$L^0 v = -\frac{\partial v}{\partial t} + \frac{1}{2} \sigma^2 \triangle v - \nabla(\sigma^2 \mathbf{b}(t, y) v).$$

Proof. For simplicity, we consider again the one-dimensional case. Then we have

$$\begin{aligned}
L^0 \mu = L^0(p\phi) &= -p \frac{\partial \phi}{\partial t} + \frac{1}{2} \sigma^2 p \phi'' - \sigma^2 b \phi' p \\
&\quad - \phi \frac{\partial p}{\partial t} + \frac{1}{2} \sigma^2 \phi p'' - \sigma^2 b \phi p' - (\sigma^2 b)' \phi p + \sigma^2 p' \phi' \\
&= -p \left(\frac{\partial \phi}{\partial t} + \frac{1}{2} \sigma^2 \phi'' + \sigma^2 b \phi' \right) \\
&\quad + \phi \left(-\frac{\partial p}{\partial t} + \frac{1}{2} \sigma^2 p'' - \sigma^2 b p' - (\sigma^2 b)' p \right) + \sigma^2 p' \phi' + \sigma^2 p \phi'',
\end{aligned}$$

where we added the correction term $\sigma^2 p \phi''$, because we changed the sign in the first parentheses. Moreover,

$$\sigma^2 p' \phi' + \sigma^2 p \phi'' = \left[\left(\sigma^2 \frac{1}{\phi} \phi' \right) \mu \right]'.$$

Therefore,

$$L^0 \mu = -p(L\phi) + \phi(L^0 p) + \left[\left(\sigma^2 \frac{1}{\phi} \phi' \right) \mu \right]',$$

and hence

$$L^0 \mu - \left[\left(\sigma^2 \frac{1}{\phi} \phi' \right) \mu \right]' = -p(L\phi + \bar{c}\phi) + \phi(L^0 p + \bar{c}p),$$

which is equation (1.5.7). $\qquad \square$

Now let us prove assertion (ii) of the theorem. Let $\phi(t, y)$ the evolution function. Then

$$L\phi + \bar{c}(t, y)\,\phi = 0.$$

Therefore, the first term on the right-hand side of equation (1.5.7) vanishes. If $p(s, x, t, y)$ is the fundamental solution of the equations of motion (1.5.6), then as a function of (t, y), it satisfies the space-time dual equation

$$L^0 p + \bar{c}(t, y)\,p = 0.$$

Hence, the second term on the right-hand side of (1.5.7) also vanishes. Therefore, by (1.5.7), the function $\mu(t, y) = p(t, y)\,\phi(t, y)$ satisfies

$$L^0 \mu - \nabla \cdot \left[\left(\sigma^2 \frac{1}{\phi}\,\nabla\,\phi\right)\mu\right] = 0,$$

i.e., equation (1.5.4*). Thus, the function $q(s, x, t, y)$ defined by formula (1.5.1) satisfies the equation (1.5.4*) as a function of the variables (t, y), as claimed. $\square$

One of the most important tasks in mechanics is **to compute the paths of the motion under external forces and to obtain solutions**. Quantum mechanics could not deal with this task, because it lacks kinematics.

In our new quantum theory, i.e., the mechanics of random motion, we can compute paths of the motion of a particle under external forces and analyze these paths (trajectories) rigorously. Namely, the following theorem holds.

Theorem 1.5.4 (Path Equation for Random Motion).

(i) *The paths $X_t(\omega)$ of the stochastic process $\{X_t(\omega); t \in [a, b], \mathbf{Q}\}$ determined by solutions $\phi(t, x) = e^{R(t,x)+S(t,x)}$, $\widehat{\phi}(t, x) = e^{R(t,x)-S(t,x)}$ of the Schrödinger-Nagasawa equations (1.4.1) are described by the stochastic differential equation*

$$X_t = X_a + \sigma B_{t-a} + \int_a^t \left(\sigma^2 \mathbf{b}(s, X_s) + \mathbf{a}(s, X_s)\right) ds, \tag{1.5.8}$$

with the evolution drift $\mathbf{a}(t, x) = \sigma^2(\nabla R(t, x) + \nabla S(t, x))$, where B_t is the Brownian motion and the starting point X_a is a random variable that is independent of B_t.

(ii) *The distribution density $\mu_t(x)$ of the stochastic process X_t is given by*

$$\mu_t(x) = \phi(t, x)\,\widehat{\phi}(t, x) = e^{2R(t,x)}, \quad t \in [a, b].$$

Equation (1.5.8) will be referred to as the **"path equation"**.

The path equation (1.5.8) is equivalent to the kinematics equation (1.5.4). This was shown in Theorem 1.3.1, as the equivalence of the equations (1.3.21) and (1.3.22).

Theorem 1.5.3 (on the kinematics equation) and Theorem 1.5.4 (on the path equation) are **fundamental theorems of kinematics in the new quantum theory, that is, in the mechanics of random motion** (Nagasawa (1989, a)).

Remark. When we apply Theorem 1.5.4 in different coordinate systems, the kinematics equation takes on the form of equation (1.3.20) in Theorem 1.3.1. In this case we can state

Theorem 1.5.5 (General coordinate systems). *The* **kinematics equation** *depends on the adopted system of coordinates* $(x^1, \ldots, x^d)$ *and has the form*

$$\frac{\partial u}{\partial t} + \frac{1}{2} \sum_{i,j=1}^{d} \sigma^2(t,x)^{ij} \frac{\partial^2 u}{\partial x^i \partial x^j} + \sum_{i=1}^{d} \mathbf{b}(t,x)^i \frac{\partial u}{\partial x^i} = 0, \qquad (1.5.9)$$

where

$$\sigma^2(t,x)^{ij} = \sum_{k=1}^{d} \sigma(t,x)_k^i \, \sigma(t,x)_k^j, \quad \sigma(t,x) = \{\sigma(t,x)_j^i\},$$

$$i,j = 1,2,\ldots,d.$$

The coordinate form of the corresponding **path equation**

$$X_t = X_a + \int_a^t \sigma(r, X_r) \, dB_{r-a} + \int_a^t \mathbf{b}(r, X_r) dr \qquad (1.5.10)$$

is

$$X_t{}^i = X_a{}^i + \int_a^t \sum_{j=1}^{d} \sigma(r, X_r)_j^i dB_{r-a}^j + \int_a^t \mathbf{b}(r, X_r)^i dr, \qquad (1.5.11)$$

$$i = 1,2,\ldots,d.$$

Thus, since the kinematics equation (1.5.9) depends on the coordinate system adopted, to get the path equation (1.5.10), (1.5.11), we must first fix the coordinate system.

Let us summarize the foregoing discussion.

We showed that our theory involves two steps. In the first step we introduced the equations of motion (1.4.1), and we used a triplet $\{p(s,x,t,y), \hat{\phi}_a(x), \phi_b(y)\}$ to construct a stochastic process. Here we should emphasize that:

The exit function $\phi_b(y)$ is prepared and settled. It depends on what we count at the end of the experiment.

An experiment which allows a good understanding of how to settle the exit function will be provided in Section 1.13.

Then we showed that the evolution function $\phi(t,x)$ yields the evolution drift vector $\mathbf{a}(t,x) = \sigma^2 \nabla \log \phi(t,x) = \sigma^2(\nabla R(t,x) + \nabla S(t,x))$, in much the same way as solving the equation of motion in classical mechanics yields the velocity.

In the second step, the kinematics equation (1.5.4) is defined by means of the evolution drift vector $\mathbf{a}(t,x) = \sigma^2(\nabla R(t,x) + \nabla S(t,x))$. The path $X_t(\omega)$ can be computed by the path equation (1.5.8) in Theorem 1.5.4. Simple examples will be provided in the following sections.

1.6   Free Random Motion of a Particle

As a simple application of the mechanics of random motion, we consider the free random motion of a particle on the line. Since "free motion" means absence of external forces, the equations of motion are

$$\begin{cases} \dfrac{\partial \phi}{\partial t} + \dfrac{1}{2}\sigma^2\dfrac{\partial^2 \phi}{\partial x^2} = 0, \\[2mm] -\dfrac{\partial \widehat{\phi}}{\partial t} + \dfrac{1}{2}\sigma^2\dfrac{\partial^2 \widehat{\phi}}{\partial x^2} = 0. \end{cases}$$

Their fundamental solution is given by

$$p\left(s,x,t,y\right) = \frac{1}{\sqrt{2\pi\sigma^2\left(t-s\right)}}\exp\left(-\frac{\left(y-x\right)^2}{2\sigma^2\left(t-s\right)}\right).$$

This is the transition probability density for the Einstein-Smoluchowski Brownian motion. Therefore, it may seem reasonable to assume that the free random motion of a particle is the Einstein-Smoluchowski Brownian motion. However, this is not the case, because to describe the random motion we must first of all construct an evolution function, which depends on an exit function.

Example 1. Let us take $\phi_b(y) = \alpha\exp\left(-\frac{\kappa^2}{2}b + \frac{\kappa}{\sigma}y\right)$ as exit function, where κ is a positive constant that we can choose freely. Then the corresponding evolution function is

$$\phi\left(t,x\right) = \alpha\exp\left(-\frac{\kappa^2}{2}t + \frac{\kappa}{\sigma}x\right).$$

Since this $\phi\left(t,x\right)$ is not bounded, we must proceed with caution.

The evolution function $\phi\left(t,x\right)$ determines the evolution drift. By formula (1.4.7),

$$a\left(t,x\right) = \sigma^2\frac{1}{\phi\left(t,x\right)}\frac{\partial\phi\left(t,x\right)}{\partial x} = \sigma\kappa. \tag{1.6.1}$$

Therefore, by (1.5.4), the kinematics equation is

$$\frac{\partial u}{\partial t} + \frac{1}{2}\sigma^2\frac{\partial^2 u}{\partial x^2} + \sigma\kappa\frac{\partial u}{\partial x} = 0, \tag{1.6.2}$$

and the paths of the motion are given by the stochastic differential equation (1.5.8) of Theorem 1.5.4 as

$$X_t = X_a + \sigma B_{t-a} + \sigma\kappa\left(t-a\right). \tag{1.6.3}$$

This can be taken as a model of the motion of an electron shot from an electron gun.

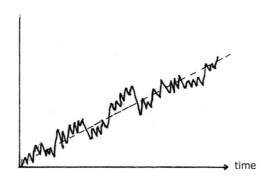

Figure 1.4

If κ is large enough and the Brownian motion $\sigma B_{t-a}(\omega)$ can be neglected, equation (1.6.3) reduces to

$$X_t = X_a + \sigma\kappa\,(t-a)\,, \tag{1.6.4}$$

i.e., it describes the classical uniform motion. Therefore, the path equation (1.6.3) describes the classical uniform motion to which a small Brownian noise $\sigma B_{t-a}(\omega)$ is added.

Let us compute the distribution density $\mu(t,x)$. If we take the backward evolution function

$$\widehat{\phi}(t,x) = \alpha\exp\left(\frac{\kappa^2}{2}t + \frac{\kappa}{\sigma}x\right),$$

then

$$\mu(t,x) = \phi(t,x)\,\widehat{\phi}(t,x) = \alpha^2\exp\left(\frac{2\kappa}{\sigma}x\right). \tag{1.6.5}$$

Consider a finite interval $[\ell_1,\ell_2]$ and observe the motion on this interval. Denote

$$\beta = \int_{\ell_1}^{\ell_2}\exp\left(\frac{2\kappa}{\sigma}x\right)dx$$

and set $\alpha^2 = 1/\beta$. Then

$$\int_{\ell_1}^{\ell_2}\mu(t,x)\,dx = 1,$$

i.e., $\mu(t,x)$ is a distribution density on the finite interval $[\ell_1,\ell_2]$. The corresponding probability distribution of the random motion $X_t(\omega)$ is given by the formula

$$\mathbf{Q}\left[X_t \in dx\right] = \alpha^2\exp\left(\frac{2\kappa}{\sigma}x\right)dx, \quad x \in [\ell_1,\ell_2],$$

and so does not depend on time.

Example 2. If we choose $\phi_b(y) \equiv 1$ as the exit function, then the evolution function is simply

$$\phi(t, x) \equiv 1.$$

By formula (1.4.7), the corresponding evolution drift $a(t, x)$ is

$$a(t, x) = \sigma^2 \frac{1}{\phi(t, x)} \frac{\partial \phi(t, x)}{\partial x} \equiv 0.$$

Therefore, by formula (1.5.4) of Theorem 1.5.3, the kinematics equation is

$$\frac{\partial u}{\partial t} + \frac{1}{2} \sigma^2 \frac{\partial^2 u}{\partial x^2} = 0.$$

According to equation (1.5.8) of Theorem 1.5.4, the paths of the motion of our particle are given by

$$X_t = X_a + \sigma B_{t-a}. \tag{1.6.6}$$

If we take the density function $\mu(x)$ of the probability distribution as an entrance function $\widehat{\phi}_a(x)$, then

$$\widehat{\phi}(t, x) = \int \mu(z) \, dz \, p(a, z, t, x)$$

$$= \int \mu(z) \, dz \frac{1}{\sqrt{2\pi \sigma^2 (t - a)}} \exp\left(-\frac{(x - z)^2}{2\sigma^2 (t - a)}\right),$$

and since in this case $\phi(t, x) \equiv 1$, the same formula gives the distribution density of the random motion $X_t(\omega)$. Thus, in this example the random motion is the well-known Einstein-Smoluchowski Brownian motion.

In the classical theory of Einstein-Smoluchowski Brownian motion, one prescribes an initial distribution density. The fact that the free random motion is the random uniform motion $X_t = X_a + \sigma B_{t-a} + \sigma \kappa (t - a)$ was first clarified by our mechanics of random motion. The classical Brownian motion is the special case "$\kappa = 0$".

When we discuss the classical motion of a particle, we specify an initial position and an initial velocity. In the conventional theory of stochastic processes, we specify an initial distribution, which corresponds to specifying an initial position. However, a mathematical procedure allowing to specify an initial velocity was not known in the conventional theory of stochastic processes. The way to specify an "initial velocity" is now clarified by the mechanics of random motion.

For example, in the mechanics of random motion the constant $\sigma \kappa$ figuring in the equation of motion (1.6.2) comes from specifying the initial evolution drift

$$\sigma^2 \frac{1}{\phi(a, x)} \frac{\partial \phi(a, x)}{\partial x} = \sigma \kappa.$$

at the initial time $t = a$. This corresponds to specifying an initial velocity in classical mechanics.

1.7 Hooke's Force

Let us consider the random motion under Hooke's force.

(i) We treat one-dimensional motion, that is, the particle moves on a straight line. Hooke's potential is $c(x) = -\frac{1}{2}\kappa^2 x^2$ and the Schrödinger-Nagasawa equations of motion are

$$
\begin{cases}
\dfrac{\partial \phi}{\partial t} + \dfrac{1}{2}\sigma^2 \dfrac{\partial^2 \phi}{\partial x^2} - \dfrac{1}{2}\kappa^2 x^2 \phi = 0, \\[2mm]
-\dfrac{\partial \widehat{\phi}}{\partial t} + \dfrac{1}{2}\sigma^2 \dfrac{\partial^2 \widehat{\phi}}{\partial x^2} - \dfrac{1}{2}\kappa^2 x^2 \widehat{\phi} = 0.
\end{cases}
$$

Consider the case of stationary motion. Assume that the evolution and backward evolution functions are given by

$$
\phi(t, x) = e^{\lambda t}\varphi(x) \quad \text{and} \quad \widehat{\phi}(t, x) = e^{-\lambda t}\varphi(x),
$$

respectively, and substitute them in the equations of motion, which yields the equation

$$
-\frac{1}{2}\sigma^2 \frac{\partial^2 \varphi}{\partial x^2} + \frac{1}{2}\kappa^2 x^2 \varphi = \lambda\varphi.
$$

This is an eigenvalue problem with eigenvalues $\lambda_n = \left(n + \frac{1}{2}\right)\sigma\kappa$.

Let us look at the random motion determined by the lowest eigenvalue $\lambda_0 = \sigma\kappa/2$. The associated eigenfunction is

$$
\varphi_0(x) = \beta \exp\left(-\frac{\kappa x^2}{2\sigma}\right).
$$

Therefore, the corresponding evolution function is

$$
\phi(t, x) = \exp\left(\frac{\sigma\kappa t}{2}\right)\varphi_0(x) = \beta \exp\left(\frac{\sigma\kappa t}{2} - \frac{\kappa x^2}{2\sigma}\right),
$$

and it generates, by formula (1.5.4), the evolution drift

$$
a(x) = \sigma^2 \frac{1}{\phi(t, x)}\frac{\partial \phi(t, x)}{\partial x} = -\sigma\kappa x.
$$

Then, by formula (1.5.4) in Theorem 1.5.3, the corresponding kinematics equation is

$$
\frac{\partial u}{\partial t} + \frac{1}{2}\sigma^2 \frac{\partial^2 u}{\partial x^2} - \sigma\kappa x \frac{\partial u}{\partial x} = 0.
$$

Finally, by (1.5.8) in Theorem 1.5.4, the path equation under Hooke's force reads

$$x_t = x_a + \sigma B_{t-a} - \sigma\kappa \int_a^t ds\, x_s. \tag{1.7.1}$$

The integral term in (1.7.1) takes negative values if $x_s > 0$, and represents a drift toward the origin. If $x_s < 0$, it takes positive values, and represents a drift toward the origin from the negative side. Therefore, the random motion given by (1.7.1) oscillates with the Brownian noise σB_{t-a}, namely, it is a random harmonic oscillation.

The distribution of the random motion x_t is given by

$$\mathbf{Q}\left[x_t \in dx\right] = \phi\left(t, x\right) \widehat{\phi}\left(t, x\right) dx = \beta^2 \exp\left(-\frac{\kappa x^2}{\sigma}\right) dx,$$

i.e., it is the normal distribution centered at the origin.

(ii) Now let us consider harmonic oscillations in the (x, y)-plane. With the harmonic oscillation x_t given by (1.7.1) we define a random motion on the x-axis of the (x, y)-plane by

$$X_t = (x_t, 0); \tag{1.7.2}$$

this is an x-polarized random harmonic oscillation. The random motion orthogonal to it is given by

$$X_t^\perp = (0, x_t), \tag{1.7.3}$$

and is a y-polarized random harmonic oscillation.

More generally, we consider random motion on a line $y = \alpha x$ of the (x, y)-plane, where α is a constant. Using the process given by (1.7.1), we set

$$X_t = (x_t, \alpha x_t). \tag{1.7.4}$$

This is a polarized random harmonic oscillation on the line $y = \alpha x$ in the (x, y)-plane. The random motion orthogonal to (1.7.4) is the polarized random harmonic oscillation

$$X_t^\perp = (-\alpha x_t, x_t), \tag{1.7.5}$$

on the line $-\alpha y = x$. Thus, we have a "combined x and y random motion".

(iii) If a random motion in the plane takes place under Hooke's force

$$c\left(x, y\right) = -\frac{1}{2}\kappa^2\left(x^2 + y^2\right),$$

then the Schrödinger-Nagasawa equations of motion are

$$\begin{cases} \dfrac{\partial \phi}{\partial t} + \dfrac{1}{2}\sigma^2\left(\dfrac{\partial^2 \phi}{\partial x^2} + \dfrac{\partial^2 \phi}{\partial y^2}\right) - \dfrac{1}{2}\kappa^2\left(x^2 + y^2\right)\phi = 0, \\[4mm] -\dfrac{\partial \widehat{\phi}}{\partial t} + \dfrac{1}{2}\sigma^2\left(\dfrac{\partial^2 \widehat{\phi}}{\partial x^2} + \dfrac{\partial^2 \widehat{\phi}}{\partial y^2}\right) - \dfrac{1}{2}\kappa^2\left(x^2 + y^2\right)\widehat{\phi} = 0. \end{cases}$$

Let us consider the case of stationary motion. Then the evolution and backward evolution functions are given by

$$\phi(t,x,y) = e^{\lambda t}\varphi(x,y) \quad \text{and} \quad \widehat{\phi}(t,x,y) = e^{-\lambda t}\varphi(x,y),$$

respectively. Substituting these expressions in the equations of motion, we obtain the equation

$$-\frac{1}{2}\sigma^2\left(\frac{\partial^2\varphi}{\partial x^2} + \frac{\partial^2\varphi}{\partial y^2}\right) + \frac{1}{2}\kappa^2\left(x^2 + y^2\right)\varphi = \lambda\varphi. \tag{1.7.6}$$

This eigenvalue problem is well known in quantum mechanics (see, for example, Section 11.c of Pauling and Wilson (1935)). Writing n as $n = n_x + n_y$, the eigenvalues of problem (1.7.6) are

$$\lambda_n = \sigma\kappa\left(n_x + n_y + 1\right), \quad n_x, n_y = 0, 1, 2, \ldots .$$

Consider the case $n = 0$. Then $\lambda_0 = \sigma\kappa$, and the associated eigenfunction is

$$\varphi_0(x,y) = \beta\exp\left(-\frac{\kappa\left(x^2 + y^2\right)}{2\sigma}\right).$$

Hence, the corresponding evolution function is given by the formula

$$\phi(t,x,y) = \exp(\sigma\kappa t)\varphi_0(x,y) = \beta\exp\left(\sigma\kappa t - \frac{\kappa\left(x^2 + y^2\right)}{2\sigma}\right).$$

In this case the distribution of the random motion $\mathbf{X}_t = (x_t, y_t)$ is

$$\mathbf{Q}\left[\mathbf{X}_t \in dxdy\right] = \phi(t,x,y)\,\widehat{\phi}(t,x,y)\,dxdy$$
$$= \beta^2\exp\left(-\frac{\kappa\left(x^2 + y^2\right)}{\sigma}\right)dxdy,$$

i.e., the two-dimensional normal distribution centered at the origin. As explained above, in quantum mechanics there is no problem computing the distribution.

To compute the paths of the random motion we resort to the evolution function. Quantum mechanics could not approach this problem.

The two-dimensional evolution drift vector $\mathbf{a}(x,y)$ determined by the evolution function $\phi(t,x,y)$ is given in components by

$$\mathbf{a}(x,y) = \sigma^2\frac{1}{\phi(t,x,y)}\left(\frac{\partial\phi(t,x,y)}{\partial x}, \frac{\partial\phi(t,x,y)}{\partial y}\right) = -\left(\sigma\kappa x, \sigma\kappa y\right).$$

Therefore, in the present case case the kinematics equation (1.5.4) takes on the form

$$\frac{\partial u}{\partial t} + \frac{1}{2}\sigma^2\left(\frac{\partial^2 u}{\partial x^2} + \frac{\partial^2 u}{\partial y^2}\right) - \left(\sigma\kappa x\frac{\partial u}{\partial x} + \sigma\kappa y\frac{\partial u}{\partial y}\right) = 0,$$

and the path equation for the stochastic process $\mathbf{X}_t = (x_t, y_t)$, i.e., equation (1.5.8) of Theorem 1.5.4, becomes

$$(x_t, y_t) = \left(x_a + \sigma B^1_{t-a} - \sigma\kappa \int_a^t ds\, x_s,\ y_a + \sigma B^2_{t-a} - \sigma\kappa \int_a^t ds\, y_s\right),$$

where B^1_t and B^2_t are independent one-dimensional Brownian motions.

1.8 Hooke's Force and an Additional Potential

Let us rewrite the eigenvalue problem (1.7.6) in the two-dimensional polar coordinates (r, η):

$$x = r\cos\eta, \quad y = r\sin\eta, \quad r \in (0, \infty), \quad \eta \in [0, 2\pi].$$

Then,

$$\operatorname{grad}\varphi = \left(\frac{\partial\varphi}{\partial r}, \frac{1}{r}\frac{\partial\varphi}{\partial\eta}\right),$$

$$\operatorname{div}\mathbf{X} = \frac{1}{r}\frac{\partial(rX_1)}{\partial r} + \frac{1}{r}\frac{\partial X_2}{\partial\eta},$$

for $\mathbf{X} = (X_1, X_2)$, and the Laplacian has the expression

$$\triangle\varphi = \operatorname{div}\operatorname{grad}\varphi = \frac{1}{r}\frac{\partial}{\partial r}\left(r\frac{\partial\varphi}{\partial r}\right) + \frac{1}{r^2}\frac{\partial^2\varphi}{\partial\eta^2}.$$

Therefore, the Schrödinger-Nagasawa equations of motion take on the form

$$\begin{cases}
\dfrac{\partial\phi}{\partial t} + \dfrac{1}{2}\sigma^2\left(\dfrac{1}{r}\dfrac{\partial}{\partial r}\left(r\dfrac{\partial\phi}{\partial r}\right) + \dfrac{1}{r^2}\dfrac{\partial^2\phi}{\partial\eta^2}\right) - \dfrac{1}{2}\kappa^2 r^2\phi = 0, \\[2ex]
-\dfrac{\partial\widehat{\phi}}{\partial t} + \dfrac{1}{2}\sigma^2\left(\dfrac{1}{r}\dfrac{\partial}{\partial r}\left(r\dfrac{\partial\widehat{\phi}}{\partial r}\right) + \dfrac{1}{r^2}\dfrac{\partial^2\widehat{\phi}}{\partial\eta^2}\right) - \dfrac{1}{2}\kappa^2 r^2\widehat{\phi} = 0,
\end{cases} \tag{1.8.1}$$

and the corresponding eigenvalue problem (1.7.6) becomes

$$-\frac{1}{2}\sigma^2\left(\frac{1}{r}\frac{\partial}{\partial r}\left(r\frac{\partial\varphi}{\partial r}\right) + \frac{1}{r^2}\frac{\partial^2\varphi}{\partial\eta^2}\right) + \frac{1}{2}\kappa^2 r^2\varphi = \lambda\varphi.$$

Substituting here $\varphi(r, \eta) = R(r)\Phi(\eta)$ and separating the variables, we obtain the equations

$$-\frac{\partial^2\Phi}{\partial\eta^2} = m^2\Phi, \tag{1.8.2}$$

and

$$-\frac{1}{2}\sigma^2\frac{1}{r}\frac{\partial}{\partial r}\left(r\frac{\partial R}{\partial r}\right) + \left(\frac{1}{2}\kappa^2 r^2 + \frac{1}{2}\sigma^2\frac{m^2}{r^2}\right)R = \lambda R. \tag{1.8.3}$$

Equation (1.8.2) has the solutions

$$\sin m\eta, \quad \cos m\eta, \quad m = 0, \pm 1, \pm 2, \ldots . \tag{1.8.4}$$

The solutions of equation (1.8.3) are well known in quantum mechanics (see, for example, Section 12 of Pauling and Wilson (1935)).

The corresponding eigenvalues are

$$\lambda_{|m|+n} = \sigma\kappa \left(|m| + n + 1\right), \quad m = 0, \pm 1, \pm 2, \ldots, \quad n = 0, 2, 4, \ldots,$$

with associated eigenfunctions

$$R_{m,n}\left(r\right) = F_{|m|,n}\left(\sqrt{\frac{\kappa}{\sigma}}r\right)\exp\left(-\frac{1}{2}\frac{\kappa}{\sigma}r^2\right),$$

where $F_{|m|,n}\left(x\right)$ is a polynomial in x.

(i) Consider the case of the smallest eigenvalue $\lambda_0 = \sigma\kappa$ (i.e., $m = 0, n = 0$, $F_{|0|,0}\left(x\right) = 1$). The corresponding evolution function is

$$\phi\left(t, r, \eta\right) = \beta\exp\left(-\frac{\kappa}{2\sigma}r^2 + \sigma\kappa t\right),$$

where β is a normalization constant.

If we denote

$$R_0 = -\frac{\kappa}{2\sigma}r^2, \quad S_0 = \sigma\kappa t,$$

then the evolution and backward evolution functions are expressed as

$$\phi\left(t, r, \eta\right) = \beta e^{R_0 + S_0} \quad \text{and} \quad \widehat{\phi}\left(t, r, \eta\right) = \beta e^{R_0 - S_0},$$

respectively. Accordingly, the distribution of the random motion X_t is given by the formula

$$\mu(t, r, \eta) = \phi(t, r, \eta)\widehat{\phi}(t, r, \eta) = \beta^2\exp\left(-\frac{\kappa r^2}{\sigma}\right),$$

i.e., it is the normal distribution centered at the origin.

To describe the motion of a particle, we apply the kinematics of random motion.

Formula (1.4.7) gives the evolution drift vector

$$\mathbf{a}\left(r, \eta\right) = \sigma^2\frac{1}{\phi}\left(\frac{\partial\phi}{\partial r}, \frac{1}{r}\frac{\partial\phi}{\partial\eta}\right) = \sigma^2\left(\frac{\partial R_0}{\partial r}, \frac{1}{r}\frac{\partial S_0}{\partial\eta}\right) = \left(-\sigma\kappa r, 0\right).$$

Therefore, the kinematics equation reads

$$\frac{\partial u}{\partial t} + \frac{1}{2}\sigma^2\left(\frac{1}{r}\frac{\partial}{\partial r}\left(r\frac{\partial u}{\partial r}\right) + \frac{1}{r^2}\frac{\partial^2 u}{\partial\eta^2}\right) - \sigma\kappa r\frac{\partial u}{\partial r} = 0. \tag{1.8.5}$$

To get the path equation we write (1.8.5) in expanded form as

$$\frac{\partial u}{\partial t} + \frac{1}{2}\sigma^2 \left(\frac{\partial^2 u}{\partial r^2} + \frac{1}{r^2}\frac{\partial^2 u}{\partial \eta^2} \right) + \frac{1}{2}\sigma^2 \frac{1}{r}\frac{\partial u}{\partial r} - \sigma\kappa r\frac{\partial u}{\partial r} = 0.$$

Then, due to the equivalence of equations (1.5.9) and (1.5.11) of Theorem 1.5.5, the path equation takes, in the space variables (r, η), the form of the system

$$\begin{cases} r_t = r_a + \sigma B^1_{t-a} + \displaystyle\int_a^t ds\, \frac{1}{2}\sigma^2 \frac{1}{r_s} - \int_a^t ds\, \sigma\kappa r_s, \\ \eta_t = \eta_a + \displaystyle\int_a^t \sigma\frac{1}{r_s}\, dB^2_{s-a}, \end{cases}$$

where B^1_t and B^2_t are independent one-dimensional Brownian motions.

For the smallest eigenvalue $\lambda_0 = \sigma\kappa$, the radial motion r_t is repelled strongly by the origin due to the drift $\frac{1}{r}$, which diverges at the origin. The stochastic process

$$r_t = \sigma B^1_{t-a} + \int_a^t ds\, \frac{1}{2}\sigma^2 \frac{1}{r_s}$$

figuring in the radial motion is called the Bessel process.

(ii) The case of the first excited eigenvalue $\lambda_1 = 2\sigma\kappa$ $(m = \pm 1, n = 0)$.

The eigenvalue problem (1.8.2) admits the solutions

$$\sin m\eta, \quad \cos m\eta, \quad m = 0, \pm 1, \pm 2, \dots ,$$

as well as their linear combinations

$$e^{im\eta} = \cos m\eta + i\sin m\eta, \quad m = 0, \pm 1, \pm 2, \dots .$$

Since $F_{|\pm 1|, 0}(x) = 2x$, when $m = \pm 1$, we have

$$\varphi_{1,0}(r, \eta) = \beta r \exp\left(-\frac{\kappa}{2\sigma}r^2 \right) \exp(i\eta),$$

$$\varphi_{-1,0}(r, \eta) = \beta r \exp\left(-\frac{\kappa}{2\sigma}r^2 \right) \exp(-i\eta),$$

are complex-valued eigenfunctions.

We can now introduce the complex evolution evolution functions by

$$\psi_{\pm 1}(t, r, \eta) = \exp(-i2\sigma\kappa t)\varphi_{\pm 1,0}(r, \eta)$$

$$= \beta r \exp\left(-\frac{\kappa}{2\sigma}r^2 + i\left(-2\sigma\kappa t \pm \eta \right) \right).$$

(For complex evolution functions in general, see the following section.)

Denoting

$$R_1 = \log r - \frac{\kappa}{2\sigma}r^2, \quad S_{\pm 1} = -2\sigma\kappa t \pm \eta, \tag{1.8.6}$$

we have

$$\psi_{\pm 1}(t, r, \eta) = \beta \exp\left(R_1 + iS_{\pm 1}\right). \tag{1.8.7}$$

The random motion (r_t, η_t) cannot be described directly by the complex-valued functions $\psi_{\pm 1}(t, r, \eta)$. We regard R_1 as the generating function of the distribution and $S_{\pm 1}$ as the drift potential, and inroduce the pair of evolution functions

$$\phi_{\pm 1}(t, r, \eta) = \beta e^{R_1 + S_{\pm 1}}, \quad \widehat{\phi}_{\pm 1}(t, r, \eta) = \beta e^{R_1 - S_{\pm 1}}, \tag{1.8.8}$$

which will allow us to describe the random motion. The relation between the complex evolution function $\psi_{\pm 1}(t, r, \eta)$ and the pair of evolution functions

$$\{\phi_{\pm 1}(t, r, \eta), \widehat{\phi}_{\pm 1}(t, r, \eta)\}$$

will be discussed in detail in the following section.

Let us consider first the case $m = +1$, and set

$$\begin{cases} \phi(t, r, \eta) = \beta e^{R_1 + S_{+1}} = \beta r \exp\left(-\frac{\kappa}{2\sigma}r^2 + (-2\sigma\kappa t + \eta)\right), \\ \widehat{\phi}(t, r, \eta) = \beta e^{R_1 - S_{+1}} = \beta r \exp\left(-\frac{\kappa}{2\sigma}r^2 - (-2\sigma\kappa t + \eta)\right). \end{cases} \tag{1.8.9}$$

These are solutions of the Schrödinger-Nagasawa equations of motion

$$\begin{cases} \dfrac{\partial \phi}{\partial t} + \dfrac{1}{2}\sigma^2\left(\dfrac{1}{r}\dfrac{\partial}{\partial r}\left(r\dfrac{\partial \phi}{\partial r}\right) + \dfrac{1}{r^2}\dfrac{\partial^2 \phi}{\partial \eta^2}\right) - \left(\dfrac{1}{2}\kappa^2 r^2 + \widetilde{V}(r)\right)\phi = 0, \\ -\dfrac{\partial \widehat{\phi}}{\partial t} + \dfrac{1}{2}\sigma^2\left(\dfrac{1}{r}\dfrac{\partial}{\partial r}\left(r\dfrac{\partial \widehat{\phi}}{\partial r}\right) + \dfrac{1}{r^2}\dfrac{\partial^2 \widehat{\phi}}{\partial \eta^2}\right) - \left(\dfrac{1}{2}\kappa^2 r^2 + \widetilde{V}(r)\right)\widehat{\phi} = 0, \end{cases} \tag{1.8.10}$$

where

$$\widetilde{V}(r) = \sigma^2 \frac{1}{r^2} - 4\sigma\kappa.$$

Equations (1.8.10) govern the motion under the potential

$$-c(r) = \frac{1}{2}\kappa^2 r^2 + \widetilde{V}(r) = \frac{1}{2}\kappa^2 r^2 + \sigma^2 \frac{1}{r^2} - 4\sigma\kappa. \tag{1.8.11}$$

The corresponding distribution density is given by

$$\mu(t, r, \eta) = \phi(t, r, \eta)\,\widehat{\phi}(t, r, \eta) = \beta^2 r^2 \exp\left(-\frac{\kappa}{\sigma}r^2\right);$$

it vanishes at the origin.

Since
$$\frac{d\left(-c(r)\right)}{dr} = \kappa^2 r - 2\sigma^2 \frac{1}{r^3},$$
the potential $-c(r)$ attains its minimum for $\widetilde{r} = 2^{\frac{1}{4}}\sqrt{\sigma/\kappa}$. Therefore, the distribution looks like an American-style doughnut in two dimensions.

To describe the random motion of a particle in this case, we must derive the path equation. By formula (1.4.7), the evolution drift vector has the components

$$\begin{cases} a^r\left(r,\eta\right) = \sigma^2 \frac{1}{\phi}\frac{\partial\phi}{\partial r} = \sigma^2 \frac{\partial R_1}{\partial r} = \sigma^2 \frac{1}{r} - \sigma\kappa, \\[2mm] a^\eta\left(r,\eta\right) = \sigma^2 \frac{1}{\phi}\frac{1}{r}\frac{\partial\phi}{\partial\eta} = \sigma^2 \frac{1}{r}\frac{\partial S_{+1}}{\partial\eta} = \sigma^2 \frac{1}{r}, \end{cases}$$

where $\phi\left(t,r,\eta\right)$ is the evolution function in (1.8.9).

Therefore, the kinematics equation (1.5.4) is

$$\frac{\partial u}{\partial t} + \frac{1}{2}\sigma^2 \left(\frac{1}{r}\frac{\partial}{\partial r}\left(r\frac{\partial u}{\partial r}\right) + \frac{1}{r^2}\frac{\partial^2 u}{\partial\eta^2}\right)$$
$$+ \left(\sigma^2 \frac{1}{r} - \sigma\kappa r\right)\frac{\partial u}{\partial r} + \sigma^2 \frac{1}{r}\frac{1}{r}\frac{\partial u}{\partial\eta} = 0. \tag{1.8.12}$$

To get the path equation, we expand equation (1.8.12) as

$$\frac{\partial u}{\partial t} + \frac{1}{2}\sigma^2 \left(\frac{\partial^2 u}{\partial r^2} + \frac{1}{r^2}\frac{\partial^2 u}{\partial\eta^2}\right)$$
$$+ \frac{1}{2}\sigma^2 \frac{1}{r}\frac{\partial u}{\partial r} + \left(\sigma^2 \frac{1}{r} - \sigma\kappa r\right)\frac{\partial u}{\partial r} + \sigma^2 \frac{1}{r}\frac{1}{r}\frac{\partial u}{\partial\eta} = 0.$$

Then, since equations (1.5.9) and (1.5.11) are equivalent, the path equation, written in the space variables (r,η) becomes the system

$$\begin{cases} r_t = r_a + \sigma B^1_{t-a} + \displaystyle\int_a^t ds\left(\frac{3}{2}\sigma^2 \frac{1}{r_s} - \sigma\kappa r_s\right), \\[4mm] \eta_t = \eta_a + \displaystyle\int_a^t \sigma\frac{1}{r_s}\,dB^2_{s-a} + \int_a^t \sigma^2 \frac{1}{r_s^2}\,ds, \end{cases}$$

where B^1_t and B^2_t are independent one-dimensional Brownian motions.

Solving the equation
$$\frac{3}{2}\sigma^2 \frac{1}{r} - \sigma\kappa r = 0,$$
we get
$$\overline{r} = \sqrt{\frac{3}{2}\frac{\sigma}{\kappa}}.$$

Therefore, the drift function takes positive values if $r < \bar{r}$, and negative values if $r > \bar{r}$, and hence the radial motion r_t is attracted to the zero-point $\bar{r}$. Thus, we have shown that the observed small particle executes a random motion in an American-style doughnut of mean radius $\bar{r} = \sqrt{\frac{3}{2}\frac{\sigma}{\kappa}}$, with an anti-clockwise rotational drift σ^2/r_s^2.

In the case $m = -1$, we use the pair

$$\begin{cases} \phi(t,r,\eta) = \beta e^{R_1 + S_{-1}} = \beta r \exp\left(-\frac{\kappa}{2\sigma}r^2 + (-2\sigma\kappa t - \eta)\right), \\ \phi(t,r,\eta) = \beta e^{R_1 - S_{-1}} = \beta r \exp\left(-\frac{\kappa}{2\sigma}r^2 - (-2\sigma\kappa t - \eta)\right). \end{cases}$$

Hence, the observed small particle executes a random motion in an American doughnut of mean radius $\bar{r} = \sqrt{\frac{3}{2}\frac{\sigma}{\kappa}}$, with a clockwise rotational drift σ^2/r_s^2.

Comment. (i) An important role in the preceding considerations was played by the functions R and S introduced by means of formulas (1.8.6), and the corresponding complex evolution function

$$\psi = e^{R+iS}, \tag{1.8.13}$$

where we regarded $\{R, S\}$ as the generating functions of random motion. Moreover, we introduced the pair of evolution functions of exponential form

$$\begin{cases} \phi = e^{R+S}, \\ \widehat{\phi} = e^{R-S}. \end{cases} \tag{1.8.14}$$

The complex evolution function in (1.8.13) defines a random motion through the pair (1.8.14) of evolution and backward evolution functions.

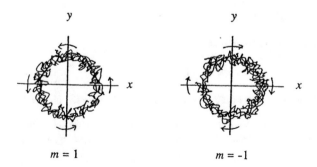

Figure 1.5. Illustration of the paths of the motion corresponding to the first excited eigenvalue

(ii) The Schrödinger-Nagasawa equations of motion (1.4.1) with an external potential were introduced by Nagasawa (1993, 2000). Schrödinger (1931) treated the case without an external potential as in (1.1.3) and stated that the case with an external potential can be treated in the same manner. But it seem that he was not aware of the existence of the functions R and S. The existence of the function S becomes clear if we represent the solutions of the equation of motion (1.4.1) in exponential form as in (1.4.6). However, the fact that the drift potential S is a fundamental physical quantity becomes clear only after the kinematics equation (1.5.4) of random motion is introduced and analyzed. Schrödinger (1932) did not treat the case with external potentials, and stopped pursuing this subject, which is a pity: intense controversies and conceptual confusion in quantum theory might have been avoided.

1.9   Complex Evolution Functions

We will prove that the well-known Schrödinger equation belongs to the theory of stochastic processes. Schrödinger himself regarded it as a "wave equation", when he introduced his theory in quantum physics as "wave mechanics". We will argue that in the context of our mechanics of random motion, the Schrödinger equation is *not a wave equation*, but a **complex evolution equation**.

Earlier, we called the pair $\{R(t, x), S(t, x)\}$ **the generating function of random motion**, and showed that it is a fundamental physical quantity. We will now use the pair $\{R(t, x), S(t, x)\}$ to introduce a **complex exponential function** and develop our theory of random motion in terms of this complex-valued physical quantity.

As promised, we regard the pair $\{R(t, x), S(t, x)\}$ as a fundamental physical quantity and associate with it

a pair of evolution functions

$$\phi(t, x) = e^{R(t,x)+S(t,x)}, \quad \widehat{\phi}(t, x) = e^{R(t,x)-S(t,x)}, \tag{1.9.1}$$

and **a complex evolution function**

$$\psi(t, x) = e^{R(t,x)+iS(t,x)}. \tag{1.9.2}$$

A key point here is that all these functions share the pair $\{R(t, x), S(t, x)\}$.

Theorem 1.9.1 (Equivalence of the Equations of Motion and the Schrödinger Equation). *The complex evolution function (1.9.2) satisfies the Schrödinger equation*

$$i\frac{\partial \psi}{\partial t} + \frac{1}{2}\sigma^2(\nabla + i\mathbf{b}(t, x))^2\psi - V(t, x)\psi = 0, \tag{1.9.3}$$

if and only if the pair (1.9.1) of evolution functions $\{\phi(t, x), \widehat{\phi}(t, x)\}$ satisfies the

Schrödinger-Nagasawa equations of motion (1.4.1)

$$\begin{cases} \dfrac{\partial \phi}{\partial t} + \dfrac{1}{2}\sigma^2(\nabla + \mathbf{b}\,(t,x))^2 \phi + c\,(t,x)\,\phi = 0, \\[3mm] -\dfrac{\partial \widehat{\phi}}{\partial t} + \dfrac{1}{2}\sigma^2(\nabla - \mathbf{b}\,(t,x))^2 \widehat{\phi} + c\,(t,x)\,\widehat{\phi} = 0. \end{cases}$$

The Schrödinger equation and the Schrödinger-Nagasawa equations determine the same random motion under the condition

$$V\,(t,x) + c\,(t,x) + \sigma^2 \mathbf{b}^2 + \widetilde{V}\,(t,x) = 0, \tag{1.9.4}$$

where

$$\widetilde{V}(t,x) = 2\frac{\partial S}{\partial t} + \sigma^2(\nabla S)^2 + 2\sigma^2 \mathbf{b} \cdot \nabla S, \tag{1.9.5}$$

Remark. As stated in Section 1.1, Schrödinger's attempt can be described as follows: "The wave theory has a defect in some sense, but it is no doubt that the Schrödinger equation (1.1.1) is effective in computing energy values in quantum theory. Therefore, if we can show that the twin equations of motion (1.4.1) are effective in computing the energy, or that the Schrödinger equation can be derived from the twin equations of motion, then we obtain a quantum theory based on Brownian motion." However, Schrödinger could not establish this, and it remained an open problem.

Theorem 1.9.1 solves Schrödinger's problem, establishing that, under the condition (1.9.4), the equations of random motion (1.4.1) and the Schrödinger equation (complex evolution equation) (1.9.3) are equivalent.

Proof of Theorem 1.9.1. Let us substitute the function $\psi\,(t,x) = e^{R(t,x)+iS(t,x)}$ in the complex evolution equation (1.9.3). This yields

$$\begin{aligned} 0 &= \left\{ i\frac{\partial \psi}{\partial t} + \frac{1}{2}\sigma^2(\nabla + i\mathbf{b}\,(t,x))^2 \psi - V\,(t,x)\,\psi \right\} \frac{1}{\psi} \\[2mm] &= -V - \frac{1}{2}\sigma^2 \mathbf{b}^2 - \frac{\partial S}{\partial t} + \frac{1}{2}\sigma^2 \triangle R \\[2mm] &\quad + \frac{1}{2}\sigma^2(\nabla R)^2 - \frac{1}{2}\sigma^2(\nabla S)^2 - \sigma^2 \mathbf{b} \cdot (\nabla S) \\[2mm] &\quad + i\left\{ \frac{\partial R}{\partial t} + \frac{1}{2}\sigma^2 \triangle S + \sigma^2 \nabla S \cdot \nabla R + \sigma^2 \mathbf{b} \cdot \nabla R + \frac{1}{2}\sigma^2\,\nabla \cdot \mathbf{b} \right\}. \end{aligned} \tag{1.9.6}$$

Separating the real and imaginary parts, we have

$$-V - \frac{1}{2}\sigma^2 \mathbf{b}^2 - \frac{\partial S}{\partial t} + \frac{1}{2}\sigma^2 \triangle R + \frac{1}{2}\sigma^2(\nabla R)^2 - \frac{1}{2}\sigma^2(\nabla S)^2 - \sigma^2 \mathbf{b} \cdot \nabla S = 0, \tag{1.9.7}$$

and

$$\frac{\partial R}{\partial t} + \frac{1}{2}\sigma^2 \triangle S + \sigma^2 \nabla S \cdot \nabla R + \sigma^2 \mathbf{b} \cdot \nabla R + \frac{1}{2}\sigma^2 \nabla \cdot \mathbf{b} = 0, \tag{1.9.8}$$

respectively.

Substituting $\phi\left(t, x\right) = e^{R(t,x)+S(t,x)}$ in the equations of motion (1.4.1), dividing the result by $\phi\left(t, x\right)$ and rearranging the terms, we obtain

$$
\begin{aligned}
0 = {}& -\frac{1}{2}\sigma^2\mathbf{b}^2 - \frac{\partial S}{\partial t} + \frac{1}{2}\sigma^2\triangle R + \frac{1}{2}\sigma^2(\nabla R)^2 - \frac{1}{2}\sigma^2(\nabla S)^2 - \sigma^2\mathbf{b}\cdot(\nabla S) \\
& + \frac{\partial R}{\partial t} + \frac{1}{2}\sigma^2\triangle S + \sigma^2\nabla S\cdot\nabla R + \sigma^2\mathbf{b}\cdot\nabla R + \frac{1}{2}\sigma^2\nabla\cdot\mathbf{b} \\
& + c + \left\{\sigma^2\mathbf{b}^2 + 2\frac{\partial S}{\partial t} + \sigma^2(\nabla S)^2 + 2\sigma^2\mathbf{b}\cdot\nabla S\right\}.
\end{aligned}
\tag{1.9.9}
$$

(i) If $\psi\left(t, x\right) = e^{R(t,x)+iS(t,x)}$ satisfies the complex evolution equation (1.9.3), then, by (1.9.7), the first line in (1.9.9) is equal to V. The second line in (1.9.9) vanishes, because of (1.9.8). Therefore, we have

$$
V + c + \left\{\sigma^2\mathbf{b}^2 + 2\frac{\partial S}{\partial t} + \sigma^2(\nabla S)^2 + 2\sigma^2\mathbf{b}\cdot\nabla S\right\} = 0,
\tag{1.9.10}
$$

and so under the condition (1.9.10) the function $\phi\left(t, x\right) = e^{R(t,x)+S(t,x)}$ satisfies the first of the equations of motion (1.4.1).

In much the same way one verifies that the function $\widehat{\phi}\left(t, x\right) = e^{R(t,x)-S(t,x)}$ satisfies the second equation in (1.4.1).

(ii) To derive the complex evolution equation (1.9.3) from the equations of motion (1.4.1), we use the following duality result.

Lemma 1.9.1. *If* $\phi\left(t, x\right) = e^{R(t,x)+S(t,x)}$ *and* $\widehat{\phi}\left(t, x\right) = e^{R(t,x)-S(t,x)}$ *satisfy the equations of motion* (1.4.1), *then*

$$
\frac{\partial R}{\partial t} + \frac{1}{2}\sigma^2\triangle S + \sigma^2\nabla S\cdot\nabla R + \sigma^2\mathbf{b}\cdot\nabla R + \frac{1}{2}\sigma^2\nabla\cdot\mathbf{b} = 0.
\tag{1.9.11}
$$

Proof. By the kinematics Theorem 1.5.3, the distribution density

$$
\mu\left(t, x\right) = \phi\left(t, x\right)\widehat{\phi}\left(t, x\right) = e^{2R(t,x)}
$$

satisfies the equation

$$
-\frac{\partial\mu}{\partial t} + \frac{1}{2}\sigma^2\triangle\mu - \nabla\cdot\left(\left(\sigma^2\mathbf{b}\left(t, x\right) + \mathbf{a}\left(t, x\right)\right)\mu\right) = 0.
$$

By the time-reversed kinematics Theorem 4.6.2, the function $\mu(t, x)$ satisfies also the equation

$$
\frac{\partial\mu}{\partial t} + \frac{1}{2}\sigma^2\triangle\mu - \nabla\cdot\left(\left(-\sigma^2\mathbf{b}\left(t, x\right) + \widehat{\mathbf{a}}\left(t, x\right)\right)\mu\right) = 0;
$$

this will be shown in Chapter 4. Substracting, we obtain

$$
\frac{\partial\mu}{\partial t} + \nabla\cdot\left\{\left(\sigma^2\mathbf{b}\left(t, x\right) + \frac{\mathbf{a}\left(t, x\right) - \widehat{\mathbf{a}}\left(t, x\right)}{2}\right)\mu\right\} = 0.
\tag{1.9.12}
$$

Since

$$\mathbf{a}(t,x) = \sigma^2 \nabla (R(t,x) + S(t,x))$$

and

$$\widehat{\mathbf{a}}(t,x) = \sigma^2 \nabla (R(t,x) - S(t,x)),$$

we have

$$\frac{\mathbf{a}(t,x) - \widehat{\mathbf{a}}(t,x)}{2} = \sigma^2 \nabla S(t,x).$$

Substituting this expression in (1.9.12), we get equation (1.9.11). This completes the proof of the lemma. □

Let us return to the proof of Theorem 1.9.1.

If $\phi(t,x) = e^{R(t,x)+S(t,x)}$ and $\widehat{\phi}(t,x) = e^{R(t,x)-S(t,x)}$ satisfy the equations of motion (1.4.1), then the equation (1.9.8) holds by (1.9.11). The second line in (1.9.9) vanishes by (1.9.11). If (1.9.10) holds, then (1.9.9) yields the equation (1.9.7).

Finally, since (1.9.7) and (1.9.8) hold, $\psi(t,x) = e^{R(t,x)+iS(t,x)}$ satisfies the complex evolution equation (1.9.3), which completes the proof of Theorem 1.9.1. □

Thus, we established the equivalence of the Schrödinger-Nagasawa equations of motion (1.4.1) and the complex evolution equation (1.9.3), i.e., the Schrödinger equation.

However, there is a major qualitative difference. The kinematics of electrons cannot be deduced directly from the complex evolution equation (1.9.3). In other words,

the paths of the motion of an electron cannot be described directly by the Schrödinger equation.

This is the reason paths of motion could not be discussed in the conventional quantum theory. However, the Schrödinger equation is of a decisive importance, because it guarantees the "superposition" of random motions.

Fortunately, Theorem 1.9.1 allows one to describe the paths of the motion through the equations of motion (1.4.1). In this way the Schrödinger equation nevertheless determines the paths of the motion of an electron, making it one of the fundamental equations in the mechanics of random motion.

The complex evolution equation (1.9.3) is a generalization of the Schrödinger wave equation (1.1.1). As already indicated, **it is not a "wave equation"**, but a **complex "evolution equation"** that describes the time evolution of the complex function $R(t,x) + iS(t,x)$ by means of the complex exponential function $\psi(t,x) = e^{R(t,x)+iS(t,x)}$. Namely,

the Schrödinger equation describes the evolution of the motion under external forces,

that is, the following theorem holds.

Theorem 1.9.2 (Schrödinger Equation and Random Motion). *Let $\psi\,(t,x)$ be a solution of the Schrödinger equation*

$$i\frac{\partial\psi}{\partial t} + \frac{1}{2}\sigma^2(\nabla + i\mathbf{b}\,(t,x))^2\psi - V\,(t,x)\,\psi = 0, \qquad (1.9.13)$$

of the form

$$\psi\,(t,x) = e^{R(t,x)+iS(t,x)}. \qquad (1.9.14)$$

Then $R\,(t,x)$ is the generating function of the distribution, $S\,(t,x)$ is the drift potential, and the pair $\{R(t,x),S(t,x)\}$ determines the evolution drift vector

$$\mathbf{a}(t,x) = \sigma^2(\nabla R(t,x) + \nabla S(t,x)), \qquad (1.9.15)$$

and the random motion

$$X_t = X_a + \ \sigma B_{t-a} + \int_a^t \left(\sigma^2\mathbf{b}\,(s,X_s) + \mathbf{a}(s,X_s)\right)ds, \qquad (1.9.16)$$

where B_t is a Brownian motion and X_a is a random variable independent of B_t. That is, the Schrödinger equation is a complex evolution equation.

Proof. By the equivalence established in Theorem 1.9.1, $R\,(t,x)$ is the generating function of distribution and $S\,(t,x)$ is the drift potential. The drift vector is given by (1.9.15). The random motion is determined by the evolution drift, as proved in Theorem 1.5.4.

The fact that the Schrödinger equation (1.9.13) is an evolution equation becomes clear once we write it in the form

$$\frac{\partial\psi}{\partial t} = i\left(\frac{1}{2}\sigma^2(\nabla + i\mathbf{b}\,(t,x))^2 - V\,(t,x)\right)\psi. \qquad (1.9.17)$$

This shows that the operator $G = i\big(\frac{1}{2}\sigma^2(\nabla + i\mathbf{b}(t,x))^2 - V(t,x)\big)$ is the generator of a semigroup of operators e^{Gt}. (See Yosida (1948) for the theory of semigroups of operators.) $\qquad\square$

Remark. We re-emphasize that Theorem 1.9.2 establishes that the Schrödinger equation is a complex evolution equation, and its solution $\psi\,(t,x) = e^{R(t,x)+iS(t,x)}$ is a complex evolution function that describes the evolution of random motion; in other words, the Schrödinger equation is not a wave equation. This is not simply a question of interpretation, it is a theorem.

Comment. 1. The pair of functions $\{R(t,x),S(t,x)\}$ figures in the complex evolution function (1.9.2) as well as in the pair of evolution functions in (1.9.1). It represents the most important physical quantity in our mechanics of random motion, and it satisfies the equations (1.9.7) and (1.9.8). However, practically it is not possible to get it by solving those equations: the pair is obtained from the complex evolution function or the pair of evolution functions. Once we have the pair $\{R(t,x), S(t,x)\}$, we can define the evolution drift vector

$$\mathbf{a}(t,x) = \sigma^2 \nabla \left(R(t,x) + S(t,x) \right).$$

The Schrödinger equation (1.9.3) itself does not really suggest how to understand the pair; its meaning becomes clear only after we derive the kinematics equation (1.5.4) or the path equation (1.5.8) from the equations of motion (1.4.1) (keep Theorem 1.9.2 in mind).

2. It appears that Feynman did regard the Schrödinger equation (1.9.17) as a complex evolution equation, with a complex evolution function $\psi(t,x)$ as its solution (see Feynman (1948)). In fact, he used the fundamental solution $\psi(s,x,t,y)$ of (1.9.17) to define a "complex-valued measure." If we fix times $a < t_1 \leq t_2 \leq \cdots \leq t_n = b$, we can define a complex-valued quantity $P^{(n)}[\Gamma]$ by the formula

$$P^{(n)}[\Gamma] = \int \mu_a(dx_0)\, \psi(a, x_0, t_1, x_1)\, dx_1 \mathbf{1}_{B_1}(x_1)\, \psi(t_1, x_1, t_2, x_2) \times \cdots \quad (1.9.18)$$

$$\times\, dx_2 \mathbf{1}_{B_2}(x_2) \cdots \psi(t_{n-1}, x_{n-1}, t_n, x_n)\, dx_n \mathbf{1}_{B_n}(x_n).$$

However, if we let $n \to \infty$, which means that we multiply infinitely many complex numbers on the right-hand side, then $P^{(n)}[\Gamma]$ does not converge. Therefore, the so-called Feynman path integral, a complex-valued measure on the path space, does not exist. Equation (1.9.18) has a meaning only when we fix times $a < t_1 \leq t_2 \leq \cdots \leq t_n = b$. In other words, Feynman did not construct a stochastic process on the path space.

1.10 Superposition Principle

In the preceding section, we introduced a complex exponential function

$$\psi(t,x) = e^{R(t,x)+iS(t,x)}, \quad (1.10.1)$$

and called it "complex evolution function." In this section we assume that $\psi(t,x)$ satisfies the Schrödinger equation, and call it "Schrödinger function." **It is the main player in this section.**

As indicated earlier, the Schrödinger function $\psi(t,x)$ does not describe the random motions directly. To that end we must look at the exponent $R(t,x) + iS(t,x)$ and interpret the real part $R(t,x)$ as the generating function of the distribution and the imaginary part $S(t,x)$ as the drift potential of a random motion. Then, as before, we introduce the pair of evolution functions

$$\phi(t,x) = e^{R(t,x)+S(t,x)}, \quad \widehat{\phi}(t,x) = e^{R(t,x)-S(t,x)}, \quad (1.10.2)$$

from which we derive the random motion

$$X_t = X_a + \sigma B_{t-a} + \int_a^t \sigma^2 \frac{\nabla\phi}{\phi}(s, X_s)ds.$$

This fact must be always kept in mind when we discuss the "superposition principle of Schrödinger functions $\psi(t,x)$" and the "entangled random motion", which are the main topics of this section.

Although the random motion cannot be described directly by means of the Schrödinger function $\psi(t,x)$, this function is useful:

Theorem 1.10.1. *The distribution function is given by*

$$\mu(t,x) = \psi(t,x)\overline{\psi(t,x)} = \phi(t,x)\widehat{\phi}(t,x) = e^{2R(t,x)}. \tag{1.10.3}$$

Now, let $\psi_1(t,x)$ and $\psi_2(t,x)$ be two solutions of the Schrödinger equation. Then, thanks to the linearity of the Schrödinger equation, any linear combination (superposition)

$$\psi^*(t,x) = \alpha_1\psi_1(t,x) + \alpha_2\psi_2(t,x)$$

is also a solution. When $\psi_1(t,x)$ and $\psi_2(t,x)$ are wave functions, this is the well-known superposition principle. However, they are not wave functions, but complex evolution functions. Therefore, we must clarify the physical meaning of the terms "superposition" and "entanglement" in our theory of random motion.

The notion of **superposed Schrödinger function** introduced by Schrödinger is well-known in quantum mechanics. However, in quantum mechanics it has been not so clear what superposition means, which led to numerous involved discussions concerning possible physical interpretations.

Things are clear in our mechanics of random motion. The superposed complex evolution function $\psi^*(t,x)$ describes a new random motion. We will prove that

the superposed complex evolution function $\psi^*(t,x)$ determines an entangled random motion.

This was first clarified in the context of the mechanics of random motion (see Nagasawa (1992)).

Let us go back for a moment to the theory of Markov processes and look at the Kolmogorov equation (1.2.21). This is also a linear equation. Therefore, a linear combination

$$u^*(t,x) = \frac{1}{2}u_1(t,x) + \frac{1}{2}u_2(t,x)$$

of two of its solutions $u_1(t,x)$ and $u_2(t,x)$ is again a solution. The sample space of the process determined by the superposition $u^*(t,x)$ is the union of the sample spaces determined by $u_1(t,x)$ and $u_2(t,x)$.

In the same way, if we take a superposed complex evolution function

$$\psi^*(t,x) = \frac{1}{\sqrt{2}}\,\psi_1(t,x) + \frac{1}{\sqrt{2}}\,\psi_2(t,x),$$

we would expect that the sample space of the random motion determined by $\psi^*(t,x)$ will be the union of the sample spaces of the random motions determined separately by $\psi_1(t,x)$ and $\psi_2(t,x)$. However, this is not the case; things are not so simple, because me must deal with the drift potentials $S_k(t,x)$ of the functions

$$\psi_k(t,x) = e^{R_k(t,x)+iS_k(t,x)}, \qquad k = 1,2.$$

In fact, the drift potential $S^*(t,x)$ of the superposed Schrödinger function

$$\psi^*(t,x) = e^{R^*(t,x)+iS^*(t,x)} = \frac{1}{\sqrt{2}}\psi_1(t,x) + \frac{1}{\sqrt{2}}\psi_2(t,x)$$

is not simply the sum of $S_1(t,x)$ and $S_2(t,x)$, but more complicated, so $\psi^*(t,x)$ induces a complicated random motion. This is one of the reasons behind the heated controversies concerning the interpretation of the concept of superposition. We will explain this, following Nagasawa (1992).

We are interested in the following question:

What kind of random motion is induced when complex exponential functions are added, that is, by the superposed complex evolution function $\psi^*(t,x)$?

The superposed Schrödinger function

$$\psi^*(t,x) = \frac{1}{\sqrt{2}}\,\psi_1(t,x) + \frac{1}{\sqrt{2}}\,\psi_2(t,x)$$

induces complicated random motions that are completely different from the random motions determined by $\psi_1(t,x)$ and by $\psi_2(t,x)$. We will call it **"entangled random motion"**.

Let us describe the entangled random motion. We prepare two random motions, called "random motion 1" and "random motion 2", determined by

the Schrödinger function $\quad \psi_1(t,x) = e^{R_1(t,x)+iS_1(t,x)}$

and

the Schrödinger function $\quad \psi_2(t,x) = e^{R_2(t,x)+iS_2(t,x)}$,

respectively. Then:

The evolution functions of the random motion k are

$$\phi_k(t,x) = e^{R_k(t,x)+S_1(t,x)}, \quad \widehat{\phi}_k(t,x) = e^{R_k(t,x)-S_k(t,x)},$$

i.e., the random motion k evolves carrying the drift potential $S_k(t,x)$, $k = 1,2$.

We want to describe the entangled random motion of the random motions 1 and 2. To this end we must use the corresponding Schrödinger functions, i.e., consider their superposition.

Thus, we consider a combination (superposition)

$$\psi^*\left(t,x\right) = \alpha_1 \psi_1\left(t,x\right) + \alpha_2 \psi_2\left(t,x\right), \tag{1.10.4}$$

of the Schrödinger functions $\psi_1(t,x)$ and $\psi_2(t,x)$ introduced above, where α_1 and α_2 are complex constants satisfying the normalization condition $\int |\psi^*\left(t,x\right)|^2 dx = 1$, and then write the function (1.10.4) in exponential form as

$$\psi^*\left(t,x\right) = e^{R^*(t,x)+iS^*(t,x)}. \tag{1.10.5}$$

Definition 1.10.1.

(i) The function $\psi^*(t,x)$ will be called the **superposition** of the Schrödinger functions $\psi_1(t,x)$ and $\psi_2(t,x)$.

(ii) The exponent $\{R^*(t,x), S^*(t,x)\}$ of $\psi^*(t,x)$ will be called the **entanglement** of the exponents $\{R_1\left(t,x\right), S_1\left(t,x\right)\}$ and $\{R_2(t,x), \ \ S_2(t,x)\}$.

(iii) Using the entangled exponent $\{R^*(t,x), S^*(t,x)\}$, we introduce the new pair of evolution functions

$$\phi^*\left(t,x\right) = e^{R^*(t,x)+S^*(t,x)}, \quad \widehat{\phi}^*\left(t,x\right) = e^{R^*(t,x)-S^*(t,x)},$$

which will be called the **entangled evolution function** of the evolution functions of the random motions 1 and 2.

(iv) The stochastic process $\{X_t(\omega); t \in [a,b], \mathbf{Q}^*\}$ determined by the entangled generating function $\{R^*(t,x), S^*(t,x)\}$ will be called the **superposed**, or **entangled** random motion of the random motions 1 and 2.

The notion of the entangled random motion $\{X_t(\omega); t \in [a,b], \mathbf{Q}^*\}$ was not known in quantum mechanics and in the conventional theory of stochastic processes.

We will prove that the **entangled random motion** just defined exists. First, a preliminary result.

Theorem 1.10.2. *Given a pair* $\{R(t,x), S(t,x)\}$, *set*

$$\phi(t,x) = e^{R(t,x)+S(t,x)}, \quad \widehat{\phi}(t,x) = e^{R(t,x)-S(t,x)}. \tag{1.10.6}$$

Then there exist a function $p(s,x,t,y)$ *and a random motion* $\{X_t(\omega)\,;\, t \in [a,b], \mathbf{Q}\}$ *such that the probability measure* $\mathbf{Q}$ *satisfies the equation* (1.4.3) *and*

$$\widehat{\phi}(t,x) = \int dz\, \widehat{\phi}_a(z) p(a,z,t,x), \quad \phi(t,x) = \int p(t,x,b,y)\phi_b(y)dy,$$

where $\widehat{\phi}_a(z) = \widehat{\phi}(a,z)$ *and* $\phi_b(y) = \phi(b,y)$.

Proof. Use $\phi(t, x)$ in (1.10.6) to define a drift $\sigma^2 \nabla \log \phi(t, x)$, and then use the solution of the stochastic differential equation

$$X_t = X_a + \sigma B_{t-a} + \int_a^t \sigma^2 \nabla \log \phi(s, X_s) ds$$

to define a Markov process $\{X_t(\omega); t \in [a, b], \mathbf{Q}\}$, where the distribution of X_a is $\widehat{\phi}_a(x)\phi(a, x)$ and $\nabla \log \phi(s, x) = \nabla R(s, x) + \nabla S(s, x)$. Let $q(s, x, t, y)$ be the transition probability density of the Markov process $\{X_t(\omega), \mathbf{Q}\}$. With $\phi(t, x)$ given by equation (1.10.6), set

$$p(s, x, t, y) = \phi(s, x)q(s, x, t, y)\frac{1}{\phi(t, y)}.$$

Then $\{X_t(\omega), \mathbf{Q}\}$ is determined by the triplet $\{p(s, x, t, y), \widehat{\phi}_a(x), \phi_b(y)\}$ and satisfies equation (1.4.3). We can see this by looking at Theorem 1.5.2 in the opposite direction. □

Now we apply Theorem 1.10.2 to the entangled generating function $\{R^*(t, x), S^*(t, x)\}$ introduced in Definition 1.10.1 to establish the existence of the entangled random motion:

Theorem 1.10.3. *There exist the entangled motion* $\{X_t(\omega); t \in [a, b], \mathbf{Q}^*\}$ *of the random motions 1 and 2, as well as the entangled density* $p^*(s, x, t, y)$. *The entangled evolution functions are given by*

$$\widehat{\phi}^*(t, x) = \int dz\, \widehat{\phi}_a^*(z)p^*(a, z, t, x), \quad \phi^*(t, x) = \int p^*(t, x, b, y)\, \phi_b^*(y) dy.$$

Theorem 1.10.4. *The entangled random motion* $\{X_t(\omega); t \in [a, b], \mathbf{Q}^*\}$ *carries the entangled drift potential* $S^*(t, x)$, *which induces the* **entangled drift vector**

$$\mathbf{a}^*(t, x) = \sigma^2 \nabla R^*(t, x) + \sigma^2 \nabla S^*(t, x).$$

The paths $X_t(\omega)$ *of the entangled random motion are given by the equation*

$$X_t = X_a^* + \sigma B_{t-a} + \int_a^t \mathbf{a}^*(s, X_s) ds, \qquad (1.10.7)$$

where the distribution density $\mu_a^*(x) = e^{2R^*(a,x)}$ *of* X_a^* *is determined by the generating function* $R^*(a, x)$. *The distribution density is given by*

$$\mu^*(t, x) = \widehat{\phi}^*(t, x)\phi^*(t, x).$$

A fact to be emphasized in Theorem 1.10.4 is the **existence of the entangled random motion** $X_t(\omega)$. Without this theorem it is difficult to understand what "superposition" means. This is a reason why "superposition" was not well understood in quantum mechanics.

Computing $R^*(t,x)$ and $S^*(t,x)$ and solving the equation (1.10.7) is not an easy task. A practical way to understand the entangled random motion is to compute its distribution density

$$\mu^*(t,x) = \phi^*(t,x)\,\widehat{\phi}^*(t,x) = |\psi^*(t,x)|^2,$$

since this can be done by means of the Schrödinger function $\psi^*(t,x)$, as described below.

Theorem 1.10.5. *The distribution density $\mu^*(t,x)$ of the entangled random motion is given by the formula*

$$\begin{aligned}
\mu^*(t,x) = {} & |\alpha_1|^2 e^{2R_1(t,x)} + |\alpha_2|^2 e^{2R_2(t,x)} \\
& + 2e^{R_1(t,x)+R_2(t,x)}\Re(\alpha_1\overline{\alpha_2})\cos(S_1(t,x) - S_2(t,x)) \qquad (1.10.8) \\
& - 2e^{R_1(t,x)+R_2(t,x)}\Im(\alpha_1\overline{\alpha_2})\sin(S_1(t,x) - S_2(t,x)).
\end{aligned}$$

The function

$$\begin{aligned}
& 2e^{R_1(t,x)+R_2(t,x)}\Re(\alpha_1\overline{\alpha_2})\cos(S_1(t,x) - S_2(t,x)) \\
& - 2e^{R_1(t,x)+R_2(t,x)}\Im(\alpha_1\overline{\alpha_2})\sin(S_1(t,x) - S_2(t,x))
\end{aligned} \qquad (1.10.9)$$

is called the **entanglement effect**.

Proof. By (1.10.3) and (1.10.4), the distribution density of the entangled random motion is

$$\begin{aligned}
\mu^*(t,x) = {} & \phi^*(t,x)\,\widehat{\phi}^*(t,x) = |\psi^*(t,x)|^2 = |\alpha_1\psi_1(t,x) + \alpha_2\psi_2(t,x)|^2 \\
= {} & |\alpha_1\psi_1(t,x)|^2 + |\alpha_2\psi_2(t,x)|^2 \\
& + \alpha_1\psi_1(t,x)\overline{\alpha_2\psi_2(t,x)} + \overline{\alpha_1\psi_1(t,x)}\alpha_2\psi_2(t,x),
\end{aligned}$$

where
$$\psi_1(t,x) = e^{R_1(t,x)+iS_1(t,x)}, \quad \psi_2(t,x) = e^{R_2(t,x)+iS_2(t,x)},$$
$$|\alpha_1\psi_1(t,x)|^2 = |\alpha_1|^2 e^{2R_1(t,x)}, \quad |\alpha_2\psi_2(t,x)|^2 = |\alpha_2|^2 e^{2R_2(t,x)}.$$

Therefore,

$$\begin{aligned}
& \alpha_1\psi_1(t,x)\overline{\alpha_2\psi_2(t,x)} + \overline{\alpha_1\psi_1(t,x)}\alpha_2\psi_2(t,x) \\
& = 2\Re\big(\alpha_1\psi_1(t,x)\overline{\alpha_2\psi_2(t,x)}\big) \\
& = 2e^{R_1(t,x)+R_2(t,x)}\Re(\alpha_1\overline{\alpha_2})\cos(S_1(t,x) - S_2(t,x)) \\
& \quad - 2e^{R_1(t,x)+R_2(t,x)}\Im(\alpha_1\overline{\alpha_2})\sin(S_1(t,x) - S_2(t,x)),
\end{aligned}$$

which yields formula (1.10.8).   □

The usefulness of Theorem 1.10.5 will be explained later. But for a while, we put superposition aside.

If a Schrödinger function $\psi(t, x) = e^{R(t,x)+iS(t,x)}$ is given, then the process $\{X_t(\omega); t \in [a, b], \mathbf{Q}\}$ is obtained by using the pair $\{R(t, x), S(t, x)\}$: the random motion $X_t(\omega)$ has the generating function of distribution $R(t, x)$ and the drift potential $S(t, x)$. The function $R(t, x)$ is connected with the distribution density $\mu(t, x)$ of the random motion $X_t(\omega)$ by the formula $\mu(t, x) = |\psi(t, x)|^2 = e^{2R(t,x)}$.

We claimed that "the random motion $X_t(\omega)$ carries the drift potential $S(t, x)$", but so far we did not provide an explicit formula to express this. The drift potential $S(t, x)$ is in a sense a hidden quantity. Nevertheless, Theorem 1.10.5 enables us to uncover the drift potential $S(t, x)$. To do this, we use the entangled random motion $\{X_t(\omega); t \in [a, b], \mathbf{Q}^*\}$, as follows.

Suppose we want to find the drift potential $S_1(t, x)$ of a Schrödinger function

$$\psi_1(t, x) = e^{R_1(t,x)+iS_1(t,x)}.$$

Then we prepare a simple Schrödinger function, for example,

$$\psi_0(t, x) = e^{R_0(t,x)+iS_0(t,x)},$$

where $S_0(t, x) = (0, 0, k)$ with a constant k, and form the superposition of $\psi_1(t, x)$ and $\psi_0(t, x)$. Then the distribution density of the entangled motion $\{X_t(\omega); t \in [a, b], \mathbf{Q}^*\}$ is given by formula (1.10.8). The sought-for drift potential $S_1(t, x)$ appears there as the difference $S_1(t, x) - (0, 0, k)$. (To "see" $S_1(t, x)$ in this way is the principle of the electron holography. Cf. Tonomura (1994).)

The superposed (entangled) stochastic process and Theorem 1.10.5 are important theoretically. Concerning experimental verification, see Sections 1.12 and 1.13 for the double-slit problem, Section 2.3 for the Bohm-Aharonov effect, and Section 2.9 for electron holography.

We close this section by stating what we have shown as a theorem.

Theorem 1.10.6. *Write the superposition* $\psi^*(t, x) = \alpha_1\psi_1(t, x) + \alpha_2\psi_2(t, x)$ *of the Schrödinger functions* $\psi_1(t, x) = e^{R_1(t,x)+iS_1(t,x)}$ *and* $\psi_2(t, x) = e^{R_2(t,x)+iS_2(t,x)}$ *in the form* $\psi^*(t, x) = e^{R^*(t,x)+iS^*(t,x)}$. *Then:*

(i) *The entangled evolution potential* $S^*(t, x)$ *induces the entangled evolution drift vector* $\mathbf{a}^*(t, x) = \sigma^2 \nabla R^*(t, x) + \sigma^2 \nabla S^*(t, x)$.

(ii) *There exists the entangled random motion* $\{X_t, \mathbf{Q}^*\}$, *determined by* $\psi^*(t, x)$, *the paths of which are given by the equation*

$$X_t = X_a^* + \sigma B_{t-a} + \int_a^t \mathbf{a}^*(s, X_s)ds.$$

(iii) *The entangled random motion* $\{X_t, \mathbf{Q}^*\}$ *carries the entangled evolution potential* $S^*(t, x)$.

(iv) *The distribution of the entangled random motion* $\{X_t, \mathbf{Q}^*\}$ *contains the effect of the entanglement* (1.10.9).

Remarks. There were intense controversies concerning the "superposition" of Schrödinger functions. To understand the physical meanings of the "superposition" we must consider the paths of the motion of an electron. But this was beyond the ability of quantum mechanics, because it does not include a kinematics. When Schrödinger discussed the "superposition", $\psi(t, x)$ was a complex wave function, and there was no problem to consider superposition of waves. However, once an electron is recognized as a particle with the electric charge e, Schrödinger's wave interpretation no longer has much meaning.

In quantum mechanics, Bohr claimed that an electron is a particle *and* a wave. But this is logically impossible, because the concepts of wave and particle belong to different categories and can not be identified. This was one of the sources of the controversies in those days.

In our mechanics of random motion there is no problem, because the Schrödinger equation is not a wave equation but a complex evolution equation.

1.11   Entangled Quantum Bit

In the "classical" computers now in use, a bit is produced by a massive amount of electrons. For example, if five volt, bit [1], and if 0 volt, bit [0]. Quantum computers, on the other hand, use a single electron or photon to produce a bit. We assume that the functions associated with those particles depend on a parameter τ.

The motion of a particle with a parameter τ_1 is given by

$$\psi\left((t, x)\,; \tau_1\right) = e^{R((t,x);\tau_1)+iS((t,x);\tau_1)}, \tag{1.11.1}$$

and the motion of a particle with a parameter τ_2 is given by

$$\psi\left((t, x)\,; \tau_2\right) = e^{R((t,x);\tau_2)+iS((t,x);\tau_2)}, \tag{1.11.2}$$

where, for example, the parameter τ_1 is spin up, $\tau_1 = +1$, and the parameter τ_2 is spin down, $\tau_2 = -1$.

We then denote the complex evolution function $\psi\left((t, x)\,; +1\right)$ with spin up by $|1\rangle$, and the complex evolution function $\psi\left((t, x)\,; -1\right)$ with spin down by $|0\rangle$, and call $|1\rangle$ and $|0\rangle$ "quantum bits". We can then use an entangled bit of the bits $|1\rangle$ and $|0\rangle$. We define the entangled complex evolution function by

$$\alpha_1\psi\left((t, x)\,; \tau_1\right) + \alpha_2\psi\left((t, x)\,; \tau_2\right), \quad |\alpha_1|^2 + |\alpha_2|^2 = 1, \tag{1.11.3}$$

where the parameter τ is the spin of an electron or the polarization of a photon.

We denote $\psi\left((t, x)\,; \tau_1\right)$ by $|1\rangle$, and $\psi\left((t, x)\,; \tau_2\right)$ by $|0\rangle$, and define an entangled bit $|\alpha_1, \alpha_2\rangle$ by

$$|\alpha_1, \alpha_2\rangle = \alpha_1|1\rangle + \alpha_2|0\rangle, \quad |\alpha_1|^2 + |\alpha_2|^2 = 1. \tag{1.11.4}$$

The entangled bit $|\alpha_1, \alpha_2\rangle = \psi((t, x); \alpha_1, \alpha_2)$ determines an entangled random motion.

If an entangled bit $|\alpha_1, \alpha_2\rangle$ occurs, then the probability that spin up is realized is $|\alpha_1|^2$ and the probability that spin down is realized is $|\alpha_2|^2$. Therefore, by a single observation we cannot determine if we have $|\alpha_1, \alpha_2\rangle$, $|1\rangle$ or $|0\rangle$. To answer this question we must repeat the observation under the same condition. This is a characteristic of quantum bits, and is not necessarily a defect.

As clarified in Theorem 1.10.5, the distribution density of the entangled random motion is given by formula (1.10.8), which contains the cross term

$$+ 2e^{R_1(t,x)+R_2(t,x)} \Re\left(\alpha_1 \overline{\alpha_2}\right) \cos(S((t, x); \tau_1) - S((t, x); \tau_2))$$

$$- 2e^{R_1(t,x)+R_2(t,x)} \Im\left(\alpha_1 \overline{\alpha_2}\right) \sin(S((t, x); \tau_1) - S((t, x); \tau_2)).$$

To construct a quantum computer one must be able to produce a practical entangled bit and preserve it for a sufficiently long time throughout the computations. This is not an easy task.

1.12 Light Emission from a Silicon Semiconductor

As an example of superposition, consider silicon (Si) crystal semi-conductor and look at the motion of an electron on the outer shell of an atom in the crystal. It is known that, due to the crystal structure of silicon, an outer-shell electron has two states of motion: one with lower energy (valence band) and one with higher energy (conduction band), separated by a gap (forbidden gap). The energy an electrons requires in order to transit from the valence band to the conduction band is called the band gap.

Let E_1 and E_2 denote the energy of an electron in the valence band and in the conduction band, respectively. The energy band gap $E_2 - E_1$ is in the range $1V < E_2 - E_1 < 1.5V$.

We look at the motion of an electron in the the the outer shell. Let ψ_1 (resp., ψ_2) be the complex evolution function with the lower energy E_1 (resp., higher energy E_2). Then the motion of the electron is described by the superposition

$$\psi^* = \alpha_1 \psi_1 + \alpha_2 \psi_2,$$

where

$$|\alpha_1|^2 + |\alpha_2|^2 = 1.$$

Let

$$\psi_1(t, x) = e^{-i\frac{E_1 t}{\hbar}} e^{R_1(x)+iS_1(x)}$$

and

$$\psi_2(t, x) = e^{-i\frac{E_2 t}{\hbar}} e^{R_2(x)+iS_2(x)},$$

where $e^{R_1(x)+iS_1(x)}$ (resp., $e^{R_2(x)+iS_2(x)}$) is the solution of the stationary Schrö-dinger equation with the energy E_1 (resp., E_2). Thus, the motion is described by the function

$$\psi^* = \alpha_1 e^{-i\frac{E_1 t}{\hbar}} e^{R_1(x)+iS_1(x)} + \alpha_2 e^{-i\frac{E_2 t}{\hbar}} e^{R_2(x)+iS_2(x)}.$$

To investigate the paths of an electron in the outer shell we must express $\psi^*(t,x)$ in the form

$$\psi^*(t,x) = e^{R^*(t,x)+iS^*(t,x)},$$

and calculate the distribution $R^*(t,x)$ and the drift potential $S^*(t,x)$.

The superposed complex evolution function $\psi^*(t,x)$ induces the entangled random motion, but the exponents $R^*(t,x)$ and $S^*(t,x)$ of $\psi^*(t,x)$ are not easy to compute, because we have to determine them from the relation

$$e^{R^*(t,x)+iS^*(t,x)} = \alpha_1 e^{-i\frac{E_1 t}{\hbar}} e^{R_1(x)+iS_1(x)} + \alpha_2 e^{-i\frac{E_2 t}{\hbar}} e^{R_2(x)+iS_2(x)}.$$

Instead, we observe the distribution density $\mu^*(t,x) = |\psi^*(t,x)|^2$ and guess the random motion. For this purpose we use formula (1.10.8) of Theorem 1.10.5 for the distribution density $\mu^*(t,x)$ of the entangled random motion:

$$\begin{aligned}
\mu^*(t,x) = {}& |\alpha_1|^2 e^{2R_1(t,x)} + |\alpha_2|^2 e^{2R_2(t,x)} \\
& + 2e^{R_1(t,x)+R_2(t,x)} \Re\left(\alpha_1\overline{\alpha_2}\right) \cos\left(S_1(t,x) - S_2(t,x)\right) \\
& - 2e^{R_1(t,x)+R_2(t,x)} \Im\left(\alpha_1\overline{\alpha_2}\right) \sin\left(S_1(t,x) - S_2(t,x)\right).
\end{aligned}$$

For simplicity, let α_1 and α_2 be real numbers. Then we get

$$\begin{aligned}
\mu^*(t,x) = {}& \alpha_1{}^2 e^{2R_1(x)} + \alpha_2{}^2 e^{2R_2(x)} \\
& + 2e^{R_1(x)+R_2(x)} \alpha_1\alpha_2 \cos\left(\frac{(E_2-E_1)t}{\hbar} + S_1(x) - S_2(x)\right);
\end{aligned} \qquad (1.12.1)$$

we are interested in the third term of the entanglement effect, which induces the oscillation $\cos\left(\frac{(E_2-E_1)t}{\hbar}\right)$.

As equation (1.12.1) indicates, there are two cases are possible for the paths of an electron depending on samples:

Case 1: At time $t = a$ the electron starts with ψ_1 of the lower energy E_1 and moves with the entangled drift potential $S^*(t,x)$ in the interval $a < t \le b$. This is the motion in the "valence band", and is relatively stable.

Case 2: At time $t = a$ the electron starts with ψ_2 of the higher energy E_2 and moves with the entangled drift potential $S^*(t,x)$ in the interval $a < t \le b$. This is the motion in the "conduction band", and induces the oscillation $\cos\left(\frac{(E_2-E_1)t}{\hbar}\right)$ as the entanglement effect, which is observed as emission of light.

Thus, we have shown that if the electron is in the state of motion ψ_2 with higher energy E_2, then it oscillates according to the law $\cos(\frac{(E_2-E_1)t}{\hbar})$ and emits light. This occurs through the entanglement of the motion ψ_2 of the higher energy E_2 and the motion ψ_1 of the lower energy E_1. Since the band gap $E_2 - E_1$ is about 1 Volt, electrons in the outer shell easily transit to the higher-energy state of motion ψ_2.

1.13   The Double-Slit Problem

A typical entangled random motion arises in the double-slit experiments. In such an experiment one sets a barrier with wo slits and one shoots electrons from an electron beam gun one by one, separated by a long time interval, and observe them as they reach a screen. Then a typical stripe (fringe) pattern is observed on the screen.

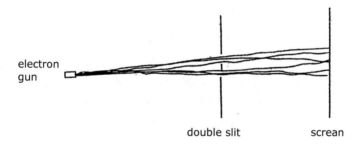

electron
gun

double slit   screan

Figure 1.6

The explanation of a double-slit experiment is an unsolved problem in quantum mechanics, first clarified theoretically by our mechanics of random motion, as we explain next. Let us consider the following experiment:

"Random motion 1": If only the first slit is open, then the random motion is described by a pair of evolution functions

$$\phi_1(t,x) = e^{R_1(t,x)+S_1(t,x)}, \quad \widehat{\phi}_1(t,x) = e^{R_1(t,x)-S_1(t,x)}.$$

"Random motion 2": If only the second slit is open, then the random motion is described by a pair of evolution functions

$$\phi_2(t,x) = e^{R_2(t,x)+S_2(t,x)}, \quad \widehat{\phi}_2(t,x) = e^{R_2(t,x)-S_2(t,x)}.$$

When both slits are open, **then which slit an electron goes through is random, because the electron performs a random motion.**

The randomness induces the entanglement of the "random motion 1" and the "random motion 2". Therefore, the behavior of the electrons is described by the entangled random motion.

The key concept here is that of the **entangled stochastic process**, introduced in Definition 1.10.1. To obtain the entangled stochastic process induced by the double-slit experiment under considerations, we first prepare

the complex evolution function 1: $\psi_1(t,x) = e^{R_1(t,x)+iS_1(t,x)}$,

and

the complex evolution function 2: $\psi_2(t,x) = e^{R_2(t,x)+iS_2(t,x)}$.

Next, we introduce the entangled complex evolution function $\psi^*(t,x) = \alpha_1\psi_1(t,x) + \alpha_2\psi_2(t,x)$, represent it in the exponential form

$$\psi^*(t,x) = e^{R^*(t,x)+iS^*(t,x)},$$

and then set

$$\phi^*(t,x) = e^{R^*(t,x)+S^*(t,x)}, \quad \widehat{\phi}^*(t,x) = e^{R^*(t,x)-S^*(t,x)}. \tag{1.13.1}$$

The entangled random motion $\{X_t(\omega), \mathbf{Q}^*\}$ determined by this pair of functions describes the random motion through the double slit. The stochastic process $\{X_t(\omega), \mathbf{Q}^*\}$ evolves carrying the entangled drift potential $S^*(t,x)$.

Suppose an electron goes through the double-slit barrier at time $t = a$. According to Theorem 1.10.4, the path X_t of the random motion after the barrier is crossed is determined by the entangled drift $\mathbf{a}^*(t,x) = \sigma^2\nabla R^*(t,x) + \sigma^2\nabla S^*(t,x)$ induced by $S^*(t,x)$ via the formula

$$X_t = X_a^* + \sigma B_{t-a} + \int_a^t \mathbf{a}^*(s, X_s)ds, \tag{1.13.2}$$

where the distribution density of X_a^* is given by $\mu_a^*(x) = e^{2R^*(a,x)}$. The key point here is that we can describe the path X_t of the motion going through the double slit with no ambiguity.

By Theorem 1.10.5, the distribution density of the entangled random motion, $\mu^*(t,x) = \phi^*(t,x)\widehat{\phi}^*(t,x)$, is calculated as

$$\mu^*(t,x) = \psi^*(t,x)\overline{\psi^*(t,x)} = |\alpha_1|^2 e^{2R_1(t,x)} + |\alpha_2|^2 e^{2R_2(t,x)}$$
$$+ 2e^{R_1(t,x)+R_2(t,x)}\Re(\alpha_1\overline{\alpha_2})\cos(S_2(t,x) - S_1(t,x))$$
$$+ 2e^{R_1(t,x)+R_2(t,x)}\Im(\alpha_1\overline{\alpha_2})\sin(S_2(t,x) - S_1(t,x)).$$

Particles arrive at the screen one by one, and as a **statistical effect** a pattern of stripes appears on the screen that depends on the difference $S_2(t,x) - S_1(t,x)$.

Thus, **the double-slit problem is explained theoretically by the mechanics of random motion.**

The double-slit problem reamined for a long time a Gedankenexperiment problem. Subsequently, experiments were performed. Among them, a double-slit experiment with electrons was carried by Tonomura (see, e.g., Tonomura (1994)). He used an electron bi-prism instead of a double-slit barrier. Electrons go through the right or left side of the bi-prism. The electron gun shoots electrons one, with a long time interval between the events. No striped pattern is observed after a few thousand shooting events, but after some ten thousand times, a typical striped pattern appears. Therefore, the striped pattern is a **statistical effect**; hence, it can not be due to "interference". (Cf. Jönssen (1961, 1974). Merli et al. (1976), Rosa (2012) incorrectly attributed it to "interference".)

Let us analyze Tonomura's experiment using the mechanics of random motion. Since an electron performs a random motion, which side (right or left) of the bi-prism it crosses is random. Let us label the motion going through the right (resp., left) side motion 1 (resp., motion 2). Then the entanglement of the two random motion occurs at the bi-prism: the electron bi-prism realizes the entanglement as a physical effect. The entangled random motion and the entanglement effect explain Tonomura's experiment theoretically.

As Feynman once stated, it is difficult to explain the double-slit experiment in quantum mechanics, because the interference effect appears even though a single electron is involved in the experiment. Therefore, it was a mystery of quantum mechanics.

We will explain this on a simple example. We set a double-slit barrier along the y-axis. A particle moves orthogonal to it. Let the drift potential of the random motion going through the slit 1 (resp., slit 2) be $S_1(t, y) = k_1 yt$ (resp., $S_2(t, y) = k_2 yt$). We assume that the time required to reach the screen after passing through a slit is t. Then, by Theorem 1.10.5, the distribution density $\mu^*(t, y)$ of particles is, with $\alpha_1 = \alpha_2 = \frac{1}{\sqrt{2}}$,

$$\mu^*(t, y) = \frac{1}{2} e^{2R_1(t,y)} + \frac{1}{2} e^{2R_2(t,y)} + e^{R_1(t,y)+R_2(t,y)} \cos(k_1 - k_2) yt. \qquad (1.13.3)$$

Formula (1.13.3) shows that if $k_1 \neq k_2$, the striped pattern appears, whereas if $k_1 = k_2$ the stripe pattern is absent. Which of the cases $k_1 \neq k_2$ or $k_1 = k_2$ occurs is determined by the drift potential when the particle goes through the double-slit barrier. In other words, the striped pattern is caused not by "interference", but by the entangled drift potential $S^*(t, x)$ of the entangled random motion $\{X_t(\omega), \mathbf{Q}^*\}$.

In the setting of our mechanics of random motion, an electron or a photon goes through one of the two slits. **Which slit the particle goes through is at random.** The randomness is defined theoretically as **entanglement at the double-slit barrier**. There is no ambiguity here. The key point is that the **striped pattern is not caused by interference**, because we are dealing with a single electron or photon. This point will be clarified in the next section.

1.14   Double-Slit Experiment with Photons

Walborn et al. (2002) reported results of double-slit experiments with photons. However, no appropriate theoretical analysis of their experiments has been given. As a matter of fact, as pointed out by Nagasawa (2012*), a theory that can analyze their experiments accurately is the mechanics of random motion. Let us describe and explain the experiments performed by Walborn et al. (2002).

In their experiment, an argon laser (351.1 nm at ~200 mW) is used to pump a 1-mm-long BBO (βBaB$_2$O$_4$, beta-barium borate) nonlinear crystal, generating 702.2 nm photons. A the polarizations of which are orthogonal of each other.

In the Figure 1.7, which describes their experimental setup, p-photons will go along the upper path, and s-photons will go along the lower path, on which the double slit is placed.

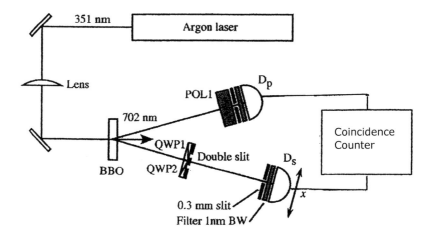

Figure 1.7

If the detector D_p catches a p-photon, it sends a click to the coincidence counter. If the detector D_s then catches an s-photon, it sends a click to the coincidence counter. **Once both clicks are detected, one registers "count 1".** Such counts will be recorded for 400 seconds. Then detector D_s is moved a millimeter and the number of counts in another 400 second interval is recorded for the new detector position. This is repeated until D_s has scanned across a region equivalent to Figure 1.7.

By using the coincidence counter one can thus select the couple of photons which should be counted.

We use s-photon which goes along the lower path for the double slit experiment. Let the s-photon be caught by the detector D_s. If we insert a filter in front of the detector D_p, then the p-photon is not necessarily detected by D_p. Therefore, a p-photon that goes along the upper path **can decide whether the s-photon detected by D_s should be counted or not.**

Walborn et al. (2002) carried out four experiments. We will call them Experiment 1, 2, 3, and 4.

Experiment 1. On the lower path of the s-photon we set a double-slit barrier. In Figure 1.7, the linear polarizer POL1 is not placed in front of the detector D_p of p-photon, and the quarter-wave plates QWP1 and QWP2 are not placed in front of the double slit along the lower path. Then a stripe like distribution pattern is observed, as shown in the left graph in Figure 1.8.

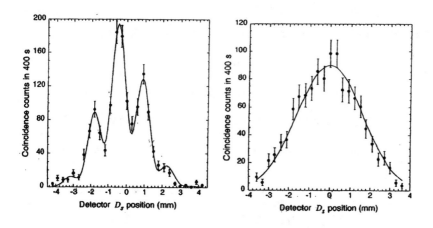

Figure 1.8

Experiment 2. The quarter-wave plates QWP1 and QWP2 are placed in front of double-slit in the above experimental setup, but the linear polarizer POL 1 is not placed in front of detector D_p . Instead, the y polarization prism is placed in front of the detector D_p of p-photon, for simplicity. Then no stripe-like pattern is observed, as shown in the right graph in Figure 1.8.

Experiment 3. The setup for the s-photon is the same as in Experiment 2; in addition, the linear polarizer POL 1 is inserted in front of detector D_p. Then a stripe-like pattern is observed similar to the figure shown as shown in the left graph in Figure 1.8.

Experiment 4. The setup is the same as in the Experiment 3, but now the polarizer POL 1 and the detector D_p are placed farther away from the BBO crystal, so that the path of the p-photon is lengthened. In this experiment, if the detector D_s catches an s-photon first and the detector D_p catches a p-photon, we register "count 1". Then a stripe-like pattern is observed, as in Experiment 3.

We will now present a mathematical (statistical) model due to Nagasawa (2012*) for the Experiments 1, 2, 3, and 4 of Walborn et al. (2002). What is important in the following analysis are the notions of **the total population and its sub-populations**.

Definition 1.14.1. The total population $\{\textbf{TP}\}$ is defined as

$$\{\textbf{TP}\} = \{\text{all photons generated by the source}\}.$$

We adopt the total population $\{\textbf{TP}\}$ for Experiment 1. For Experiments 2, 3, and 4 we introduce sub-populations.

Definition 1.14.2. The sub-population adopted for Experiment 2 is

$$\{\textbf{SP}\}_2 = \{y\text{-polarized } p\text{-photons}\}.$$

Definition 1.14.3. The sub-population adopted for Experiment 3 is

$$\{\textbf{SP}\}_3 = \{p\text{-photons with combined } x\text{- and } y \text{ - polarizations}\}.$$

Definition 1.14.4. The sub-population $\{\textbf{SP}\}_4$ adopted for Experiment 4 is the same as $\{\textbf{SP}\}_3$.

To analyze theoretically the experiments of Walborn et al. (2002) we need a mathematical tool that will allow us to describe the event that we select a sub-population $\{\textbf{SP}\}_2$, $\{\textbf{SP}\}_3$ from the total population $\{\textbf{TP}\}$.

Such a mathematical tool is the **"exit function"** of the mechanics of random motion: selecting an appropriate "exit function" corresponds to selecting a sub-population. Let us explain this.

Remark. We claim that quantum mechanics is not suitable for analyzing the experiments of Walborn et al. (2002) theoretically, because it does not include the notion of the exit function.

We will now apply our mechanics of random motion to the experiments considered above. It is not self-evident that this is appropriate for photons, because photons have no mass. However, things work out. For this issue we refer to the next section.

To guarantee the existence of the stochastic processes that describe the four experiments, we resort to results established in Section 1.4. We proved there that if we are given a triplet $\{p(s, x, t, y), \widehat{\phi}_a(x), \phi_b(y)\}$, then we can construct a stochastic process. We recall formula (1.5.2) for the probability measure $\textbf{Q}$, and write it for the entangled random motion $\{X_t(\omega); t \in [a, b], \mathbf{Q}^*\}$ at the double-slit barrier. For times $a \le t_1 \le t_2 \le \cdots \le t_n \le b$,

$$
\begin{aligned}
&\mathbf{Q}^*[f(X_{t_1}, X_{t_2}, \ldots, X_{t_n})] \\
&= \int dz\, \widehat{\phi}_a^*(z) p^*(a, z, t_1, x_1) dx_1\, p^*(t_1, x_1, t_2, x_2) dx_2 \times \cdots \quad (1.14.1) \\
&\quad \times p^*(t_n, x_n, b, y) \phi_b^*(y) dy\, f(x_1, \ldots, x_n).
\end{aligned}
$$

We then consider the pair of evolution functions

$$
\begin{cases}
\widehat{\phi}^*(t, x) = \displaystyle\int dz\, \widehat{\phi}_a^*(z) p^*(a, z, t, x), \\[2mm]
\phi^*(t, x) = \displaystyle\int p^*(t, x, b, y) \phi_b^*(y) dy,
\end{cases}
$$

where a is the time at the double-slit barrier and b is the time at which the s-photons are counted. What is essential here is that formula (1.14.1) contains the exit function $\phi_b^*(y)$.

Experiment 1. The entangled evolution functions of an s-photon are

$$
\widehat{\phi}^*(t, x; \tau_p) \quad \text{and} \quad \phi^*(t, x; \tau_p),
$$

where τ_p denotes the polarization of the p-photon. The distribution density of s-photons is therefore

$$
\mu_t^*(x) = \widehat{\phi}^*(t, x; \tau_p)\, \phi^*(t, x; \tau_p),
$$

where x is the space variable of the s-photon.

In Experiment 1 we use all photons generated by the source. Therefore, the distribution density at the detector D_s $(t = b)$ is

$$
\mu_b^*(x) = \widehat{\phi}^*(b, x; \tau_s) \phi_b^*(x).
$$

This case is the same as Tonomura's double-slit experiments for electrons.

Experiment 2. In this experiment we count selectively y-polarized p-photons, that is, we use the sub-population $\{\mathbf{SP}\}_2$.

The entangled evolution functions of an s-photon are

$$
\widehat{\phi}^*(t, x; \tau_p) \quad \text{and} \quad \phi^*(t, x; \tau_p),
$$

where τ_p denotes the polarization of the p-photon. The distribution density of s-photons is

$$
\mu_t^*(x) = \widehat{\phi}^*(t, x; \tau_p)\, \phi^*(t, x; \tau_p),
$$

where x is the space variable of the s-photon.

We specify the exit function by the polarization prism in front of the detector D_p and the coincidence counter; namely, exit function is

$$1_{\{\mathbf{SP}\}_2}(\tau_p)\phi_b^*(x),$$

where $1_{\{\mathbf{SP}\}_2}(\tau_p) = 1$ if τ_p is y-polarized, $1_{\{\mathbf{SP}\}_2}(\tau_p) = 0$ otherwise. Hence, the s-photon is counted only when its partner p-photon is in the sub-population $\{\mathbf{SP}\}_2$.

Therefore, the distribution density at the detector D_s $(t = b)$ is

$$\mu_b^*(x) = \widehat{\phi}^*(b, z; \tau_s)1_{\{\mathbf{SP}\}_2}(\tau_p)\phi_b^*(x),$$

where $\widehat{\phi}^*(b, z; \tau_s)$ is the entangled evolution function.

Let the evolution function of an s-photon through slit 1 be $\widehat{\phi}_1(t, x; \tau_1) = \widehat{\phi}_1(t, x) \otimes (\tau_1)$, and the evolution function of an s-photon through slit 2 be $\widehat{\phi}_2(t, x; \tau_2) = \widehat{\phi}_2(t, x) \otimes (\tau_2)$.

Then the cross terms of the evolution functions $\widehat{\phi}_1(t, x; \tau_s)$ and $\widehat{\phi}_2(t, x; \tau_s)$ are orthogonal, because

$$\langle (\tau_1), (\tau_2) \rangle = 0, \tag{1.14.2}$$

and hence no strip-like pattern is observed.

Moreover, we know which slit the s-photon came through in this experiment. Since the s-photon is x-polarized, it came through the first (resp., second) slit if it is left (resp., right) circularly polarized at the screen. We remark that we do not look at s-photon at the slits, even though we can say which slit it came through.

Remark. In the experiment of Walborn et al. (2002) the y polarization prism is not placed in front of the detector D_p of p-photon, that is, the sub-population $\{\mathbf{SP}\}_2$ is used instead of $\{\mathbf{TP}\}$.

Experiment 3. The setup for s-photons is the same as in Experiment 2. In addition, a linear polarizer POL 1 is inserted in front of the detector D_p. Then stripe a pattern was observed, similar to the figure shown in the left graph in Figure 1.8.

However, the sub-population $\{\mathbf{SP}\}_3$ is used. Therefore, the exit function is $1_{\{\mathbf{SP}\}_3}(\tau_p)\phi_b^*(x)$, and the distribution density at the detector D_s $(t = b)$ is

$$\mu_b^*(x) = \widehat{\phi}^*(b, z; \tau_p)1_{\{\mathbf{SP}\}_3}(\tau_p)\phi_b^*(x)$$

Since the polarization of counted s-photons is a linear combination of x and y polarizations, the plates QWP1 and QWP2 in front of the double-slit barrier do not transfer s-photons to circularly polarized ones. Therefore, the cross term (the effect of entanglement) of the evolution function does not vanish and a stripe- like pattern appears in the distribution.

In Experiment 2 and Experiment 3 the experimental setup for s-photon is the same. Nevertheless, we get different results. This is caused by the fact that we use the two different sub-populations, $\{\mathbf{SP}\}_2$ and $\{\mathbf{SP}\}_3$.

We note that in mechanics of random motion, different exit functions are adopted for the above two cases.

Remark. Experiment 2 and Experiment 3 reveal that the stripe-like pattern is not caused by interference, since the experimental setup for s-photons is the same for both experiments.

Experiment 4. The sub-population used in this experiment is $\{SP\}_4$, the same as $\{SP\}_3$ for Experiment 3. Therefore, we get the same result.

1.15 Theory of Photons

Maxwell's theory is not applicable to a single photon. It is a theory for electromagnetic waves, that is, for groups consisting of a huge number of photons. We need a theory of motion for a single photon to which our mechanics of random motion is applicable.

The intensity σ^2 of the random motion of an electron or atom is determined by its mass. Since photons have no mass, we assume that the intensity σ^2 of random motion of a photon is determined by its energy. Then we can apply our theory to a photon, as described below.

Suppose that a photon moves with the velocity c along the z-axis, and in the plane orthogonal to the z-axis the photon performs a random motion under Hooke's force with the intensity determined by its energy.

We assume that the energy of the photon in an electromagnetic wave of frequency ν is $h\nu$, and that the intensity of random motion is $\sigma^2 = h\nu$. Then the paired equations of motion of the photon are, with $\sigma^2 = h\nu$,

$$
\begin{cases}
\dfrac{\partial \phi}{\partial t} + \dfrac{1}{2}\sigma^2 \left(\dfrac{\partial^2 \phi}{\partial x^2} + \dfrac{\partial^2 \phi}{\partial y^2} \right) + c\dfrac{\partial \phi}{\partial z} - \dfrac{1}{2}\kappa^2 \left(x^2 + y^2 \right) \phi = 0, \\[2mm]
-\dfrac{\partial \widehat{\phi}}{\partial t} + \dfrac{1}{2}\sigma^2 \left(\dfrac{\partial^2 \widehat{\phi}}{\partial x^2} + \dfrac{\partial^2 \widehat{\phi}}{\partial y^2} \right) + c\dfrac{\partial \widehat{\phi}}{\partial z} - \dfrac{1}{2}\kappa^2 \left(x^2 + y^2 \right) \widehat{\phi} = 0.
\end{cases}
$$

Separating variables, the equations of motion along the z-axis are

$$
\begin{cases}
\dfrac{\partial \phi}{\partial t} + c\dfrac{\partial \phi}{\partial z} = 0, \\[2mm]
-\dfrac{\partial \widehat{\phi}}{\partial t} + c\dfrac{\partial \widehat{\phi}}{\partial z} = 0.
\end{cases}
$$

Therefore, they describe a uniform motion with velocity c.

The equations of motion in the (x, y)-plane are, with $\sigma^2 = h\nu$,

$$
\begin{cases}
\dfrac{\partial \phi}{\partial t} + \dfrac{1}{2}\sigma^2 \left(\dfrac{\partial^2 \phi}{\partial x^2} + \dfrac{\partial^2 \phi}{\partial y^2} \right) - \dfrac{1}{2}\kappa^2 \left(x^2 + y^2 \right) \phi = 0, \\[2mm]
-\dfrac{\partial \widehat{\phi}}{\partial t} + \dfrac{1}{2}\sigma^2 \left(\dfrac{\partial^2 \widehat{\phi}}{\partial x^2} + \dfrac{\partial^2 \widehat{\phi}}{\partial y^2} \right) - \dfrac{1}{2}\kappa^2 \left(x^2 + y^2 \right) \widehat{\phi} = 0.
\end{cases}
$$

The random motion in the (x, y)-plane was discussed in Sections 1.7 and 1.8. Applying the results obtained therein to the motion of a photon, we can state

Proposition 1.15.1. *Let x_t be determined by the equation*

$$
x_t = x_a + \sigma B_{t-a} - \sigma \kappa \int_a^t ds\, x_s,
$$

where $\sigma^2 = h\nu$ and B_t is one-dimensional Brownian motion.

Then there are three possible modes of motion:

(i) *$X_t = (x_t, 0)$ is an x-polarized random harmonic oscillation, and $X_t^\perp = (0, x_t)$ is an y-polarized random harmonic oscillation.*

(ii) *More generally, let $X_t = (x_t, \alpha x_t)$, where α is a real constant. This is a random harmonic oscillation along the line $y = \alpha x$ in the (x, y)-plane. The motion orthogonal to it is $X_t^\perp = (-\alpha x_t, x_t)$.*

(iii) *In polar coordinates (r, η), the random left and right rotations are given by the equations*

$$
\begin{cases}
r_t = r_a + \sigma B_{t-a}^1 + \displaystyle\int_a^t ds \left(\dfrac{3}{2}\sigma^2 \dfrac{1}{r_s} - \sigma \kappa r_s \right), \\[3mm]
\eta_t = \eta_a + \displaystyle\int_a^t \sigma \dfrac{1}{r_s}\, dB_{s-a}^2 \pm \int_a^t \sigma^2 \dfrac{1}{r_s^2}\, ds,
\end{cases}
$$

where the sign $+$ (resp., $-$) in the expression of η_t corresponds to left (resp., right) rotation . We interpret these as the polarization of the photon.

For instance, if the photon performs a random harmonic oscillation along the x-axis, it is an x-polarized photon. If it performs a random harmonic oscillation along a line $y = x$, then it is a linear combination of an x-polarized photon and an y-polarized photon. Finally, if the photon performs a left (resp., right) random rotation on a circle, then it is a left (resp., right) polarized photon.

Assertions (i) and (ii) were dealt with in Section 1.7; assertion (iii) was proved in Section 1.8. Figure 1.9 illustrates the circular random motions.

The electromagnetic wave has a frequency ν, whereas a photon has no frequency, but has the energy $h\nu$.

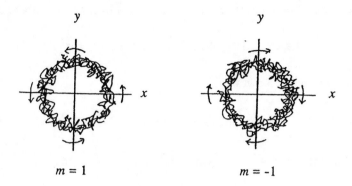

Figure 1.9

In the preceding section we applied the mechanics of random motion to a photon, but it was not clear whether this is correct or not. In this section we have shown that it is indeed appropriate to do so by presenting a particle model of a photon (Nagasawa (2012*)).

1.16 Principle of Least Action

The principle of least action describes the motion of a particle from a fixed initial time to a fixed terminal time. This is well known in classical mechanics. The principle of least action holds also in the mechanics of random motion. Since in the mechanics of random motion the entrance function at the initial time and the exit function at the terminal time are fixed, the principle of least action is a suitable way of describing the motion.

Now, suppose the vector potential $\mathbf{b}(t, x)$ is given, and add an arbitrary drift $\mathbf{a}(t, \omega)$, which is independent of events after time t. Let the stochastic process $\{Y_t, t \in [a, b], \mathbf{P}\}$ be given by a solution of the stochastic differential equation

$$Y_t = X_a + \sigma B_{t-a} + \int_a^t \left(\sigma^2 \mathbf{b}(s, Y_s) + \mathbf{a}(s, \omega)\right) ds. \qquad (1.16.1)$$

The paths Y_t depend on the additional drift $\mathbf{a}(t, \omega)$.

We denote

$$\mathbf{p}(t, \omega) = \sigma^2 \mathbf{b}(t, Y_t) + \mathbf{a}(t, \omega)$$

and define the Lagrangian $\mathcal{L}(Y_t)$ of the stochastic process Y_t as the expectation

$$L(Y_t) = \mathbf{P}\left[\frac{1}{2}\sigma^{-2}\left(\mathbf{p}(t) - \sigma^2 \mathbf{b}(t, Y_t)\right)^2 - \overline{c}(t, Y_t)\right],$$

which is a generalization of the classical Lagrangian to stochastic processes, with

$$\bar{c}\left(t,x\right) = c\left(t,x\right) + \frac{1}{2}\sigma^2\left(\nabla\cdot\mathbf{b}\left(t,x\right) + \mathbf{b}\left(t,x\right)^2\right).$$

The associated action integral,

$$\int_a^b \mathcal{L}\left(Y_t\right)dt = \int_a^b \mathbf{P}\left[\frac{1}{2}\sigma^{-2}\left(\mathbf{p}\left(t\right) - \sigma^2\mathbf{b}\left(t,Y_t\right)\right)^2 - \bar{c}\left(t,Y_t\right)\right]dt, \qquad (1.16.2)$$

is a generalization of the classical action integral to random motion.

The principle of least action asserts that the path that minimizes the action integral is the motion path that is actually realized.

To prove it, we apply Itô's formula (1.3.12) to the solution Y_t of equation (1.16.1) and the function $f\left(t,x\right) = \log\phi\left(t,x\right)$. Then, after some manipulations , we get

$$\log\phi\left(b,Y_b\right) - \log\phi\left(a,Y_a\right) = \int_a^b \sigma\frac{\nabla\phi}{\phi}\,dB_t$$
$$+ \int_a^b\left(\frac{L\phi}{\phi} + \frac{1}{2}\sigma^{-2}(\mathbf{a})^2\right)dt - \int_a^b\frac{1}{2}\left(\sigma^{-1}\mathbf{a} - \sigma\frac{\nabla\phi}{\phi}\right)^2 dt,$$

where L denotes the differential operator

$$L = \frac{\partial}{\partial t} + \frac{1}{2}\sigma^2\triangle + \sigma^2\mathbf{b}\left(t,x\right)\cdot\nabla.$$

In the above formula, $L\phi + \bar{c}\phi = 0$, since the function ϕ satisfies the equation of motion. Therefore, $-L\phi/\phi = \bar{c}$. We substitute this in the integral on the right-hand side of the formula and then take the expectation of both sides. Since the expectation of the stochastic integral vanishes, we deduce that

$$\mathbf{P}\left[\int_a^b\left(\frac{1}{2}\sigma^{-2}(\mathbf{a})^2 - \bar{c}\right)dt\right] = \mathbf{P}\left[\log\phi\left(b,Y_b\right) - \log\phi\left(a,Y_a\right)\right]$$
$$+ \mathbf{P}\left[\int_a^b\frac{1}{2}\left(\sigma^{-1}\mathbf{a} - \sigma\frac{\nabla\phi}{\phi}\right)^2 dt\right]. \qquad (1.16.3)$$

As, by definition, $\mathbf{a}\left(t,\omega\right) = \mathbf{p}\left(t,\omega\right) - \sigma^2\mathbf{b}\left(t,Y_t\right)$, the left-hand side of equation (1.16.3) is equal to the action integral in (1.16.2), and the right-hand side attains its minimum when the second term vanishes.

Namely, when

$$\mathbf{a}(t, \omega) = \sigma^2 \frac{\nabla \phi(t, Y_t)}{\phi(t, Y_t)},$$

the action integral attains its minimum, i.e., the choice $\mathbf{a}(t, x) = \sigma^2 \nabla \phi(t, x)/\phi(t, x)$ for the evolution drift vector minimizes the action integral. Therefore, the stochastic process determined by the principle of least action is the random motion determined by the equations of motion (1.4.1).

Theorem 1.16.1. *The principle of least action*

$$\int_a^b \mathcal{L}(X_t) \, dt = \inf_Y \int_a^b \mathcal{L}(Y_t) \, dt$$

determines the random motion $\{X_t, t \in [a, b], \mathbf{P}\}$, *which satisfies the path equation*

$$X_t = X_a + \sigma B_{t-a} + \int_a^t \left(\sigma^2 \mathbf{b}(s, X_s) + \sigma^2 \frac{\nabla \phi}{\phi}(s, X_s) \right) ds,$$

where the distribution density of X_a *is* $\phi(a, x) \, \widehat{\phi}(a, x)$. *That is, the random motion with the evolution drift* $\mathbf{a}(t, x) = \sigma^2 \nabla \phi(t, x)/\phi(t, x)$ *satisfies the principle of least action. In other words, the random motion determined by the equations of motion* (1.4.1) *minimizes the action integral.*

Comment to Theorem 1.16.1. Fényes (1952) was the first to discuss the principle of least action for stochastic processes. The existence of stochastic processes which satisfy the principle of least action will be proved in Section 5.2.

1.17 Transformation of Probability Measures

There are three ways to derive a description of the random motion of a particle under external forces from the equations of motion (1.4.1).

(i) The orthodox way. This is the combination of the equations of motion in Section 1.4 and the kinematics developed in Section 1.5.

(ii) The principle of least action. This is explained in the preceding section.

(iii) The third way, based on a transformation of probability measures on the path space Ω. We will explain it next.

First, we prepare a Markov process $\{X_t; \mathbf{P}_{(s,x)}, (s, x) \in [a, b] \times \mathbf{R}^d\}$, given by the stochastic differential equation

$$X_t = x + \sigma B_{t-a} + \int_a^t \sigma^2 \mathbf{b}(s, X_s) \, ds, \tag{1.17.1}$$

where σ is a positive constant and $\mathbf{b}(s, x)$ is a given vector potential.

Definition 1.17.1. Given a solution $\phi(s, x)$ of the equations of motion, define the functional N_s^t by the formula

$$N_s^t = \exp\left(\int_s^t \overline{c}\,(r, X_r)\,dr\right)\frac{\phi\,(t, X_t)}{\phi\,(s, X_s)}, \qquad (1.17.2)$$

where

$$\overline{c}\,(t, x) = -\frac{L\phi(t, x)}{\phi(t, x)} = c\,(t, x) + \frac{1}{2}\sigma^2(\nabla \cdot \mathbf{b}\,(t, x) + \mathbf{b}\,(t, x)^2).$$

Theorem 1.17.1 (Nagasawa (1989, a)).

(i) *The functional N_s^t defined by formula (1.17.2) satisfies the normalization condition*

$$\mathbf{P}_{(s,x)}\left[N_s^t\right] = 1. \qquad (1.17.3)$$

(ii) *The transformed probability measure $\mathbf{Q}_{(s,x)}$ given by*

$$\mathbf{Q}_{(s,x)} = \mathbf{P}_{(s,x)}N_s^b \qquad (1.17.4)$$

defines a new Markov process $\{X_t; \mathbf{Q}_{(s,x)}, (s, x) \in [a, b] \times \mathbf{R}^d\}$, where the functional N_s^t serves as the density of $\mathbf{Q}_{(s,x)}$ with respect to $\mathbf{P}_{(s,x)}$.

(iii) *The density N_s^t induces the evolution drift*

$$\mathbf{a}\,(t, x) = \sigma^2\frac{\nabla\phi\,(t, x)}{\phi\,(t, x)},$$

and the transformed Markov process $\{X_t; \mathbf{Q}_{(s,x)}, (s, x) \in [a, b] \times \mathbf{R}^d\}$ with the transformed probability measure $\mathbf{Q}_{(s,x)} = \mathbf{P}_{(s,x)}N_s^b$ satisfies the stochastic differential equation

$$X_t = x + \sigma\widetilde{B}_{t-a} + \int_a^t \left(\sigma^2\mathbf{b}\,(s, X_s) + \mathbf{a}\,(s, X_s)\right) ds, \qquad (1.17.5)$$

where $\widetilde{B}_t$ is the Brownian under the probability measure $\mathbf{Q}_{(s,x)} = \mathbf{P}_{(s,x)}N_s^b$.

Theorem 1.17.2 (Nagasawa (1989, a)). *The transformed Markov process $\{X_t; \mathbf{Q}\}$ with the initial distribution density given by $\mu_a(x) = \phi(a, x)\widehat{\phi}_a(x)$, where*

$$\mathbf{Q} = \int \mu_a\,(x)\,dx\,\mathbf{Q}_{(a,x)},$$

coincides with the random motion constructed by using the triplet

$$\{p(a, z, b, y), \widehat{\phi}_a(z), \phi_b(y)\}$$

in Section 1.4. In other words, if we transform the basic Markov process $\{X_t; \mathbf{P}_{(s,x)}\}$ *by using the functional*

$$N_s^t = \exp\left(\int_s^t \overline{c}\,(r, X_r)\,dr\right) \frac{\phi(t, X_t)}{\phi(s, X_s)},$$

we get the random motion determined by the equations of motion (1.4.1).

We call Theorem 1.17.2 the *method of transformation of probability measures* on the path space Ω by using the functional N_s^t.

Since the proofs of the above theorems require the theory of transformation of Markov processes, we will give them in Section 4.2.

1.18 Schrödinger Equation and Path Equation

Recall that a solution of the Schrödinger equation of the form

$$\psi\,(t, x) = e^{R(t,x)+iS(t,x)}$$

is a complex evolution function, with $R\,(t, x)$ the generating function of distribution and $S\,(t, x)$ the drift potential. As already remarked,

the Schrödinger equation does not yield a stochastic process or the path equation.

To obtain a stochastic process or the path equation, we need the equations of motion (1.4.1). For the sake of convenience, we state a theorem that will be referred to in applications discussed in the next chapter.

Theorem 1.18.1. *Consider the Schrödinger equation*

$$i\frac{\partial \psi}{\partial t} + \frac{1}{2}\sigma^2(\nabla + i\mathbf{b}\,(t, x))^2\psi - V\,(t, x)\,\psi = 0, \tag{1.18.1}$$

where $\mathbf{b}(t, x)$ *and* $V(t, x)$ *are a vector and a scalar potential, respectively, and represent its solutions in the exponential form*

$$\psi\,(t, x) = e^{R(t,x)+iS(t,x)}. \tag{1.18.2}$$

Then:

(i) *The complex evolution function* $\psi(t, x)$ *and its complex conjugate* $\overline{\psi}(t, x)$ *give the distribution density*

$$\mu(t, x) = \psi(t, x)\overline{\psi}(t, x) = e^{2R(t,x)}. \tag{1.18.3}$$

(This is well known in quantum mechanics.)

(ii) *By the equivalence established in Theorem 1.9.1, the equations of motion*
   (1.4.1) yield a stochastic process $\{X_t, t \in [a, b] \,; \mathbf{Q}\}$ *via formula (1.4.3):*

$$\mathbf{Q}[\{\omega : \omega(t_1) \in \Gamma_1, \omega(t_2) \in \Gamma_2, \dots, \omega(t_n) \in \Gamma_n\}]$$

$$= \int dx_0 \, \widehat{\phi}_a(x_0) p(a, x_0, t_1, x_1) \mathbf{1}_{\Gamma_1}(x_1) dx_1 \, p(t_1, x_1, t_2, x_2) \mathbf{1}_{\Gamma_2}(x_2) dx_2 \times \cdots$$

$$\times \cdots p(t_{n-1}, x_{n-1}, t_n, x_n) \mathbf{1}_{\Gamma_n}(x_n) dx_n \, p(t_n, x_n, b, y) \phi_b(y) dy,$$

where $\widehat{\phi}_a(x)$ *is an entrance function and* $\phi_b(y)$ *is an exit function.*

   The solution of the equations of motion (1.4.1) is a pair of evolution
functions

$$\phi(t, x) = e^{R(t,x) + S(t,x)}, \quad \widehat{\phi}(t, x) = e^{R(t,x) - S(t,x)}, \tag{1.18.4}$$

where $R(t, x)$ *is the generating function of distribution and* $S(t, x)$ *is the drift*
potential.

   The evolution drift is given by the formula

$$\mathbf{a}(t, x) = \sigma^2 \frac{\nabla \phi(t, Y_t)}{\phi(t, Y_t)} = \sigma^2 (\nabla R(a, x) + \nabla S(a, x)). \tag{1.18.5}$$

   The kinematics equation of the random motion $\{X_t, t \in [a, b] \,; \mathbf{Q}\}$ *is*

$$\frac{\partial u}{\partial t} + \frac{1}{2} \sigma^2 \triangle u + \left(\sigma^2 \mathbf{b}(t, x) + \mathbf{a}(t, x)\right) \cdot \nabla u = 0, \tag{1.18.6}$$

where the evolution drift is given by (1.18.5).

   The random motion $\{X_t, t \in [a, b] \,; \mathbf{Q}\}$ *satisfies the path equation*

$$X_t = X_a + \sigma B_{t-a} + \int_a^t \left(\sigma^2 \mathbf{b}(s, X_s) + \mathbf{a}(s, X_s)\right) ds, \tag{1.18.7}$$

where the distribution density of X_a *is* $\psi(a, x) \overline{\psi}(a, x) = e^{2R(a,x)}$ *and the evo-*
lution drift is $\mathbf{a}(t, x) = \sigma^2(\nabla R(a, x) + \nabla S(a, x))$. *The distribution density of*
the stochastic process $\{X_t, t \in [a, b] \,; \mathbf{Q}\}$ *is given by* $\mu(t, x) = \phi(t, x) \widehat{\phi}(t, x) =$
$e^{2R(t,x)}$.

(iii) *The superposition*

$$\psi^*(t, x) = \alpha_1 \psi_1(t, x) + \alpha_2 \psi_2(t, x)$$

of the Schrödinger functions

$$\psi_1(t, x) = e^{R_1(t,x) + iS_1(t,x)} \quad and \quad \psi_2(t, x) = e^{R_2(t,x) + iS_2(t,x)}$$

induces an entangled random motion $\{X_t, \mathbf{Q}^*\}$ *with*

$$\psi^*(t,x) = e^{R^*(t,x)+iS^*(t,x)} = \alpha_1\psi_1(t,x) + \alpha_2\psi_2(t,x),$$

and paths given by the equation

$$X_t = X_a^* + \sigma B_{t-a} + \int_a^t \mathbf{a}^*(s, X_s)ds,$$

where $\mathbf{a}^*(t,x) = \sigma^2 \nabla R^*(t,x) + \sigma^2 \nabla S^*(t,x)$.

For example, let

$$\psi_1 = e^{R_1(x)}e^{-i\frac{E_1 t}{\hbar}}, \quad \psi_2 = e^{R_2(x)}e^{-i\frac{E_2 t}{\hbar}}.$$

Then the superposed Schrödinger function is

$$\psi^* = \alpha_1 e^{R_1(x)}e^{-i\frac{E_1 t}{\hbar}} + \alpha_2 e^{R_2(x)}e^{-i\frac{E_2 t}{\hbar}},$$

and, by formula (1.10.8) in Theorem 1.10.5, the corresponding distribution density is

$$\mu^*(t,x) = \alpha_1{}^2 e^{2R_1(x)} + \alpha_2{}^2 e^{2R_2(x)} + 2e^{R_1(x)+R_2(x)}\alpha_1\alpha_2 \cos\left(\frac{(E_2 - E_1)t}{\hbar}\right),$$

where the third term on the right-hand side represents the effect of entanglement, which induces the oscillation $\cos\left(\frac{(E_2-E_1)t}{\hbar}\right)$ of electrons. This is observed as emission of light.

Chapter 2

Applications

The mechanics of random motion developed in Chapter 1 can be applied to various problems in quantum theory, as quantum theory itself is the theory of random motion. The Schrödinger equation determines stochastic processes through the equations of motion, and then we can derive the kinematics equation and the path equation. We will first discuss the random motion induced by the Coulomb potential. Then the motion of charged particles in a magnetic field, the Aharonov-Bohm effect, the tunnel effect, and the Bose-Einstein condensation will be treated. In Section 2.6 we will explain that the relation between the light-cone and random motion is different from the one for smooth motion, namely, a randomly moving particle can leave the light-cone in the short space-time scale. In Section 2.7 the theory of random motion will be applied to discuss a model of the origin of the Universe.

Section 2.9 is devoted to a particle theory of electron holography. In Section 2.10, as an application of the distribution density of random particles, septation of mutant Escherichia coli and the mass spectrum of mesons are discussed. Finally, Section 2.11 deals with high-temperature superconductivity.

2.1 Motion induced by the Coulomb Potential

In this section we will discuss the hydrogen atom. Schrödinger's wave mechanics and Born-Heisenberg's quantum mechanics established their fame through this problem, namely, the computation of the spectrum of the hydrogen atom.

We will show that not only spectra, but also the paths of the motion of an electron in the hydrogen atom can be computed.

The equation of motion is the Schrödinger equation (complex evolution equation)

$$i\frac{\partial \psi}{\partial t} + \frac{1}{2}\sigma^2 \triangle \psi - V(r)\psi = 0,$$

M. Nagasawa, *Markov Processes and Quantum Theory*,
Monographs in Mathematics 109, https://doi.org/10.1007/978-3-030-62688-4_2

where $V(r)$ is the Coulomb potential:

$$V(r) = -\frac{\alpha}{r}, \quad r = \sqrt{x^2 + y^2 + z^2}.$$

In the case of an electron in a hydrogen atom, $\alpha = \frac{e^2}{\hbar}$, $\sigma^2 = \frac{\hbar}{m}$, e is the charge of the electron, m is its mass and h is the Planck constant.

Since the potential is a function of r, it is appropriate to work in the spherical coordinates (r, θ, η),

$$x = r \sin \theta \cos \eta, \quad y = r \sin \theta \sin \eta, \quad z = r \cos \theta.$$

Then the volume element is

$$dx\, dy\, dz = r^2 \sin \theta dr\, d\theta\, d\eta,$$

and the differential operators involved are

$$\nabla u = \operatorname{grad} u = \left(\frac{\partial u}{\partial r}, \; \frac{1}{r} \frac{\partial u}{\partial \theta}, \; \frac{1}{r \sin \theta} \frac{\partial u}{\partial \eta} \right),$$

and

$$\nabla \cdot \mathbf{X} = \operatorname{div} \mathbf{X} = \frac{1}{r^2} \frac{\partial \left(r^2 X_1 \right)}{\partial r} + \frac{1}{r \sin \theta} \frac{\partial \left(\sin \theta X_2 \right)}{\partial \theta} + \frac{1}{r \sin \theta} \frac{\partial X_3}{\partial \eta},$$

where $\mathbf{X} = (X_1, X_2, X_3)$.

The Laplacian has the form

$$\triangle u = \nabla \cdot \nabla u = \frac{1}{r^2} \frac{\partial}{\partial r} \left(r^2 \frac{\partial u}{\partial r} \right) + \frac{1}{r^2 \sin \theta} \frac{\partial}{\partial \theta} \left(\sin \theta \frac{\partial u}{\partial \theta} \right) + \frac{1}{r^2 \sin^2 \theta} \frac{\partial^2 u}{\partial \eta^2}.$$

Therefore, in spherical coordinates the Schrödinger equation (complex evolution equation) reads

$$i \frac{\partial \psi}{\partial t} + \frac{1}{2} \sigma^2 \left\{ \frac{1}{r^2} \frac{\partial}{\partial r} \left(r^2 \frac{\partial \psi}{\partial r} \right) + \frac{1}{r^2 \sin \theta} \frac{\partial}{\partial \theta} \left(\sin \theta \frac{\partial \psi}{\partial \theta} \right) + \frac{1}{r^2 \sin^2 \theta} \frac{\partial^2 \psi}{\partial \eta^2} \right\} - V(r)\, \psi = 0.$$

We consider stationary motions and set $\psi = e^{-i\lambda t} \varphi(r, \theta, \eta)$. Then

$$-\frac{1}{2} \sigma^2 \left\{ \frac{1}{r^2} \frac{\partial}{\partial r} \left(r^2 \frac{\partial \varphi}{\partial r} \right) + \frac{1}{r^2 \sin \theta} \frac{\partial}{\partial \theta} \left(\sin \theta \frac{\partial \varphi}{\partial \theta} \right) + \frac{1}{r^2 \sin^2 \theta} \frac{\partial^2 \varphi}{\partial \eta^2} \right\} + V(r)\, \varphi = \lambda \varphi.$$

This is an eigenvalue problem; its eigenvalues λ represent the values of the energy. (For the energy of random motion, refer to Section 3.1.)

Separating variables by substituting $\varphi = R(r)\, \Theta(\theta)\, \Phi(\eta)$, we obtain the ordinary differential equations

$$-\frac{d^2 \Phi}{d\eta^2} = m^2 \Phi,$$

$$-\frac{1}{\sin \theta} \frac{d}{d\theta} \left(\sin \theta \frac{d\Theta}{d\theta} \right) + \frac{m^2}{\sin^2 \theta} \Theta = \beta \Theta,$$

$$-\frac{1}{2} \sigma^2 \left\{ \frac{1}{r^2} \frac{d}{dr} \left(r^2 \frac{dR}{dr} \right) - \frac{\beta}{r^2} R \right\} + V(r)\, R = \lambda_n R,$$

with the corresponding eigenvalues

$$\beta = \ell\,(\ell+1), \quad \ell = |m|\,, |m|+1, \ldots$$
$$\lambda_n = -\frac{\alpha^2}{2\sigma^2}\frac{1}{n^2}, \quad n = \ell+1, \ell+2, \ell+3, \ldots \tag{2.1.1}$$
$$\ell = 0, 1, 2, \ldots, n-1, \quad m = 0, \pm 1, \pm 2, \ldots, \pm \ell$$

(see, for instance, Chapter 5 of Pauling and Wilson (1935)).

Let us describe the motion of the electron for each choice of the numbers (n, ℓ, m).

Let $n = 1$. By (2.1.1), $\ell = 0$, $m = 0$. We will label this case as $\{n, \ell, m\} = \{1, 0, 0\}$. Then the eigenvalue (energy) is

$$\lambda_1 = -\frac{\alpha^2}{2\sigma^2},$$

with associated eigenfunction

$$\varphi\,(r, \theta, \eta) = \gamma e^{-\frac{\alpha r}{\sigma^2}}.$$

Therefore, the complex evolution function is

$$\psi\,(t, (r, \theta, \eta)) = \gamma e^{-\frac{\alpha r}{\sigma^2} - i\lambda_1 t},$$

where γ a normalization constant.

In Born's quantum-mechanical statistical interpretation (1926),

$$\mu_t(d\,(r, \theta, \eta)) = \psi\,(t, (r, \theta, \eta))\,\overline{\psi}\,(t, (r, \theta, \eta))\,r^2 \sin\theta dr d\theta d\eta$$

represents the distribution density of an electron.

In our mechanics of random motion we established Theorem 1.10.1, which in spherical coordinates reads

Theorem 2.1.1. *The distribution density*

$$\mu_t(d\,(r, \theta, \eta)) = \mathbf{Q}[\mathbf{X}_t \in d\,(r, \theta, \eta)]$$

of the random motion $\mathbf{X}_t$ of an electron is given by the formula

$$\mathbf{Q}[\mathbf{X}_t \in d\,(r, \theta, \eta)] = \psi\,(t, (r, \theta, \eta))\,\overline{\psi}\,(t, (r, \theta, \eta))\,r^2 \sin\theta\,dr\,d\theta\,d\eta$$

$$= \gamma^2 e^{-\frac{2\alpha r}{\sigma^2}}\,r^2 \sin\theta\,dr\,d\theta\,d\eta,$$

i.e., $\psi\,(t, (r, \theta, \eta))\,\overline{\psi}\,(t, (r, \theta, \eta))$ is the distribution density of the motion $\mathbf{X}_t$ of an electron per volume element $r^2 \sin\theta dr d\theta d\eta$.

Proof. By (1.4.3), in spherical coordinates the distribution of a path $\mathbf{X}_t = (r_t, \theta_t, \eta_t)$ of the random motion of an electron is

$$\mathbf{Q}[\mathbf{X}_t \in d\,(r, \theta, \eta)] = \phi(t, (r, \theta, \eta))\widehat{\phi}(t, (r, \theta, \eta))r^2 \sin\theta\, dr\, d\theta\, d\eta,$$

where

$$\phi(t, (r, \theta, \eta))\widehat{\phi}(t, (r, \theta, \eta)) = \psi\,(t, (r, \theta, \eta))\,\overline{\psi}\,(t, (r, \theta, \eta)) = \gamma^2 e^{-\frac{2\alpha r}{\sigma^2}},$$

as we needed to show.   □

Let us we emphasize that the above formula is not an "interpretation", but an actual "theorem" in our mechanics of random motion.

To investigate the paths of the motion of an electron we apply the theory of random motion, specifically, Theorem 1.17.1.

The complex evolution function is

$$\psi\,(t, (r, \theta, \eta)) = \gamma e^{-\frac{\alpha r}{\sigma^2} - i\lambda_1 t},$$

so the corresponding generating function and drift potential are

$$R\,(r) = \log\gamma - \frac{\alpha r}{\sigma^2} \quad \text{and} \quad S\,(t) = -\lambda_1 t. \tag{2.1.2}$$

Therefore, the evolution functions are, with the normalization constant γ,

$$\phi\,(t, (r, \theta, \eta)) = e^{R(r)+S(t)} = \gamma e^{-\frac{\alpha r}{\sigma^2} - \lambda_1 t},$$
$$\widehat{\phi}\,(t, (r, \theta, \eta)) = e^{R(r)-S(t)} = \gamma e^{-\frac{\alpha r}{\sigma^2} + \lambda_1 t}.$$

Then, by formula (1.4.7), the evolution drift is

$$\mathbf{a} = \sigma^2\,\nabla\log\phi = \sigma^2\,(\nabla R\,(r) + \nabla S\,(t)) = (-\alpha, 0, 0).$$

Hence, the equation of motion reads

$$\frac{\partial u}{\partial t} + \frac{1}{2}\sigma^2 \left\{ \frac{1}{r^2}\frac{\partial}{\partial r}\left(r^2\frac{\partial u}{\partial r}\right) + \frac{1}{r^2\sin\theta}\frac{\partial}{\partial\theta}\left(\sin\theta\frac{\partial u}{\partial\theta}\right) + \frac{1}{r^2\sin^2\theta}\frac{\partial^2 u}{\partial\eta^2} \right\} - \alpha\frac{\partial u}{\partial r} = 0.$$

Expanding, we have

$$\frac{\partial u}{\partial t} + \frac{1}{2}\sigma^2\left(\frac{\partial^2 u}{\partial r^2} + \frac{1}{r^2}\frac{\partial^2 u}{\partial\theta^2} + \frac{1}{r^2\sin^2\theta}\frac{\partial^2 u}{\partial\eta^2}\right) + \left(\frac{\sigma^2}{r} - \alpha\right)\frac{\partial u}{\partial r} + \frac{1}{2}\frac{\sigma^2}{r^2}\cot\theta\frac{\partial u}{\partial\theta} = 0,$$

and the corresponding path equation is given by (1.5.10). Expressed in terms of the three-dimensional Brownian motion $\mathbf{B}_t = \left(B_t^1, B_t^2, B_t^3\right)$, it takes the form of the system

$$
\begin{cases}
r_t = r_a + \sigma B^1_{t-a} + \displaystyle\int_a^t (\sigma^2 \frac{1}{r_s} - \alpha) ds, \quad r_t > 0, \\[3mm]
\theta_t = \theta_a + \displaystyle\int_a^t \sigma \frac{1}{r_s} dB^2_{s-a} + \frac{1}{2} \int_a^t \sigma^2 \frac{1}{r_s{}^2} \cot \theta_s \, ds, \quad \theta_t \in (0, \pi), \\[3mm]
\eta_t = \eta_a + \displaystyle\int_a^t \sigma \frac{1}{r_s \sin \theta_s} dB^3_{s-a}.
\end{cases}
$$

The radial component r_t of the random motion is attracted to the origin by the drift $-\alpha$, but also strongly repelled by the drift $\sigma^2 \frac{1}{r}$, and so it does not reach the origin.

Let $\bar{r} = \frac{\sigma^2}{\alpha}$ be the zero of the radial drift, i.e., the solution of

$$
\sigma^2 \frac{1}{r} - \alpha = 0.
$$

If $r < \bar{r}$ (resp., $r > \bar{r}$), then the radial drift is positive (resp., negative). Therefore, the zero $\bar{r} = \frac{\sigma^2}{\alpha}$ of the radial drift is the equilibrium point of the motion.

In the case of the hydrogen atom,

$$
\alpha = \frac{e^2}{\hbar} = 2.4341349 \times 10^{-11}, \quad \sigma^2 = \frac{\hbar}{m} = 1.28809 \times 10^{-21},
$$

and

$$
\bar{r} = \frac{\sigma^2}{\alpha} = \frac{\hbar^2}{me^2} = 0.529177 \text{ Å} \quad (1 \text{ Å} = 10^{-8} \text{ cm}),
$$

which coincides with the classical **Bohr radius**.

Bohr proposed a model in which an electron performs circular motion with the Bohr radius. He had no idea of random motion, but his model was a good approximation.

Next, we look at the θ-component of the random motion. The drift

$$
\frac{1}{2} \frac{\sigma^2}{r^2} \cot \theta
$$

diverges for $r > 0$ at $\theta = 0$ and $\theta = \pi$, hence the θ-component cannot reach the z-axis: the particle remains near by (x, y)-plane. Since there is no drift in the η-direction, the η-component of the motion does not perform a "circular motion".

By Theorem 2.1.1 the distribution density $\gamma^2 e^{-\frac{2\alpha r}{\sigma^2}} r^2 \sin \theta$ of the random motion $\mathbf{X}_t = (r_t, \theta_t, \eta_t)$ with respect to the measure $dr \, d\theta \, d\eta$ vanishes as $r \downarrow 0$.

In this treatment, the z-axis is artificially fixed. However, for a free hydrogen atom we can use Cartesian coordinates and take the evolution function

$$
\phi(t, x, y, z) = \gamma e^{-\frac{\alpha r}{\sigma^2} - \lambda_1 t}, \quad r = (x^2 + y^2 + z^2)^{1/2}.
$$

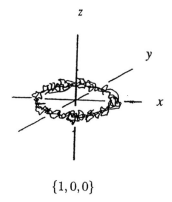

$$\{1,0,0\}$$

Figure 2.1

In this case the evolution drift vector is

$$\mathbf{a}(x,y,z) = \sigma^2 \nabla \log \phi(x,y,z) = \sigma^2 \nabla \left(-\frac{\alpha}{\sigma^2} r \right)(x,y,z)$$

$$= (-\alpha \frac{x}{r}, -\alpha \frac{y}{r}, -\alpha \frac{z}{r}),$$

where $\alpha > 0$.

Therefore the path equation of the random motion $\mathbf{X}_t = (x_t, y_t, z_t)$ is, in components,

$$\begin{cases} x_t = x_a + \sigma B^1_{t-a} - \alpha \int_a^t \frac{x_s}{r_s} ds, \\ \\ , y_t = y_a + \sigma B^1_{t-a} - \alpha \int_a^t \frac{y_s}{r_s} ds, \\ \\ z_t = z_a + \sigma B^1_{t-a} - \alpha \int_a^t \frac{z_s}{r_s} ds, \end{cases}$$

where $\mathbf{B}_t = (B^1_t, B^2_t, B^3_t)$ is the three-dimensional Brownian motion. This represents a random oscillation about at the origin, but the probability that paths go through the origin is zero.

We first take a look of the x-component x_t. The drift on the x-axis is

$$-\alpha \frac{x_t}{r_t} = -\alpha \operatorname{sign}(x_t).$$

Therefore, if $x_t > 0$, then x_t performs a Brownian motion with drift $-\alpha$, i.e., it is pulled towards the origin, goes through the origin, and if it becomes $x_t < 0$, then drift switches from $-\alpha$ to α and x_t performs a Brownian motion pulled also toward the origin. Thus, x_t performs a random oscillation about the origin, as claimed.

The y- and z-components behave in the same manner as the x-component. The distribution of the path $\mathbf{X}_t = (x_t, y_t, z_t)$ of the electron is, by the definition (1.4.3) of the probability law $\mathbf{Q}$,

$$\mu_t(d(x, y, z)) = \phi(t, x, y, z)\widehat{\phi}(t, x, y, z)dx\,dy\,dz = \gamma^2 e^{-\frac{2\alpha r}{\sigma^2}}dx\,dy\,dz.$$

In Cartesian coordinates we cannot discuss physical quantities such as the Bohr radius. Figure 2.2 illustrates the paths of the motion.

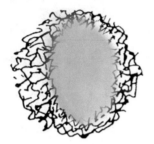

Figure 2.2

A free hydrogen atom looks like a golf ball. It contains the paths of the electron with drift $\mathbf{a}(x, y, z) = (-\alpha\,\text{sign}(x), -\alpha\,\text{sign}(y), -\alpha\,\text{sign}(z))$ toward the origin.

Now let us consider the case $n = 2$. We choose (ℓ, m) in (2.1.1).

(i) The case $\{n, \ell, m\} = \{2, 0, 0\}$. The eigenvalue is

$$\lambda_2 = -\frac{\alpha^2}{2\sigma^2}\frac{1}{2^2},$$

with associated eigenfunction

$$\varphi(r, \theta, \eta) = \gamma\left(1 - \frac{\alpha r}{2\sigma^2}\right)e^{-\frac{\alpha r}{2\sigma^2}}.$$

(See, for instance, Chapter 5 of Pauling and Wilson (1935).)

Therefore, the complex evolution function is given by

$$\psi(t, (r, \theta, \eta)) = \gamma\left(1 - \frac{\alpha r}{2\sigma^2}\right)e^{-\frac{\alpha r}{2\sigma^2} - i\lambda_2 t},$$

where γ is a normalization constant.

Again according to Born's quantum-mechanical statistical interpretation,

$$\mu_t(d\,(r,\theta,\eta)) = \psi\,(t,(r,\theta,\eta))\,\overline{\psi}\,(t,(r,\theta,\eta))\,r^2\sin\theta\,dr\,d\theta\,d\eta$$

$$= \gamma^2\left(1 - \frac{\alpha r}{2\sigma^2}\right)^2 e^{-\frac{\alpha r}{\sigma^2}}r^2\sin\theta\,dr\,d\theta\,d\eta \tag{2.1.3}$$

is the distribution of an electron. This is all that quantum mechanics can do.

In the mechanics of random motion we can compute the paths $\mathbf{X}_t$ of the motion of an electron; specifically, we apply Theorem 1.17.1.

Consider the pair of evolution functions

$$\phi\,(t,(r,\theta,\eta)) = \gamma\left(1 - \frac{\alpha r}{2\sigma^2}\right)e^{-\frac{\alpha r}{2\sigma^2}-\lambda_2 t},$$

$$\widehat{\phi}\,(t,(r,\theta,\eta)) = \gamma\left(1 - \frac{\alpha r}{2\sigma^2}\right)e^{-\frac{\alpha r}{2\sigma^2}+\lambda_2 t},$$

and then use formula (1.4.7) and compute the corresponding evolution drift as

$$\mathbf{a} = \sigma^2\,\nabla\log\phi = \left(-\frac{\alpha\sigma^2}{2\sigma^2 - \alpha r} - \frac{\alpha}{2}, 0, 0\right).$$

In spherical coordinates, the kinematics equation is

$$\frac{\partial u}{\partial t} + \frac{1}{2}\sigma^2\left\{\frac{1}{r^2}\frac{\partial}{\partial r}\left(r^2\frac{\partial u}{\partial r}\right) + \frac{1}{r^2\sin\theta}\frac{\partial}{\partial\theta}\left(\sin\theta\frac{\partial u}{\partial\theta}\right) + \frac{1}{r^2\sin^2\theta}\frac{\partial^2 u}{\partial\eta^2}\right\}$$
$$-\left(\frac{\alpha\sigma^2}{2\sigma^2 - \alpha r} + \frac{\alpha}{2}\right)\frac{\partial u}{\partial r} = 0.$$

Expanding, we have

$$\frac{\partial u}{\partial t} + \frac{1}{2}\sigma^2\left(\frac{\partial^2 u}{\partial r^2} + \frac{1}{r^2}\frac{\partial^2 u}{\partial\theta^2} + \frac{1}{r^2\sin^2\theta}\frac{\partial^2 u}{\partial\eta^2}\right)$$
$$+\left(\frac{\sigma^2}{r} - \frac{\alpha\sigma^2}{2\sigma^2 - \alpha r} - \frac{\alpha}{2}\right)\frac{\partial u}{\partial r} + \frac{1}{2}\frac{\sigma^2}{r^2}\cot\theta\frac{\partial u}{\partial\theta} = 0.$$

Writing $\mathbf{X}_t = (r_t,\theta_t,\eta_t)$, the path equation is (1.5.11), so, in coordinates

$$\begin{cases} r_t = r_a + \sigma B_{t-a}^1 + \displaystyle\int_a^t\left(\sigma^2\frac{1}{r_s} - \frac{\alpha\sigma^2}{2\sigma^2 - \alpha r_s} - \frac{\alpha}{2}\right)ds, \\[2mm] \theta_t = \theta_a + \displaystyle\int_a^t\sigma\frac{1}{r_s}dB_{s-a}^2 + \frac{1}{2}\int_a^t\sigma^2\frac{1}{r_s^2}\cot\theta_s\,ds, \\[2mm] \eta_t = \eta_a + \displaystyle\int_a^t\sigma\frac{1}{r_s\sin\theta_s}dB_{s-a}^3, \end{cases}$$

where $\mathbf{B}_t = \left(B_t^1, B_t^2, B_t^3\right)$ is the three-dimensional Brownian motion.

The equation for the radial component shows that the random motion r_t does not reach the origin, as it is attracted to it by the drift $-\frac{\alpha}{2}$, but is strongly repelled by the drift $\sigma^2 \frac{1}{r}$. Moreover, the drift

$$-\frac{\alpha\sigma^2}{2\sigma^2 - \alpha r}$$

diverges at $\bar{r} = \frac{2\sigma^2}{\alpha}$: it goes to $-\infty$ if $r \uparrow \bar{r}$, and to ∞ if $r \downarrow \bar{r}$. Therefore, the random motion is divided into the regions $r < \bar{r}$ and $r > \bar{r}$.

The path equation for the θ-component shows that it cannot reach the z-axis, because the drift $\frac{1}{2}\frac{\sigma^2}{r^2}\cot\theta$ diverges at $\theta = 0$ and $\theta = \pi$. Therefore, the particle remains near the (x, y)-plane. Moreover, the η-component does not perform a "circular motion", because there is no drift.

Figure 2.3, which illustrates the paths, uses spherical coordinates with the fixed z-axis fixed.

In Cartesian coordinates, the evolution function

$$\phi(t, x, y, z) = \gamma \left(1 - \frac{\alpha r}{2\sigma^2}\right) e^{-\frac{\alpha r}{2\sigma^2} - \lambda_2 t}, \quad r = (x^2 + y^2 + z^2)^{1/2}$$

with $\alpha > 0$ determines the evolution drift

$$\mathbf{a}(x, y, z) = \sigma^2 \nabla \log \phi(x, y, z) = \sigma^2 \nabla \left(\log(1 - \frac{\alpha}{2\sigma^2}r) - \frac{\alpha}{2\sigma^2}r\right)(x, y, z)$$

$$= \left(-\frac{\alpha}{2}\frac{1}{1 - \frac{\alpha}{2\sigma^2}r}\frac{x}{r} - \frac{\alpha}{2}\frac{x}{r}, -\frac{\alpha}{2}\frac{1}{1 - \frac{\alpha}{2\sigma^2}r}\frac{y}{r} - \frac{\alpha}{2}\frac{y}{r}, -\frac{\alpha}{2}\frac{1}{1 - \frac{\alpha}{2\sigma^2}r}\frac{z}{r} - \frac{\alpha}{2}\frac{z}{r}\right),$$

which diverges at $\bar{r} = \frac{2\sigma^2}{\alpha}$: it goes to $-\infty$ if $r \uparrow \bar{r}$, and ∞ if $r \downarrow \bar{r}$. Therefore, r_t cannot reach the point $\bar{r} = \frac{2\sigma^2}{\alpha}$.

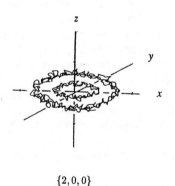

$\{2, 0, 0\}$

Figure 2.3

The drift on the x-axis (also on the y- and z-axes) is

$$-\frac{\alpha}{2}\left(\frac{1}{1-\frac{\alpha}{2\sigma^2}|x|}+1\right)\text{sign}(x).$$

By formula (1.4.3), the distribution of the path of the electron $\mathbf{X}_t = (x_t, y_t, z_t)$ is

$$\mu_t(d(x,y,z)) = \phi(t,x,y,z)\widehat{\phi}(t,x,y,z))dx\,dy\,dz$$
$$= \gamma^2\left(1-\frac{\alpha r}{2\sigma^2}\right)^2 e^{-\frac{\alpha r}{\sigma^2}}\,dx\,dy\,dz. \tag{2.1.3'}$$

Therefore, graphically it looks like a tennis ball containing a golf ball inside.

Figure 2.4. illustrates a section of the model of the paths of the random motion of an electron.

Figure 2.4

There is only one electron in a hydrogen atom. If the electron is in the inner doughnut or golf ball, it performs a random motion there and does not pass to the outer doughnut or to the golf ball, or vice versa.

Therefore, the paths of an electron in the hydrogen atom cannot realize the whole distribution given in equation (2.1.3) or (2.1.3'), namely, this distribution cannot be determined by a single hydrogen atom; rather, it represents a **statistical property of many hydrogen atoms.**

When $n \geq 2$, then as we have seen the separation of distribution occurs . However, the distribution itself does not tell us that the separation occurs for the path of an electron. We have detected it through the detailed analysis of paths of the random motion.

If a hydrogen atom is excited, it emits light and returns to the ground state. Let us analyze this process.

If we denote the energy of the ground state by E_1, then

$$\frac{E_1}{\hbar} = -\frac{\alpha^2}{2\sigma^2},$$

where $\alpha = e^2/\hbar$, $\sigma^2 = \hbar/m$, e is the charge of the electron, m is its mass, $\hbar = h/(2\pi)$, and h is the Planck constant.

The complex evolution function of the ground state is given by the formula

$$\psi_1(t,(r,\theta,\eta)) = \gamma\, e^{-\frac{\alpha r}{\sigma^2} - i\frac{E_1 t}{\hbar}},$$

where γ is a normalization constant.

Now consider an excited state $\{n,\ell,m\} = \{2,0,0\}$ and denote its energy by E_2. Then

$$\frac{E_2}{\hbar} = -\frac{\alpha^2}{2\sigma^2}\frac{1}{2^2}.$$

The complex evolution function of the excited state is given by

$$\psi_2(t,(r,\theta,\eta)) = \gamma\left(1 - \frac{\alpha r}{2\sigma^2}\right) e^{-\frac{\alpha r}{\sigma^2} - i\frac{E_2 t}{\hbar}},$$

where γ is the normalization constant.

Let us show that the emission of light is induced by the superposition (entanglement) of the complex evolution functions ψ_1 and ψ_2,

$$\psi^* = a_1\psi_1 + a_2\psi_2,$$

where a_1, a_2 are real constants satisfying the normalization condition $a_1^2 + a_2^2 = 1$, and

$$\psi_1 = e^{R_1(x)}e^{-i\frac{E_1 t}{\hbar}}, \qquad \psi_2 = e^{R_2(x)}e^{-i\frac{E_2 t}{\hbar}}.$$

Hence,

$$\psi^* = a_1 e^{R_1(x)}e^{-i\frac{E_1 t}{\hbar}} + a_2 e^{R_2(x)}e^{-i\frac{E_2 t}{\hbar}}.$$

To analyze the paths of the superposed random motion of an electron we must represent $\psi^*(t,x)$ as

$$\psi^*(t,x) = e^{R^*(t,x)+iS^*(t,x)},$$

and compute the distribution $R^*(t,x)$ and drift-potential $S^*(t,x)$ in the exponent.

However, this is not an easy task, because we must solve

$$e^{R^*(t,x)+iS^*(t,x)} = a_1 e^{R_1(x)}e^{-i\frac{E_1 t}{\hbar}} + a_2 e^{R_2(x)}e^{-i\frac{E_2 t}{\hbar}}.$$

To avoid this difficulty, we analyze the distribution $\mu^*(t,x) = |\psi^*(t,x)|^2$, and guess the form of the random motion. For this purpose, we use formula (1.10.8) in Theorem 1.10.5. Since a_1, a_2 are real constants, we obtain

$$\mu^*(t,x) = a_1{}^2 e^{2R_1(t,x)} + a_2{}^2 e^{2R_2(t,x)}$$
$$+ 2e^{R_1(t,x)+R_2(t,x)} a_1 a_2 \cos(S_1(t,x) - S_2(t,x)).$$

In our case the distribution density is

$$\mu^*(t,x) = a_1{}^2 e^{2R_1(x)} + a_2{}^2 e^{2R_2(x)}$$
$$+ 2e^{R_1(x)+R_2(x)} a_1 a_2 \cos\left(\frac{(E_2 - E_1)t}{\hbar}\right).$$

The first term $|\psi_1|^2 = e^{2R_1(t,x)}$ on the right-hand side is the contribution of the random motion of the ground state ψ_1. The second term $|\psi_2|^2 = e^{2R_2(t,x)}$ is the contribution of the random motion of the excited state ψ_2.

The third term shows the effect of the entanglement, which induces the oscillation $\cos\left(\frac{(E_2-E_1)t}{\hbar}\right)$. This oscillation is emitted as light, and the electron drops from the excited state to the ground state.

Summarizing the above, we state

Theorem 2.1.2. *If the complex evolution function ψ_2 of the excited state is entangled with the complex evolution function ψ_1 of the ground state, the entanglement induces the oscillation $\cos\left(\frac{(E_2-E_1)t}{\hbar}\right)$ of the electron. This is emitted as light.*

(ii) The case $\{n, \ell, m\} = \{2, 1, 0\}$. The eigenvalue is

$$\lambda_2 = -\frac{\alpha^2}{2\sigma^2}\frac{1}{2^2},$$

with the associated eigenfunction

$$\varphi(r, \theta, \eta) = \gamma \frac{\alpha r}{2\sigma^2} e^{-\frac{\alpha r}{2\sigma^2}} \cos\theta.$$

(See, for instance, Chapter 5 of Pauling and Wilson (1935).)

Therefore, the complex evolution function is

$$\psi(t, (r, \theta, \eta)) = \gamma\left(\frac{\alpha r}{2\sigma}\right) e^{-\frac{\alpha r}{2\sigma^2} - i\lambda_2 t} \cos\theta.$$

Again, according to Born's quantum-mechanical statistical interpretation,

$$\mu_t(d(r, \theta, \eta)) = \psi(t, (r, \theta, \eta)) \overline{\psi}(t, (r, \theta, \eta)) r^2 \sin\theta \, dr \, d\theta \, d\eta$$

$$= \gamma^2 \left(\frac{\alpha r}{2\sigma^2}\right)^2 e^{-\frac{\alpha r}{\sigma^2}} \cos^2\theta \, r^2 \sin\theta \, dr \, d\theta \, d\eta$$

is the distribution of an electron.

To compute the paths of an electron we apply the mechanics of random motion instead of quantum mechanics. In short, we use Theorem 1.17.1.

First of all, if we apply formula (1.4.7) to the evolution function $\phi = \varphi e^{-\lambda_2 t}$, then the evolution drift vector is given by

$$\mathbf{a} = \sigma^2 \nabla \log \phi = \left(\frac{\sigma^2}{r} - \frac{\alpha}{2}, -\frac{\sigma^2}{r}\tan\theta, 0\right).$$

Therefore, in spherical coordinates the kinematics equation reads

$$\frac{\partial u}{\partial t} + \frac{1}{2}\sigma^2 \left\{\frac{1}{r^2}\frac{\partial}{\partial r}\left(r^2\frac{\partial u}{\partial r}\right) + \frac{1}{r^2\sin\theta}\frac{\partial}{\partial\theta}\left(\sin\theta\frac{\partial u}{\partial\theta}\right) + \frac{1}{r^2\sin^2\theta}\frac{\partial^2 u}{\partial\eta^2}\right\}$$

$$+ \left(\frac{\sigma^2}{r} - \frac{\alpha}{2}\right)\frac{\partial u}{\partial r} - \frac{\sigma^2}{r}\tan\theta\frac{1}{r}\frac{\partial u}{\partial\theta} = 0.$$

Expanding, we have

$$\frac{\partial u}{\partial t} + \frac{1}{2}\sigma^2 \left(\frac{\partial^2 u}{\partial r^2} + \frac{1}{r^2}\frac{\partial^2 u}{\partial \theta^2} + \frac{1}{r^2\sin^2\theta}\frac{\partial^2 u}{\partial \eta^2} \right)$$
$$+ \left(\frac{2\sigma^2}{r} - \frac{\alpha}{2} \right)\frac{\partial u}{\partial r} + \frac{1}{2}\frac{\sigma^2}{r^2}\left(\cot\theta - 2\tan\theta \right)\frac{\partial u}{\partial \theta} = 0,$$

which has the form of equation (1.5.9) in Theorem 1.5.5. Therefore, the path equation is given by (1.5.10). If we write it in coordinates, then, by (1.5.11), we have

$$\begin{cases} r_t = r_a + \sigma B^1_{t-a} + \int_a^t \left(2\sigma^2\frac{1}{r_s} - \frac{\alpha}{2} \right) ds, \\[2mm] \theta_t = \theta_a + \int_a^t \sigma\frac{1}{r_s}dB^2_{s-a} + \frac{1}{2}\int_a^t \sigma^2\frac{1}{r_s{}^2}\left(\cot\theta_s - 2\tan\theta_s \right) ds, \\[2mm] \eta_t = \eta_a + \int_a^t \sigma\frac{1}{r_s\sin\theta_s}dB^3_{s-a}, \end{cases}$$

where $\mathbf{B}_t = \left(B^1_t, B^2_t, B^3_t \right)$ is the three-dimensional Brownian motion.

The equation for the radial component tells us that the random motion r_t is attracted to the origin by the constant drift $-\frac{\alpha}{2}$, but it cannot reach the origin, because it is repelled strongly near the origin by the drift $2\sigma^2\frac{1}{r}$.

Solving

$$2\sigma^2\frac{1}{r} - \frac{\alpha}{2} = 0,$$

we see that the drift vanishes at the point

$$\bar{r} = \frac{2^2\sigma^2}{\alpha}.$$

Therefore, if $r < \bar{r}$ (resp., $r > \bar{r}$), then the drift is positive (resp., negative), namely, $\bar{r} = 2^2\sigma^2/\alpha$ is an equilibrium point of the random motion. For the hydrogen atom, $\alpha = e^2/\hbar$ and $\sigma^2 = \hbar/m$. Therefore,

$$\bar{r} = \frac{\hbar^2}{me^2}2^2,$$

which coincides with the classical Bohr radius for $n = 2$. If we look at the θ-component, the drift

$$\frac{1}{2}\frac{\sigma^2}{r^2}\left(\cot\theta - 2\tan\theta \right)$$

diverges at $\theta = 0$, $\theta = \pi/2$ and $\theta = \pi$, and so the random motion θ_t is repelled strongly near the z-axis and the (x,y)-plane. Therefore, the random motion θ_t is separated into the intervals $\left(0, \frac{\pi}{2} \right)$ and $\left(\frac{\pi}{2}, \pi \right)$. Moreover, the random motion of the η-component is not circular, because there is no drift. Thus, the motion of the electron takes place into two separated American-doughnut shaped regions, above and under the (x, y)-plane. The electron moves either in the doughnut above the (x, y)-plane, or in that under that plane, but does not jump between the two doughnuts.

Figure 2.5, where the z-axis is fixed, illustrates the path of motion in this case.

In rectangular coordinates, the eigenfunction

$$\varphi(x, y, z) = \gamma \frac{\alpha r}{2\sigma^2} e^{-\frac{\alpha r}{2\sigma^2}} \frac{z}{r} = \gamma \frac{\alpha}{2\sigma^2} e^{-\frac{\alpha r}{2\sigma^2}} z, \quad r = (x^2 + y^2 + z^2)^{1/2}$$

determines the evolution drift

$$\mathbf{a}(x, y, z) = \sigma^2 \nabla \log \varphi(x, y, z) = \sigma^2 \nabla \left(-\frac{\alpha}{2\sigma^2} r + \log \frac{\alpha z}{2\sigma^2}\right)(x, y, z)$$

$$= \left(-\frac{\alpha}{2} \frac{x}{r}, -\frac{\alpha}{2} \frac{y}{r}, -\frac{\alpha}{2} \frac{z}{r} + \frac{\sigma^2}{z}\right),$$

where $\alpha > 0$. The x- and y-components of the drift vector $\mathbf{a}(x, y, z)$ point to $x = 0$ and $y = 0$, respectively.

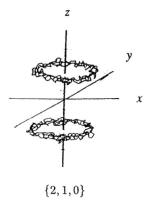

$$\{2, 1, 0\}$$

Figure 2.5

Due to the presence of the term $\frac{1}{z}$ in the z-component of the drift vector $\mathbf{a}(x, y, z)$, the electron cannot reach the (x, y)-plane. Therefore, its paths of remain above or below the (x, y)-plane, without crossing between the two regions.

(iii) The case $\{n, \ell, m\} = \{2, 1, \pm 1\}$. The eigenvalue is

$$\lambda_2 = -\frac{\alpha^2}{2\sigma^2} \frac{1}{2^2},$$

and the associated complex-valued eigenfunctions are

$$\varphi_1(r, \theta, \eta) = \gamma \frac{\alpha r}{2\sigma^2} e^{-\frac{\alpha r}{2\sigma^2}} \sin\theta e^{i\eta}, \quad m = +1,$$

and

$$\varphi_{-1}(r, \theta, \eta) = \gamma \frac{\alpha r}{2\sigma^2} e^{-\frac{\alpha r}{2\sigma^2}} \sin\theta e^{-i\eta}, \quad m = -1,$$

and so the corresponding complex evolution functions are

$$\psi_{\pm 1}(t, (r, \theta, \eta)) = \gamma \frac{\alpha r}{2\sigma^2} e^{-\frac{\alpha r}{2\sigma^2} \pm i\eta - i\lambda_2 t} \sin\theta.$$

In Born's quantum-mechanical statistical interpretation,

$$\mu_t(d(r, \theta, \eta)) = \psi_{\pm 1}(t, (r, \theta, \eta)) \overline{\psi}_{\pm 1}(t, (r, \theta, \eta)) r^2 \sin\theta \, dr \, d\theta \, d\eta$$

$$= \gamma^2 \left(\frac{\alpha r}{2\sigma^2}\right)^2 e^{-\frac{\alpha r}{\sigma^2}} \sin^2\theta \; r^2 \sin\theta \, dr \, d\theta \, d\eta$$

is the distribution of an electron.

We apply again Theorem 1.17.1 of the mechanics of random motion.
We write the complex-valued evolution function in the exponential form

$$\psi_{\pm 1}(t, (r, \theta, \eta)) = e^{R(t, (r, \theta, \eta)) + iS_{\pm 1}(t, (r, \theta, \eta))},$$

where the drift potential is

$$S_{\pm 1}(t, (r, \theta, \eta)) = \pm \eta - \lambda_2 t.$$

The complex evolution function determines a pair of real-valued evolution functions, in terms of which we compute the drift vector.

We first set $m = +1$. Then, since $S_{+1}(t, (r, \theta, \eta)) = \eta - \lambda_2 t$, the of evolution functions are given by the expressions

$$\phi_{+1}(t, (r, \theta, \eta)) = \gamma \frac{\alpha r}{2\sigma^2} e^{-\frac{\alpha r}{2\sigma^2} + \eta - \lambda_2 t} \sin\theta,$$

$$\widehat{\phi}_{+1}(t, (r, \theta, \eta)) = \gamma \frac{\alpha r}{2\sigma^2} e^{-\frac{\alpha r}{2\sigma^2} - \eta + \lambda_2 t} \sin\theta.$$

By formula (1.4.7), the drift vector determined by the evolution function $\phi_{+1}(t, (r, \theta, \eta))$ is

$$\mathbf{a}_{+1} = \sigma^2 \frac{\nabla \phi_{+1}}{\phi_{+1}} = \sigma^2 \frac{1}{\phi_{+1}} \left(\frac{\partial \phi_{+1}}{\partial r}, \frac{1}{r} \frac{\partial \phi_{+1}}{\partial \theta}, \frac{1}{r\sin\theta} \frac{\partial \phi_{+1}}{\partial \eta}\right)$$

$$= \left(\frac{\sigma^2}{r} - \frac{\alpha}{2}, \frac{\sigma^2}{r} \cot\theta, \frac{\sigma^2}{r\sin\theta}\right). \tag{2.1.4}$$

Therefore, the kinematics equation in spherical coordinates reads

$$\frac{\partial u}{\partial t} + \frac{1}{2}\sigma^2 \left\{\frac{1}{r^2}\frac{\partial}{\partial r}\left(r^2 \frac{\partial u}{\partial r}\right) + \frac{1}{r^2 \sin\theta}\frac{\partial}{\partial \theta}\left(\sin\theta \frac{\partial u}{\partial \theta}\right) + \frac{1}{r^2 \sin^2\theta}\frac{\partial^2 u}{\partial \eta^2}\right\} + \mathbf{a}_{+1} \cdot \nabla u = 0,$$

where

$$\mathbf{a}_{+1} \cdot \nabla u = \left(\frac{\sigma^2}{r} - \frac{\alpha}{2}\right)\frac{\partial u}{\partial r} + \frac{\sigma^2}{r}\cot\theta\frac{1}{r}\frac{\partial u}{\partial\theta} + \frac{\sigma^2}{r\sin\theta}\frac{1}{r\sin\theta}\frac{\partial u}{\partial\eta}.$$

In expanded form,

$$\frac{\partial u}{\partial t} + \frac{1}{2}\left(\sigma^2\frac{\partial^2 u}{\partial r^2} + \frac{\sigma^2}{r^2}\frac{\partial^2 u}{\partial\theta^2} + \frac{\sigma^2}{r^2\sin^2\theta}\frac{\partial^2 u}{\partial\eta^2}\right)$$
$$+ \frac{\sigma^2}{r}\frac{\partial u}{\partial r} + \frac{1}{2}\frac{\sigma^2}{r^2}\cot\theta\frac{\partial u}{\partial\theta} + \mathbf{a}_{+1}\cdot\nabla u = 0,$$

that is,

$$\frac{\partial u}{\partial t} + \frac{1}{2}\left(\sigma^2\frac{\partial^2 u}{\partial r^2} + \frac{\sigma^2}{r^2}\frac{\partial^2 u}{\partial\theta^2} + \frac{\sigma^2}{r^2\sin^2\theta}\frac{\partial^2 u}{\partial\eta^2}\right)$$
$$+ \left(\frac{2\sigma^2}{r} - \frac{\alpha}{2}\right)\frac{\partial u}{\partial r} + \frac{3}{2}\frac{\sigma^2}{r^2}\cot\theta\frac{\partial u}{\partial\theta} + \frac{\sigma^2}{r^2\sin^2\theta}\frac{\partial u}{\partial\eta} = 0.$$

Applying Theorem 1.5.5 in the coordinate (r, θ, η), we obtain, by formula (1.5.11), the system

$$\begin{cases} r_t = r_a + \sigma B^1_{t-a} + \int_a^t (2\sigma^2\frac{1}{r_s} - \frac{\alpha}{2})ds, \\[2mm] \theta_t = \theta_a + \int_a^t \sigma\frac{1}{r_s}dB^2_{s-a} + \frac{3}{2}\int_a^t \sigma^2\frac{1}{r_s^2}\cot\theta_s\,ds, \\[2mm] \eta_t = \eta_a + \int_a^t \sigma\frac{1}{r_s\sin\theta_s}dB^3_{s-a} + \int_a^t \sigma^2\frac{1}{r_s^2\sin^2\theta_s}\,ds, \end{cases}$$

where $\mathbf{B}_t = (B^1_t, B^2_t, B^3_t)$ is the three-dimensional Brownian motion.

The radial motion r_t is attracted by the drift $-\alpha/2$, but repelled strongly by the drift $2\sigma^2\frac{1}{r}$ near the origin, so the electron does not reach the origin.

To determine the zero point of the drift, we solve

$$2\sigma^2\frac{1}{r} - \frac{\alpha}{2} = 0,$$

and get

$$\overline{r} = \frac{2^2\sigma^2}{\alpha}.$$

If $r < \overline{r}$ (resp., $r > \overline{r}$), then the drift of the radial direction is positive (resp., negative). Therefore, $\overline{r} = 2^2\sigma^2/\alpha$ is an equilibrium point of the radial motion r_t.

In the case of a hydrogen atom,

$$\overline{r} = \frac{\hbar^2}{me^2}2^2, \tag{2.1.5}$$

where m is the mass of the electron and $\alpha = e^2/\hbar$, $\sigma^2 = \hbar/m$. This coincides with the classical Bohr radius for $n = 2$. The motion in the θ-direction does not reach the z-axis, because the drift

$$\frac{3}{2}\frac{\sigma^2}{r^2}\cot\theta$$

diverges at $\theta = 0$ and $\theta = \pi$ and is strongly repelling. Thus, the electron remains close to the (x, y)-plane. Moreover, the η-component exhibits the counter-clockwise drift

$$\frac{\sigma^2}{r^2\sin^2\theta},$$

which induces the magnetic moment of the hydrogen atom. When $m = -1$, the drift changes sign and becomes

$$-\frac{\sigma^2}{r^2\sin^2\theta},$$

so we have a clockwise drift.

Thus, the electron performs a random motion in an American doughnut with either a counter-clockwise drift, or a clockwise one.

Figure 2.6 illustrates the path of the motion.

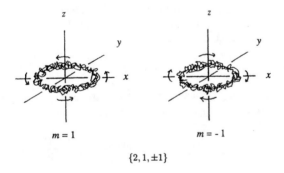

$$\{2, 1, \pm 1\}$$

Figure 2.6

Now, let ψ_1 and ψ_2 be states of motion with energy E_1 and E_2, respectively. If the excited random motion determined by ψ_2 is entangled with the random motion of the ground state ψ_1, then the entanglement induces the oscillation

$$\cos\left(\frac{(E_2 - E_1)t}{\hbar}\right),$$

and the atom emits light (see Theorem 2.1.1).

The fact that the spectrum $\cos\left(\frac{(E_n - E_m)t}{\hbar}\right)$ of hydrogen atom is determined by the energy difference $\frac{E_n - E_m}{\hbar}$ was discovered by Bohr (Bohr's law). Bohr regarded this as an instantaneous discontinuous phenomenon and called it "transition".

If the "transition" is an instantaneous discontinuous phenomena, it cannot be described by the Schrödinger equation. However, the "transition" is in fact described by the Schrödinger equation as the superposition (entanglement) of the states of motion ψ_n and ψ_m. This was pointed out by Y. Suematsu and K. Iga (2014).

For the case of the random motion with higher energy, we refer to Nagasawa (2000).

2.2   Charged Particle in a Magnetic Field

The complex evolution equation governing the motion of a charged particle in an electromagnetic field is

$$i\frac{\partial\psi}{\partial t} + \sigma^2(\nabla + i\mathbf{A})^2\psi - V\psi = 0,$$

where $\mathbf{A}$ is a vector potential and V is a scalar potential. Let us consider the case when there is a uniform magnetic field H in the direction of the z-axis, but no electric field, i.e., $V \equiv 0$, and look at the motion of the charged particle.

Then the vector potential $\mathbf{A}$ is

$$\mathbf{A} = (-\alpha Hy, \alpha Hx, 0),$$

where α is a constant that depends on the mass and charge of the particle. Therefore, in this case the complex evolution equation reduces to

$$i\frac{\partial\psi}{\partial t} + \frac{1}{2}\sigma^2(\nabla_2 + i\mathbf{A}_2\,(x,y))^2\psi + \frac{1}{2}\sigma^2\frac{\partial^2\psi}{\partial z^2} = 0, \qquad (2.2.1)$$

where

$$\mathbf{A}_2\,(x,y) = (-\alpha Hy, \alpha Hx),$$

and ∇_2 denotes the gradient in the (x,y)-plane.

Substituting $\psi(t,(x,y,z)) = \psi_1(t,(x,y))\psi_2(t,z)$ in equation (2.2.1) and separating variables, we obtain the equations

$$i\frac{\partial\psi_1}{\partial t} + \frac{1}{2}\sigma^2(\nabla_2 + i\mathbf{A}_2\,(x,y))^2\psi_1 = 0, \qquad (2.2.2)$$

$$i\frac{\partial\psi_2}{\partial t} + \frac{1}{2}\sigma^2\frac{\partial^2\psi_2}{\partial z^2} = 0. \qquad (2.2.3)$$

According to Theorem 1.9.1, the motion along the z-axis described by equation (2.2.3) is the one-dimensional free motion discussed in Section 1.6. The paired equations of motion are

$$\begin{cases} \dfrac{\partial \phi}{\partial t} + \dfrac{1}{2}\sigma^2 \dfrac{\partial^2 \phi}{\partial z^2} = 0, \\[2mm] -\dfrac{\partial \widehat{\phi}}{\partial t} + \dfrac{1}{2}\sigma^2 \dfrac{\partial^2 \widehat{\phi}}{\partial z^2} = 0. \end{cases}$$

If we take

$$\phi\left(t, z\right) = \alpha \exp\left(-\frac{\kappa^2}{2}t + \frac{\kappa}{\sigma}z\right),$$

with an arbitrary positive constant, κ as the evolution function, then paths of the random motion along the z-axis are given by the formula

$$Z_t = Z_a + \sigma B_{t-a} + \sigma\kappa\left(t - a\right).$$

Let the solution of equation (2.2.2) in the (x, y)-plane, that is, the complex evolution function, be $\psi(t, (x,y)) = e^{R(t,(x,y))+iS(t,(x,y))}$. By Theorem 1.9.1, the evolution function $\phi\left(t, (x,y)\right) = e^{R(t,(x,y))+S(t,(x,y))}$ satisfies the equation

$$\frac{\partial \phi}{\partial t} + \frac{1}{2}\sigma^2\left(\nabla_2 + \mathbf{A}_2\left(x, y\right)\right)^2 \phi - \left(\sigma^2 \mathbf{A}_2(x, y)^2 + \widetilde{V}\right)\phi = 0, \qquad (2.2.4)$$

which contains the self-potential

$$\widetilde{V} = 2\frac{\partial S}{\partial t} + \sigma^2(\nabla_2 S)^2 + \sigma^2 \mathbf{A}_2 \cdot \nabla_2 S.$$

Expanding equation (2.2.4), we get

$$\frac{\partial \phi}{\partial t} + \frac{1}{2}\sigma^2 \triangle_2 \phi + \sigma^2 \mathbf{A}_2 \cdot \nabla_2 \phi + \frac{1}{2}\sigma^2\left(\nabla_2 \cdot \mathbf{A}_2\right)\phi - \frac{1}{2}\sigma^2 \mathbf{A}_2^2 \phi - \widetilde{V}\phi = 0,$$

where

$$\mathbf{A}_2\left(x, y\right) = \left(-\alpha H y, \alpha H x\right), \quad \nabla_2 \cdot \mathbf{A}_2 = 0, \quad \mathbf{A}_2^2 = \left(\alpha H\right)^2\left(x^2 + y^2\right).$$

Therefore, equation (2.2.4) takes on the form

$$\frac{\partial \phi}{\partial t} + \frac{1}{2}\sigma^2\left(\frac{\partial^2}{\partial x^2} + \frac{\partial^2}{\partial y^2}\right)\phi + \sigma^2 \alpha H\left(-y\frac{\partial \phi}{\partial x} + x\frac{\partial \phi}{\partial y}\right)$$
$$-\frac{1}{2}\sigma^2(\alpha H)^2\left(x^2 + y^2\right)\phi - \widetilde{V}\phi = 0, \qquad (2.2.5)$$

where we should emphasize the presence of the Hooke potential

$$-\frac{1}{2}\sigma^2(\alpha H)^2(x^2 + y^2).$$

Moreover, we remark that, mathematically speaking, the drift

$$\sigma^2 \alpha H \left(-y \frac{\partial \phi}{\partial x} + x \frac{\partial \phi}{\partial y} \right) \tag{2.2.6}$$

can be easily added, as will be discussed later on.

The random motion in the (x, y)-plane determined by equation (2.2.5) stays close to the origin due to the presence of the Hooke potential.

The dynamic equation determined by the evolution equation (2.2.5) is, therefore,

$$\begin{aligned}
&\frac{\partial u}{\partial t} + \frac{1}{2}\sigma^2 \left(\frac{\partial^2 u}{\partial x^2} + \frac{\partial^2 u}{\partial y^2} \right) + \sigma^2 \left(\mathbf{b}_x \frac{\partial u}{\partial x} + \mathbf{b}_y \frac{\partial u}{\partial y} \right) \\
&+ \sigma^2 \alpha H \left(-y \frac{\partial u}{\partial x} + x \frac{\partial u}{\partial y} \right) = 0,
\end{aligned} \tag{2.2.7}$$

where

$$\mathbf{b} = \sigma^2 \frac{\nabla_2 \phi_0}{\phi_0}$$

is the evolution drift determined by the solution ϕ_0 of the evolution equation. The drift term (2.2.6) entrains the charged particle into a "counter-clockwise rotational motion" in the (x, y)-plane. For the so-called drift transformation, see Theorem 4.2.7 and its applications.

2.3   Aharonov-Bohm Effect

We continue our discussion on the motion of a charged particle in an electromagnetic field. The electromagnetic field $(\mathbf{E}, \mathbf{H})$ is expressed in terms of a vector potential $\mathbf{A}$ and a scalar potential V as

$$\mathbf{H} = \operatorname{rot} \mathbf{A}, \quad \mathbf{E} = -\operatorname{grad} V - \frac{\partial \mathbf{A}}{\partial t}.$$

As a matter of fact, the motion of a charged particle in the field $(\mathbf{E}, \mathbf{H})$ is determined not by the vectors $\mathbf{E}, \mathbf{H}$, but by the vector potential $\mathbf{A}$ and the scalar potential V: the Schrödinger equation

$$i\frac{\partial \psi}{\partial t} + \frac{1}{2}\sigma^2 (\nabla + i\mathbf{A})^2 \psi - V\psi = 0,$$

does not involve $\mathbf{E}$ and $\mathbf{H}$ directly.

Aharonov and Bohm (1959) pointed out this fact and gave the following interesting example.

Write the vector potential as $\mathbf{A} = (A_x, A_y, A_z)$ and suppose that the components are

$$A_x = 0, \quad A_y = 0, \quad A_z = \beta$$

in a cylinder of radius r and height ℓ with base in the (x, y)-plane, and zero otherwise, where β is a constant. Moreover, suppose that the scalar potential V vanishes everywhere. Then there is no electromagnetic field, because

$$\mathbf{H} = \operatorname{rot}\mathbf{A} = 0 \quad \text{and} \quad \mathbf{E} = -\operatorname{grad}V - \frac{\partial \mathbf{A}}{\partial t} = 0.$$

It would therefore be quite reasonable to guess that an electron performs a free motion. But Aharonov and Bohm (1959) discovered that this is not the case.

Thus, there is no electromagnetic field in the cylinder, but the z-component of the vector potential is different from zero. Therefore, the Schrödinger equation in the z-direction reads

$$i\frac{\partial \psi}{\partial t} + \frac{1}{2}\sigma^2 \left(\frac{\partial}{\partial z} + i\beta\right)^2 \psi = 0, \tag{2.3.1}$$

where $\sigma^2 = \hbar/m$ and m is the mass of the electron.

Aharonov and Bohm (1959) claimed that therefore the effect of the vector potential must be felt. This is called the Aharonov-Bohm effect (see Tonomura (1990, 1994) for details).

Let us apply the mechanics of random motion to this problem. Substituting $\psi = e^{-i\lambda\hbar t}\varphi(z)$ in equation (2.3.1), we obtain the equation

$$\frac{1}{2}\frac{d^2\varphi}{dz^2} + i\beta\frac{d\varphi}{dz} + \frac{1}{2}\left(\alpha^2 - \beta^2\right)\varphi = 0, \tag{2.3.2}$$

where λ is an eigenvalue and $\alpha^2 = 2m\lambda/\hbar^2$.

Equation (2.3.2) admits two solutions, and in the cylinder we have

$$\varphi_+ = e^{-i(\beta+\alpha)z} \quad \text{and} \quad \varphi_- = e^{-i(\beta-\alpha)z}.$$

Therefore, the corresponding complex evolution functions are

$$\psi_+ = e^{-i(\lambda\hbar t + (\beta+\alpha)z)} \quad \text{and} \quad \psi_- = e^{-i(\lambda\hbar t + (\beta-\alpha)z)}, \tag{2.3.3}$$

and the generating function and the drift potential are

$$R(t, z) = 0 \quad \text{and} \quad S_\pm(t, z) = -(\lambda\hbar t + (\beta\pm\alpha)z).$$

Then in the cylinder we have random motions of two types along the z-axis, which are determined by $\psi_+ = e^{-i(\lambda\hbar t + (\beta+\alpha)z)}$ and $\psi_- = e^{-i(\lambda\hbar t + (\beta-\alpha)z)}$. But the random motion is actually determined by the superposition

$$\psi^*(t, x) = \frac{1}{\sqrt{2}}\psi_+(t, x) + \frac{1}{\sqrt{2}}\psi_-(t, x)$$

The entangled random motion determined by the superposed Schrödinger function $\psi^*(t, x)$ was discussed in Section 1.10.

By Theorem 1.10.5, the function $S_+(t, z) - S_-(t, z)$ appears in the distribution density of the entangled motion. Therefore, along the z-axis there arises the stripe pattern

$$\cos\left(2\frac{\sqrt{2m\lambda}}{\hbar}z\right).$$

Tonomura (1990, 1994) demonstrated this experimentally.

Comments.

(i) The conventional Maxwell theory contains no vector potential, and consequently it cannot treat the Aharonov-Bohm effect. Aharonov and Bohm (1959) proved that the vector potential is absolutely necessary for the analysis of electromagnetic field experiments.

(ii) In the context of the uncertainty principle, Heisenberg claimed "since the paths of the motion of an electron are not observable, it should not be included in the theory". Einstein criticized this belief, asserting that "we do not build a theory with experimental data, but what should be observed is determined by a theory". The Aharonov-Bohm effect supports Einstein's comment.

2.4   Tunnel Effect

For simplicity, we consider the problem in one dimension and a particle moving to the right of the origin. If there is a barrier of finite height, a finite drift $0 < a(t, x) < \infty$ arises near the origin. Nevertheless, since the path equation is

$$X_t = X_a + \sigma B_{t-a} + \int_a^t a(s, X_s)\, ds,$$

because of presence of the Brownian motion σB_t, it turns out that $X_t < 0$ with a positive probability, that is, the particle crosses the barrier. This is called "tunnel effect."

If the height of the barrier is infinite, a total reflection occurs. In this case, the random motion is described by

$$Y_t = |X_t|, \quad \text{where } X_t \text{ is the motion without barrier}$$

and

$$
\begin{aligned}
P_t f(x) &= \mathbf{P}_x\left[f(Y_t)\right] = \mathbf{P}_x\left[f(|B_t|)\right] \\
&= \mathbf{P}_x\left[f(B_t) : B_t \geq 0\right] + \mathbf{P}_x\left[f(-B_t) : B_t < 0\right] \\
&= \int_{\mathbf{R}^+} p_t(x, y)\, dy f(y) + \int_{\mathbf{R}^-} p_t(x, y)\, dy f(-y) \\
&= \int_{\mathbf{R}^+} (p_t(x, y) + p_t(x, -y))\, dy f(y),
\end{aligned}
$$

where $p_t(x, y)$ is the density function of the transition probability

$$p_t(x, y) = \frac{1}{\sqrt{2\pi t}} \exp\left(-\frac{|x - y|^2}{2t}\right).$$

If we set

$$p_t^+(x, y) = p_t(x, y) + p_t(x, -y)$$

$$= \frac{1}{\sqrt{2\pi t}} \left[\exp\left(-\frac{|x - y|^2}{2t}\right) + \exp\left(-\frac{|x + y|^2}{2t}\right)\right],$$

then $p_t^+(x, y)$ satisfies the reflecting boundary condition

$$\frac{\partial}{\partial x} p_t^+(x, y)\Big|_{x=0} = 0.$$

Hence, $p_t^+(x, y)$ is the fundamental solution of the boundary problem

$$\frac{\partial u}{\partial t} = \frac{1}{2} \frac{\partial^2 u}{\partial x^2} \quad \text{for } x \geq 0, \quad \frac{\partial u}{\partial x} = 0 \quad \text{for } x = 0.$$

We will call $\{Y_t = |B_t|, t \geq 0; \mathbf{P}_x, x \geq 0\}$ a reflecting Brownian motion.

The reflecting Brownian motion reaches the origin and is reflected there; however, there is another case, when the random motion cannot reach the origin. For example, the case (ii) of the radial motion

$$r_t = r_a + \sigma B_{t-a}^1 + \int_a^t ds\left(\frac{3}{2}\sigma^2 \frac{1}{r_s} - \sigma \kappa r_s\right) \tag{2.4.1}$$

is of this kind. In this example, although the term σB_{t-a}^1 is present, the random motion cannot reach the origin due to the drift $\frac{3}{2}\sigma^2 \frac{1}{r_s}$, which diverges at the origin.

In contrast to the above example, the reflection at the origin is not so strong; nevertheless, the random motion cannot reach the origin because it becomes too weak. For example, if stocks take a sharp drop and trades become very thin for some time, then the market will recover slowly; an endangered species will recover slowly; and so on. In those cases, equation (2.4.1) is replaced, for example, by

$$X_t = X_a + \int_a^t \sigma\sqrt{X_s}\, dB_s + \int_a^t ds\left(\frac{3}{2}\sigma^2 - \sigma \kappa X_s^2\right). \tag{2.4.2}$$

Since $\sigma\sqrt{x}$ vanishes near the origin and the drift $\frac{3}{2}\sigma^2$ is the dominating effect, X_t does not become zero.

Equations (2.4.1) and (2.4.2) describe different phenomena, but the former can be derived from latter, and vice versa, through the so-called "time change", which will be treated in Section 4.3.

2.5   Bose-Einstein Distribution

Bose computed the distribution of a population of photons (Bose (1924)). In his computation he used

$$\frac{(N + M - 1)!}{(M - 1)!N!},$$

as the number of ways to distribute n particles among m states. This is called "Bose statistics". Then he showed that its distribution coincided with the Planck distribution of heat radiation.

Einstein (Einstein (1924, 1925)) analyzed Bose's computation and clarified that it represented the statistics of indistinguishable particles (called Bose particles). He then applied it to the distribution of an ideal gas, and deduced that the mean number of particles with energy ϵ obeys

$$\langle n_\epsilon \rangle = \frac{1}{\exp((\epsilon - \mu)/k_\mathrm{B}T) - 1}, \qquad (2.5.1)$$

where μ is the chemical potential, k_B is the Boltzmann constant, and T is the temperature. This is called the Bose-Einstein distribution.

Let us prove formula (2.5.1).

Consider a population of Bose particles of energy ϵ. The number of ways to distribute indistinguishable particles N_ϵ into M_ϵ states is

$$W_\epsilon = \frac{(N_\epsilon + M_\epsilon - 1)!}{(M_\epsilon - 1)!N_\epsilon!}$$

Since N_ϵ and M_ϵ are sufficiently large, we may assume that

$$W_\epsilon = \frac{(M_\epsilon + N_\epsilon)!}{M_\epsilon!N_\epsilon!}.$$

Then the number of states of the population of Bose particles is

$$W = \prod_\epsilon W_\epsilon.$$

Now define the entropy S by

$$S = k_\mathrm{B} \log W.$$

Then

$$S = k_\mathrm{B} \sum_\epsilon \big[\log((M_\epsilon + N_\epsilon)!) - \log(M_\epsilon!) - \log(N_\epsilon!) \big].$$

Applying Stirling's formula

$$n! = \sqrt{2\pi} n^{n + \frac{1}{2}} e^{-n},$$

we have

$$S = k_{\mathrm{B}} \sum_{\epsilon} \left[\left(1 + \frac{N_\epsilon}{M_\epsilon}\right) \log\left(1 + \frac{N_\epsilon}{M_\epsilon}\right) - \frac{N_\epsilon}{M_\epsilon} \log\left(\frac{N_\epsilon}{M_\epsilon}\right) \right] M_\epsilon.$$

If we denote the mean number of particles in a state of energy ϵ by

$$\langle n_\epsilon \rangle = \frac{N_\epsilon}{M_\epsilon},$$

then the entropy is

$$S = k_{\mathrm{B}} \sum_{\epsilon} \left[(1 + \langle n_\epsilon \rangle) \log(1 + \langle n_\epsilon \rangle) - \langle n_\epsilon \rangle \log(\langle n_\epsilon \rangle) \right] M_\epsilon. \tag{2.5.2}$$

Now the total number N and the total energy E are

$$N = \sum_{\epsilon} N_\epsilon \quad \text{and} \quad E = \sum_{\epsilon} \epsilon N_\epsilon. \tag{2.5.3}$$

To compute the distribution $\langle n_\epsilon \rangle$ of particles in thermodynamic equilibrium, it is enough to find the distribution that maximizes S for fixed N and E. According to Lagrange's method of indeterminate coefficients (Lagrange multipliers), the distribution that maximizes

$$\widetilde{S} = S - \alpha N - \beta E = S - \alpha \sum_{\epsilon} \langle n_\epsilon \rangle M_\epsilon - \beta \sum_{\epsilon} \epsilon \langle n_\epsilon \rangle M_\epsilon$$

maximum is given by

$$\frac{\partial \widetilde{S}}{\partial \langle n_\epsilon \rangle} = \left[\log \frac{1 + \langle n_\epsilon \rangle}{\langle n_\epsilon \rangle} - \alpha - \beta\epsilon \right] M_\epsilon = 0,$$

and so

$$\langle n_\epsilon \rangle = \frac{1}{\exp(\alpha + \beta\epsilon) - 1}. \tag{2.5.4}$$

Now let us compute the Lagrange multipliers α, β. We first recast equation (2.5.2) as

$$S = k_{\mathrm{B}} \sum_{\epsilon} \left[\log(1 + \langle n_\epsilon \rangle) + \langle n_\epsilon \rangle \log\left(1 + \frac{1}{\langle n_\epsilon \rangle}\right) \right] M_\epsilon,$$

where we apply equation (2.5.4). Since $\log\left(1 + \frac{1}{\langle n_\epsilon \rangle}\right) = \alpha + \beta\epsilon$, we have

$$S = k_{\mathrm{B}} \sum_{\epsilon} \left[\log(1 + \langle n_\epsilon \rangle) + \langle n_\epsilon \rangle (\alpha + \beta\epsilon) \right] M_\epsilon$$

$$= k_{\mathrm{B}} \sum_{\epsilon} M_\epsilon \log(1 + \langle n_\epsilon \rangle) + k_{\mathrm{B}} \sum_{\epsilon} N_\epsilon (\alpha + \beta\epsilon).$$

Hence, with N and E given in (2.5.3),

$$S = k_{\mathrm{B}} \sum_\epsilon M_\epsilon \log(1 + \langle n_\epsilon \rangle) + k_{\mathrm{B}}(\alpha N + \beta E).$$

Therefore,

$$\frac{\partial S}{\partial N} = k_{\mathrm{B}}\alpha, \quad \frac{\partial S}{\partial E} = k_{\mathrm{B}}\beta. \tag{2.5.5}$$

On the other hand, we have the thermodynamic relations

$$\frac{\partial S}{\partial N} = -\frac{\mu}{T}, \quad \frac{\partial S}{\partial E} = \frac{1}{T}, \tag{2.5.6}$$

where μ is the chemical potential. If we combine equation (2.5.5) and equation (2.5.6), we have

$$\alpha = -\frac{\mu}{k_{\mathrm{B}}T}, \quad \beta = \frac{1}{k_{\mathrm{B}}T},$$

and substituting these expressions in equation (2.5.4), we get the Bose-Einstein distribution

$$\langle n_\epsilon \rangle = \frac{1}{\exp((\epsilon - \mu)/k_{\mathrm{B}}T) - 1}.$$

Next, let us consider the condensation phenomenon. Let the total number of particles be N. The mean number of excited state is

$$N_{\mathrm{ex}} = \sum_\epsilon \langle n_\epsilon \rangle.$$

To compute it we replace the sum in N_{ex} by an integration. Then we see that N attains its maximum at $\mu = 0$, where its value is

$$N(0) = \sum_{p \neq 0} \frac{1}{\exp\left(\frac{p^2}{2m}\frac{1}{k_{\mathrm{B}}T}\right) - 1} = \frac{V^3}{2\pi\hbar} \int_0^\infty \frac{4\pi p^2}{\exp\left(\frac{p^2}{2m}\frac{1}{k_{\mathrm{B}}T}\right) - 1} \, dp = N\left(\frac{T}{T_{\mathrm{c}}}\right)^{3/2},$$

where T_{c} is the critical temperature,

$$T_{\mathrm{c}} = \frac{1}{k_{\mathrm{B}}} \left(\frac{2\pi\hbar^2}{m}\right) \left(\frac{N}{\zeta(3/2)V}\right)^{3/2},$$

ζ is the Riemann ζ-function, and $\zeta(3/2) = \displaystyle\sum_{k=1}^\infty \frac{1}{k^{3/2}} = 2.612\ldots$.

According to Einstein (1924, 1925), the ideal gas is saturated at the temperature $T = T_{\mathrm{c}}$, and for $T < T_c$ the remaining particles

$$N_0 = N - N(0) = N\left(1 - \left(\frac{T}{T_{\mathrm{c}}}\right)^{3/2}\right)$$

are condensed into the lowest energy state; the ratio

$$\frac{N_0}{N} = 1 - \left(\frac{T}{T_c}\right)^{3/2}$$

becomes $\frac{N_0}{N} = 1$ at $T = 0$, i.e., all particles drop to the lowest energy state.

Thus, at very low temperatures the gas condenses. This phenomenon is called "Bose-Einstein condensation". It will play a decisive role in our discussion of high-temperature superconductivity in Section 2.11.

The Bose-Einstein condensation has been considered simply as a problem of theoretical interest, because Einstein (1924, 1925) discussed the ideal gas. However, it could be actually verified experimentally problem when it became possible to suitably lower the temperature. For example, Mewes et al. (1996) carried out such an experiment with an gas of sodium atoms achieving condensates that contained up to 5×10^6 atoms, and D. S. Jin (1996) performed it with a ^{87}Rb gas with total number of atoms 4500 ± 300.

We will further on show that

"If the temperature is lowered, then the Bose-Einstein progresses with the decrease in the temperature, but the process has a limit."

If we start lowering the temperature, then the kinetic energy will decrease and approach the ground energy, and the variance of the position of gas particles will become small. However, we cannot make the variance as small as we wish (Nagasawa (2012)).

To show this we will apply the inequality (3.6.1):

$$\langle \psi_t, (x - \langle \psi_t, x\psi_t \rangle)^2 \psi_t \rangle \, \langle \psi_t, \frac{1}{2m}\mathbf{p}^2 \psi_t \rangle \geq \frac{1}{4}\frac{1}{2m} \, \hbar^2 \qquad (2.5.7)$$

(which should not be confused with Robertson's inequality).

As a matter of fact, the left-hand side of inequality (2.5.7) contains the expectation of the kinetic energy $\langle \psi_t, \frac{1}{2m}\mathbf{p}^2 \psi_t \rangle$ instead of the variance of the momentum. Namely

"The product of the dispersion on the axis x-axis (and similarly for the y and z axes) and the expectation of the kinetic energy is bounded from below by $\frac{1}{4}\frac{1}{2m} \, \hbar^2$."

Inequality (2.5.7) shows that, in the ground state of small kinetic energy, there exists a limit for gas particles beyond which the variance of the position cannot be smaller.

Consequently, the Bose-Einstein condensation has a lower limit. That is, if the condensation goes on, the variance of the position of particles must become large, and finally the condensation blows up (as reported in Cornish et al. (2001), and Ueda and Saito (2002)).

This implies that as the kinetic energy approaches the ground energy, the dispersion of position must become large so as to obey inequality (2.5.7). Therefore, the fluidity of the condensed population increases at extremely low temperatures, even though the Bose-Einstein condensation occurs unless gas particles are confined with a potential. This fact is adopted as the theoretical foundation of superfluidity at extremely low temperatures.

Remark. In some papers the Heisenberg's inequality (uncertainty principle) is used in discussions of Bose-Einstein condensation. We claim that this is inadequate, because the kinetic energy becoming small is different from the dispersion of the momentum becoming small. The Heisenberg's inequality (uncertainty principle) will be discussed in Section 3.6.

2.6   Random Motion and the Light Cone

Classical particles that do not perform random motion cannot leave the light cone. This is one of the conclusions of relativity theory (Einstein (1905)). However, particles that perform random motion may leave immediately the light cone. This means that we might be able to communicate faster than the speed of light! However, things are not so simple.

For simplicity, let us assume that $X(t) = \sigma B_t$, where B_t is the one-dimensional Brownian motion. Then there is no upper limit for $X(t) = \sigma B_t$, but

$$\limsup_{t \to \infty} \frac{X(t)}{t} = \bar{a} < \infty \quad \text{a.s..}$$

Therefore, if we take $a > \bar{a}$ and set

$$X^{(a)}(t) = X(t) - at,$$

then $X^{(a)}(t)$ has an upper limit for. We denote it by

$$M^{(a)} = \sup_{t>0} X^{(a)}(t),$$

and denote the first time at which the stochastic process $X^{(a)}(t)$ attains $M^{(a)}$ as

$$T^{(a)} = \sup\left\{t : X^{(a)}(t) = M^{(a)}\right\}.$$

Set

$$H^{(a)} = X\big(T^{(a)}\big),$$

then the stochastic process $X(t)$ attains the value $H^{(a)} = M^{(a)} + aT^{(a)}$ at $T^{(a)}$. With c denoting the speed of light, we consider the set of paths

$$\Omega^{(a)} = \left\{\omega : H^{(a)}(\omega) \le cT^{(a)}\right\}.$$

If $\omega \in \Omega^{(a)}$, then the path $X_t(\omega) = \omega(t)$ leaves the light cone for $t \le T^{(a)}$, but it does not for $t \ge T^{(a)}$.

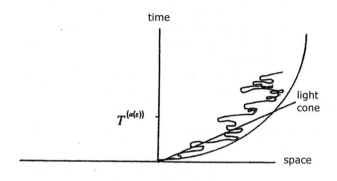

Figure 2.7

If we set

$$a(\varepsilon) = \sup\left\{a : \mathbf{P}[\Omega^{(a)}] \ge 1 - \varepsilon\right\}$$

and

$$\Omega^{(a(\varepsilon))} = \left\{\omega : H^{(a(\varepsilon))}(\omega) \le cT^{(a(\varepsilon))}\right\},$$

then we can show that

$$\mathbf{P}[\Omega^{(a(\varepsilon))}] \ge 1 - \varepsilon.$$

Namely,

$$\mathbf{P}\left[\{X_t \text{ does not leave the light cone for } t \ge T^{(a(\varepsilon))}\}\right] \ge 1 - \varepsilon,$$

where the parameter ε can be taken arbitrarily small.

The expected value of the time $T^{(a(\varepsilon))}$ is

$$\mathbf{P}[T^{(a(\varepsilon))}] = \int_0^\infty dt \int_{a(\varepsilon)t}^\infty dx \frac{1}{\sqrt{2\pi\sigma^2 t}} \exp\left(-\frac{x^2}{2\sigma^2 t}\right) \le \frac{2}{\sqrt{\pi}} \frac{\sigma^2}{a(\varepsilon)^2},$$

where $\sigma^2 = \frac{\hbar}{m}$. This result is due to Nagasawa and Tanaka. For details, we refer to Chapter X of Nagasawa (2000).

Therefore, Einstein's classical statement that "particles cannot leave the light cone" is modified as follows:

Proposition 2.6.1. *In the long time range the above classical proposition of relativity theory holds, namely, except for a short space-time scale, the probability that one can communicate faster than the speed of light is very small.*

Here the restriction "for a short space-time scale" is very important, because it implies that we have the following proposition holds.

Proposition 2.6.2. *It is expected that in a short space-time scale one can communicate faster than the speed of light velocity with a high probability.*

Proposition 2.6.2 is one of the important consequences of the theory of random motion. Thus, it is not appropriate to assume a priori that relativity theory holds in a short space-time scale.

The experimental verification of the theoretical prediction of Proposition 2.6.2 is an important task, hopefully to be carried out.

2.7   Origin of the Universe

According to Hubble's law, the Universe expands at a constant speed. Therefore, if we reverse time and follow the development of the Universe backward, then the Universe must converge to a single point. Was the Universe born spontaneously from a point? We will investigate the origin of the Universe by introducing Hubble's law into our mechanics of random motion.

We adopt a boundaryless model of the Universe in one dimension, in which the Universe is a circle of radius $R(t)$. In cylindrical coordinates in three dimensions, we identify $\theta \in \left[-\frac{\pi}{2}, \frac{\pi}{2}\right]$ with $\pi - \theta$, and introduce a new variable

$$x = 2R(t)\theta/\pi \in [-R(t), R(t)].$$

Then the space-time domain that describes our one-dimensional Universe is

$$D = \{(t, x) ; t > 0, \ x \in [-R(t), R(t)]\}.$$

When the Universe started, it was a collection of radiation particles. According to Hubble's law, the Universe is continuously expanding. Therefore, a particle at a space-time point (t, x) is subject to the Hubble flow

$$a(t, x) = \frac{dR(t)}{dt}\frac{x}{R(t)}, \quad -1 \le \frac{x}{R(t)} \le 1, \quad t > 0. \tag{2.7.1}$$

The Hubble flow becomes stronger proportionally to x: the farther a particle is, the faster in moves away. Moreover, radiation particles are reflected at the boundary points $-R(t)$ and $R(t)$. In the three-dimensional cylindrical coordinates, they go behind.

We assume that the boundary function $R(t)$ is given by the formula

$$R(t) = (\alpha t)^\gamma, \quad 0 < \gamma < 1, \tag{2.7.2}$$

for sufficiently small $t > 0$. Then, by (2.7.1), the Hubble flow is

$$a(t, x) = \gamma\frac{x}{t}. \tag{2.7.3}$$

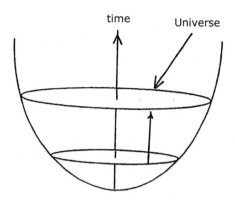

Figure 2.8

Therefore, the random motion of radiation particles is governed by the evolution equation

$$\frac{\partial u}{\partial t} + \frac{1}{2}\tilde{\sigma}^2\frac{\partial^2 u}{\partial x^2} + \gamma\frac{x}{t}\frac{\partial u}{\partial x} = 0,$$

or, after changing the time scale, by the equation

$$\frac{\partial u}{\partial t} + \frac{1}{2}\sigma^2\frac{\partial^2 u}{\partial x^2} + \frac{x}{t}\frac{\partial u}{\partial x} = 0, \qquad (2.7.4)$$

where $\sigma = \tilde{\sigma}/\sqrt{\gamma}$.

Thus, the random motion of radiation particles is described by the evolution equation (2.7.4) and the time-dependent reflecting boundary condition

$$\frac{\partial}{\partial x}u(x) = 0, \qquad x = -R(t) \quad \text{and} \quad x = R(t).$$

This is not the equation of motion, but the kinematics equation of particles of the Universe.

Mathematically, the evolution equation (2.7.4) with a time-dependent boundary condition and with a singular point at the origin $(t,x) = (0,0)$ is not easy to solve. This problem is equivalent to the Skorokhod problem with two moving boundaries

$$X(t) = \sigma\beta(t) + \int_0^t ds\,\frac{X(s)}{s} + \Phi(t), \qquad |X(t)| \le R(t), \qquad (2.7.5)$$

where $\beta(t)$ is the one-dimensional Brownian motion and the function $\Phi(t)$ satisfies the following conditions:

$\Phi(t)$, $t > 0$, is continuous   and $\Phi(0) = 0$;

$\Phi(t)$ is of bounded variation in the interval $[\varepsilon, 1/\varepsilon]$, $\varepsilon > 0$;

$\Phi(t)$ is increasing only on $\{s : X(s) = -R(s)\}$;

$\Phi(t)$ is decreasing only on $\{s : X(s) = R(s)\}$;

$\Phi(t)$ is a constant elsewhere.

We do not discuss further this difficult problem, and refer the reader to Chapter 14 of Nagasawa (2000). Based on theorems proved therein, we can state the following result due to Nagasawa and Domenig (1996):

Theorem 2.7.1 (Existence of the Universe).

(i) *Radiation particles perform a random motion, i.e., the Skorokhod problem (2.7.5) with two moving boundaries admits a solution $X(t)$, $t \geq 0$. In other words, our model establishes the existence of a Universe starting at time $t = 0$.*

(ii) *Uniqueness of the solution is determined by the exponent γ of the boundary function $R(t) = (\alpha t)^{\gamma}$ (i.e., the radius of the Universe). Specifically, if $0 < \gamma < 1/2$, then the Universe is not unique, whereas if $1/2 \leq \gamma < 1$ the Universe is unique.*

(iii) *Assume $0 < \gamma < \frac{1}{2}$. If for sufficiently large t*

$$R(t) = (\alpha t)^{\gamma_2}, \quad with \quad 0 < \gamma_2 \leq 1/2,$$

then the Universe is asymptotically unique. If for sufficiently large t

$$R(t) = (\alpha t)^{\gamma_2}, \quad with \quad 1/2 < \gamma_2 < 1,$$

then the Universe is not asymptotically unique.

As already remarked, the evolution equation (2.7.4) is the kinematics equation of the Universe. We now consider the paired equations of motion of the Universe

$$\begin{cases} \dfrac{\partial \phi}{\partial t} + \dfrac{1}{2}\sigma^2 \dfrac{\partial^2 \phi}{\partial x^2} + c(t,x)\,\phi = 0, \\[2mm] -\dfrac{\partial \widehat{\phi}}{\partial t} + \dfrac{1}{2}\sigma^2 \dfrac{\partial^2 \widehat{\phi}}{\partial x^2} + c(t,x)\,\widehat{\phi} = 0, \end{cases} \tag{2.7.6}$$

close to the birth of the Universe, i.e., for small t.

First of all we solve the equation

$$\sigma^2 \frac{\partial \log \phi(t,x)}{\partial x} = \frac{x}{t},$$

which expresses the relation between the evolution function $\phi(t,x)$ and the drift x/t.

For an appropriately chosen integration constant we have

$$\phi(t,x) = \frac{1}{\sqrt{2\pi\sigma^2 t}} \exp\left(\frac{x^2}{2\sigma^2 t}\right). \tag{2.7.7}$$

Substituting this expression in equation (2.7.6), we obtain

$$-c(t,x)\phi = \frac{\partial\phi}{\partial t} + \frac{1}{2}\sigma^2\frac{\partial^2\phi}{\partial x^2} = 0,$$

whence $c(t,x) \equiv 0$. Therefore, the equations of motion close to the birth of the Universe, i.e., for small values of t, are

$$\begin{cases} \dfrac{\partial\phi}{\partial t} + \dfrac{1}{2}\sigma^2\dfrac{\partial^2\phi}{\partial x^2} = 0, \\[2mm] -\dfrac{\partial\widehat{\phi}}{\partial t} + \dfrac{1}{2}\sigma^2\dfrac{\partial^2\widehat{\phi}}{\partial x^2} = 0. \end{cases} \tag{2.7.8}$$

We state this conclusion as a theorem:

Theorem 2.7.2. *At the birth of the Universe there was no (external) potential: the Universe was born spontaneously.*

This is conclusion is quite reasonable, because our model assumed that there was nothing before or "external" to the Universe.

Recall that the evolution equation (2.7.4), or equation (2.7.5) of the Skorokhod problem with two moving boundaries, were derived based on Hubble's law. This means that we predict the evolution of the Universe backwards, with the *time reversed* $t \downarrow 0$ near its birth. Then, as we saw, the singular term $\frac{x}{t}$ appears in equation (2.7.4). Therefore, we should write equation (2.7.5) as

$$X(t) = \sigma\beta(t) + \int_0^t ds\frac{X(s)}{s} + \Phi(t), \quad |X(t)| \le R(t), \quad t \downarrow 0, \tag{2.7.9}$$

where we look at the Universe time reversed, towards the moment of its birth.

One can ask how does the Universe $X(t)$ evolve in the natural direction of time, $0 \uparrow t$. One might think that

"there would be not much difference between the two directions of time,"

but in fact this is not the case. To see this, we apply the following result (refer to Theorem 14.6.1 of Nagasawa (2000)).

Theorem 2.7.3. *Suppose the boundary function $R(t)$ satisfies*

$$\lim_{t\to\infty}\frac{R(t)}{t} = 0.$$

*Let $\beta(t)$ be the one-dimensional Brownian motion. Then $X(t)$ satisfies equation
(2.7.9) if and only if it satisfies the equation*

$$X(t) = \sigma B(t) - t \int_t^\infty d\Phi(s) \frac{1}{s}, \quad 0 \uparrow t, \qquad (2.7.10)$$

where $B(t)$ is the one-dimensional Brownian motion defined by

$$B(t) = -t \int_t^\infty \frac{d\beta(s)}{s}.$$

Theorem 2.7.3 claims that equation (2.7.9) with time reversed is equivalent
to the equation (2.7.10) with the natural direction of time.

The singularity appearing in equation (2.7.9) is not present in equation
(2.7.10) with the natural direction of time $0 \uparrow t$ of the Universe. Thus, according to (2.7.10), the Universe starts naturally at its birth as a Brownian motion.
We state this fact as a theorem.

Theorem 2.7.4. *The Universe was born naturally from "nothing." Namely, with the
natural time evolution $0 \uparrow t$, the Universe started naturally as a Brownian motion.*

Comment to Theorem 2.7.4. If we look at equation (2.7.9), then because of of
the presence of the singular Hubble flow $\frac{x}{t}$, it looks as if the Universe started at
$t = 0$ like an explosion (i.e., big bang). However, equation (2.7.10) shows that this
actually is not the case: the Universe expanded as a Brownian motion at its birth.

It was accepted that Universe started with a big bang, the reason being that
one looks at the Universe backward in time towards the birth time, $t \downarrow 0$, and
thought that this observation is independent of the direction of time. Theorem
2.7.4 shows that when one analyzes the birth of the Universe, the direction of
time must be treated carefully.

If we express the equation of motion in complex numbers, then, as shown in
Theorem 1.9.1, we get the complex evolution equation

$$i \frac{\partial \psi}{\partial t} + \frac{1}{2} \sigma^2 \frac{\partial^2 \psi}{\partial x^2} - V(t, x) \psi = 0. \qquad (2.7.11)$$

Therefore, the complex evolution equation (2.7.11) can be used to discuss the
evolution of the Universe.

The potential $V(t, x)$ that appears in equation (2.7.11) can be computed by
means of the condition (1.9.4) of Theorem 1.9.1 and formula (1.9.5). In the present
case, since $c(t, x) \equiv 0$ and $b(t, x) \equiv 0$, we have

$$V(t, x) = -2 \frac{\partial S}{\partial t} - \sigma^2 \left(\frac{\partial S}{\partial x} \right)^2.$$

Since

$$\phi(t, x) = e^{R(t,x) + S(t,x)} \quad \text{and} quad \mu(t, x) = e^{2R(t,x)},$$

we have

$$S\left(t, x\right) = \log \phi\left(t, x\right) - \frac{1}{2} \log \mu\left(t, x\right).$$

We assume that

$$\mu\left(t, x\right) = \frac{1}{\sqrt{2\pi\sigma^2 t}} \exp\left(-\frac{x^2}{2\sigma^2 t}\right)$$

for sufficiently small t. Since $\phi\left(t, x\right)$ is given by formula (2.7.7),

$$V\left(t, x\right) = -\frac{3}{4}\frac{1}{\sigma^2}\frac{x^2}{t^2} + \frac{1}{2}\frac{1}{t} \tag{2.7.12}$$

for sufficiently small t. The potential $V(t, x)$ that appears in the complex evolution equation (2.7.11), does not correspond to external forces.

The potential $V(t, x)$ given by formula (2.7.12) satisfies

$$V\left(t, 0\right) \uparrow \infty, \quad t \downarrow 0,$$

that is, it becomes infinite at the origin $x = 0$ when $t \downarrow 0$.

Therefore, the complex evolution equation yields another description of the birth of the Universe:

Theorem 2.7.5. *The birth of the Universe can be described by means of the complex evolution equation. The potential $V(t, x)$ in equation (2.7.11) is given by formula (2.7.12) for sufficiently small $t > 0$. Moreover, $V(t, x)$ is a concave function, and it diverges as $t \downarrow 0$. Therefore, after its birth, the Universe started to expand rapidly.*

Comments to Theorem 2.7.5.

(i) Theorem 2.7.5 is based on the fact that in terms of the solution $\psi\left(t, x\right)$ of the complex evolution equation the distribution density is given by $\mu\left(t, x\right) = \psi\left(t, x\right) \overline{\psi}\left(t, x\right)$. Therefore, the assertion of the theorem is not about the motion of particles, but about the distribution density of the Universe.

(ii) Vilenkin (1982, 1983) discussed the birth of the Universe by means of the complex evolution equation (Schrödinger equation).

Theorems 2.7.2, 2.7.4 and 2.7.5 tell the story of the birth of the Universe based on the theory of stochastic processes. Summarizing,

"The Universe was born spontaneously from nothing, and started as a Brownian motion (not as as big bang). After birth, it went into an explosive inflation."

2.8 Classification of Boundary Points

Let us consider a diffusion process in an open interval (α, β). One of the main problems is whether such a processes can reach or leave the boundary points. We will present and explain Feller's theorem on the classification of boundary points.

Consider the random motion $\{X_t, t \geq 0; \mathbf{P}_x\}$ determined in an open interval (α, β) by the equation

$$\frac{\partial u}{\partial t} = Au, \tag{2.8.1}$$

where

$$A = \frac{1}{2} a\,(x)\,\frac{d^2}{dx^2} + b\,(x)\,\frac{d}{dx}, \quad a\,(x) > 0. \tag{2.8.2}$$

Fix an arbitrary point $c \in (\alpha, \beta)$, and set

$$W\,(x) = \int_c^x dy\,\frac{2b\,(y)}{a\,(y)}.$$

Then the operator A can be written in divergence form

$$A = \frac{1}{2} a\,(x)\,e^{-W(x)}\,\frac{d}{dx}\left(e^{W(x)}\,\frac{d}{dx}\right). \tag{2.8.3}$$

We then define the canonical scale $S(x)$ by

$$S\,(x) = \int_c^x dy\,e^{-W(y)},$$

and the speed measure M by

$$M(dx) = \frac{1}{a\,(x)} e^{W(x)} dx.$$

Now using the canonical scale and the speed measure, we define Feller's canonical operator by

$$\mathcal{A} = \frac{1}{2}\,\frac{d}{dM}\,\frac{d}{dS}. \tag{2.8.4}$$

Its domain $\mathcal{D}(\mathcal{A})$ is the set of all functions $f \in C\,([\alpha, \beta])$ that satisfy the following conditions:

   (i) df is absolutely continuous with respect to dS;

   (ii) $d\left(\frac{df}{dS}\right)$ is absolutely continuous with respect to dM;

   (iii) $\frac{d}{dM}\left(\frac{df}{dS}\right) \in C\,([\alpha, \beta])$.

The canonical operator $(\mathcal{A}, \mathcal{D}(\mathcal{A}))$ is the generator of an operator semigroup. The Markov process $\{X_t, t \geq 0; \mathbf{P}_x\}$ determined by this operator semigroup is called a "Feller process".

Note on operator semigroups.

(i) Let P_t be a family of operators depending on a (time) parameter $t \geq 0$ and defined on the space $C([\alpha, \beta])$ of bounded continuous functions. If

$$P_{t+s}f(x) = P_t P_s f(x), \quad t, s \geq 0 \quad \text{and} \quad P_0 f(x) = f(x),$$

P_t is called an operator semigroup on the space $C([\alpha, \beta])$.

(ii) If

$$\frac{d}{dt} P_t f(x) = \mathcal{A} P_t f(x) \quad \text{for all } f \in \mathcal{D}(\mathcal{A}),$$

then $\mathcal{A}$ is called the generator of semigroup P_t.

(iii) There exists a transition probability $P(t, x, dy)$ satisfying

$$P_t f(x) = \int P(t, x, dy) f(y) dy,$$

which can be used to construct a Markov process $\{X_t, t \geq 0; \mathbf{P}_x\}$.

(iv) The operator semigroup P_t on the space $C([\alpha, \beta])$ is represented in terms of the Markov process $\{X_t, t \geq 0; \mathbf{P}_x\}$ by the formula

$$P_t f(x) = \mathbf{P}_x[f(X_t)], \quad f \in C([\alpha, \beta]).$$

(v) The Markov process constructed by means of the semigroup of operators generated by the Feller canonical operator $\mathcal{A} = \frac{1}{2} \frac{d}{dM} \frac{d}{dS}$ defined in equation (2.8.4) is a Feller process.

Details on operator semigroups and Markov processes will be provided in Section 4.1.

Now consider a Feller process X_t and define the first hitting time of an interior point $y \in (\alpha, \beta)$ by

$$T_y(\omega) = \inf \{t : X_t(\omega) = y\}.$$

We assume that

$$\mathbf{P}_x[T_y < \infty] > 0, \quad x, y \in (\alpha, \beta).$$

Namely, our particle can come and go between any two interior points x and y.

However, we do not know whether our particle can come and go between an interior point $y \in (\alpha, \beta)$ and a boundary point α or β. To determine this, one resorts to the *Feller test*.

We state the test for a boundary point α, which we classify using the functions

$$S(\alpha, x] = S(x) - \lim_{y \downarrow \alpha} S(y) \quad \text{and} \quad M(\alpha, x] = M((\alpha, x]),$$

as follows: α is

(i) a regular boundary point, if $S(\alpha, x] < \infty$ and $M(\alpha, x] < \infty$;

(ii) an exit boundary point, if $S(\alpha, x] < \infty$ and $M(\alpha, x] = \infty$;

(iii) an entrance boundary point, if $S(\alpha, x] = \infty$ and $M(\alpha, x] < \infty$;

(iv) a natural boundary point, if $S(\alpha, x] = \infty$ and $M(\alpha, x] = \infty$.

Then the following theorem holds (Feller (1954)).

Theorem 2.8.1 (Feller's Theorem). *Let $X(t)$ be a Feller process. Then if the left boundary point α is*

(i) *a regular point, then α is accessible from (α, β), and (α, β) is accessible from α;*

(ii) *an exit point, then α is accessible from (α, β), but (α, β) is not accessible from α;*

(iii) *an entrance point, then α is not accessible from (α, β), but (α, β) is accessible from α;*

(iv) *a natural point, then α is not accessible from (α, β), and (α, β) is not accessible from α.*

The proof of this theorem is omitted, because it is rather involved. For details, we refer, e.g., to Sections 2.10 and 2.11 of Nagasawa (1993).

Example 1. Let us apply Theorem 2.8.1 to a simple example. Let κ be a constant and consider the operator

$$A = \frac{1}{2}\frac{d^2}{dx^2} + \kappa\frac{1}{x}\frac{d}{dx}$$

in the open interval $(0, \infty)$. In this case we can assume that

$$W(x) = 2\kappa \log x. \tag{2.8.5}$$

Then

$$M(\varepsilon, x] = \begin{cases} \displaystyle\int_\varepsilon^x dy\, e^{W(y)} = \int_\varepsilon^x dy\, y^{2\kappa} \\ \displaystyle\qquad = \frac{1}{1+2\kappa}\left(x^{1+2\kappa} - \varepsilon^{1+2\kappa}\right), & \text{if } \kappa \neq -\tfrac{1}{2}, \\ \log x - \log \varepsilon, & \text{if } \kappa = -\tfrac{1}{2}. \end{cases}$$

Therefore,

$$M(0, x] < \infty, \quad \text{if } \kappa > -\frac{1}{2},$$

$$M(0, x] = \infty, \quad \text{if } \kappa \leq -\frac{1}{2}.$$

For S we have

$$S\left[\varepsilon, x\right] = \begin{cases} \displaystyle\int_\varepsilon^x dy\, e^{-W(y)} = \int_\varepsilon^x dy\, y^{-2\kappa} \\[2mm] \displaystyle = \frac{1}{1-2\kappa}\left(x^{1-2\kappa} - \varepsilon^{1-2\kappa}\right), & \text{if } \kappa \neq \tfrac{1}{2}, \\[2mm] \log x - \log \varepsilon, & \text{if } \kappa = \tfrac{1}{2}. \end{cases}$$

Therefore,

$$S(0, x] < \infty, \quad \text{if } \kappa < \frac{1}{2},$$

$$S(0, x] = \infty, \quad \text{if } \kappa \geq \frac{1}{2},$$

and the behavior of the Feller process X_t depends on the constant κ. According to Theorem 2.8.1, the origin 0 is

(i) a regular boundary point, if $-\frac{1}{2} < \kappa < \frac{1}{2}$,

(ii) an exit boundary point, if $\kappa \leq -\frac{1}{2}$,

(iii) an entrance boundary point, if $\frac{1}{2} \leq \kappa$.

Example 2. In Section 2.1, in the case where $\{n, \ell, m\} = \{1, 0, 0\}$, the radial component of the random motion is given by

$$r_t = r_a + \sigma B_{t-a} + \int_a^t \left(\sigma^2 \frac{1}{r_s} - \alpha\right) ds, \quad r_t > 0,$$

where B_t is the one-dimensional Brownian motion. The generator of this Markov process is

$$A = \frac{1}{2}\sigma^2 \frac{d^2}{dx^2} + \left(\sigma^2 \frac{1}{x} - \alpha\right)\frac{d}{dx}, \quad x \in (0, \infty).$$

Since $-\alpha$ is negligible compared to $1/x$ near the origin, we can consider instead the operator

$$A = \frac{1}{2}\sigma^2 \frac{d^2}{dx^2} + \sigma^2 \frac{1}{x}\frac{d}{dx}.$$

Then

$$W(x) = 2\log x.$$

This is the case when $\kappa = 1$ in (2.8.5). Accordingly, as shown in Example 1, the origin is an entrance boundary point. Therefore, the radial random motion r_t cannot reach the origin.

This is important for the motion of an electron in the hydrogen atom. If the electron can reach the origin, it is absorbed there by the proton, and the hydrogen atom disappears.

2.9   Particle Theory of Electron Holography

In this application, an electron gun shoots electrons at a sample and its image is observed with an electron microscope or electron holography. Electron holography is theoretically an application of the entanglement of random motions discussed in Section 1.10. Akira Tonomura and his collaborators at Hitachi Research Laboratory developed an electron holography technique of practical use (see, e.g., Tonomura (1999)).

Let us show how our theory of random motion of particles applies to electron holography.

Remark. Tonomura (1990) regarded electrons as "waves", and thus considered that they have the "wave property", and talked about "electron waves". For example he regarded an electron shot from an electron gun as a plane wave. What Tonomura calls "plane electron wave" is the random motion of an electron with the drift potential $\mathbf{a}(t,x) = \sigma^2 \nabla S(t,x) = \sigma^2(0,0,k)$. Now, as explained in Sections 1.10 and 1.12, there are no theoretical grounds for asserting that electrons have the so-called wave property. Instead, an electron performs a Brownian motion with the **drift potential** $S(t,x)$. Thus, our aim is to establish a particle theory of electron holography instead of a wave theory.

The electron gun (electron source) used in an electron microscope or electron holography is a tungsten needle with a 0.1 micron tip. At some kilo voltage a regularly parallel field emission occurs.

Suppose an electron is shot from the electron gun. The motion of the electron is described by the pair of evolution functions

$$\phi_1(t,x) = e^{R_1(t,x)+S_1(t,x)}, \quad \widehat{\phi}_1(t,x) = e^{R_1(t,x)-S_1(t,x)}, \tag{2.9.1}$$

where $\{R_1(t,x), S_1(t,x)\}$ is the generating function of the random motion, or, alternatively, by the complex evolution function

$$\psi_1(t,x) = e^{R_1(t,x)+iS_1(t,x)}, \tag{2.9.2}$$

where we direct the z-axis downward, and we can take $S_1 = kz$, because we are using a regularly parallel field-emission electron source.

The electron will hit the sample or pass through near the sample, and it will be influenced by the electromagnetic field around the sample. We assume that its motion is described by the pair of evolution function

$$\phi_2(t,x) = e^{R_2(t,x)+S_2(t,x)}, \quad \widehat{\phi}_2(t,x) = e^{R_2(t,x)-S_2(t,x)}. \tag{2.9.3}$$

Namely, we assume that, due to the presence of the sample, the generating function $\{R_1(t,x), S_1(t,x)\}$ is transformed into $\{R_2(t,x), S_2(t,x)\}$. Then the distribution density of the electron is

$$\mu_2(t,x) = e^{2R_2(t,x)}, \tag{2.9.4}$$

because $\mu_2(t,x) = \phi_2(t,x)\widehat{\phi}_2(t,x)$. Therefore, electrons are shot one by one, we can see the image of the sample distributed with the density $\mu_2(t,x)$. This is the sought-for particle theory formulated in terms of our mechanics of random motion, and the operation of an electron microscope can be designed on the basis of this particle theory.

Here we must emphasize that even though the drift potential $S_1(t,x)$ is transformed into $S_2(t,x)$ due to the influenced of the sample, we cannot see this: what we can can actually observe is only the distribution density $\mu_2(t,x) = e^{2R_2(t,x)}$, which does not contain the drift potential $S_2(t,x)$. In other words, when we look at the sample with the electron microscope, we discard important information encoded in the drift potential $S_2(t,x)$.

Suppose, for instance, that the sample is a small magnet. Since electrons are influenced by the lines of magnetic force when passing near the sample, if we look at the motion of electrons, we should be able to see the lines of magnetic force. However, as we saw in Section 2.3, the influence of the lines of magnetic force manifests itself in the drift-potential $S_2(t,x)$, not in the distribution exponent $R_2(t,x)$. Therefore, only the magnet can be seen in the image of the electron microscope represented by the distribution formula (2.9.4): the lines of magnetic force around remain invisible.

Recall that if you place a glass plate on a big magnet and sprinkle iron sand on the plate, you can see the lines of magnetic force around the magnet. In the above example, the electrons play the role of the iron sand. Even so, with the electron microscope, lines of magnetic force cannot be visualized by sprinkling electrons.

If we nevertheless want to visualize the lines of magnetic force around specimens, we have find a way to see the influence of the lines of magnetic force on the drift-potential $S_2(t,x)$. So is there a way to access the information contained in $S_2(t,x)$? The answer is provided by the method of electronic holography, as we explain next.

The basis of electronic holography is the "superposition" or "entanglement" of the two random motions described in Section 1.10.

The entanglement is actually realized by an "electron biprism", such as the one used by Tonomura in the double-slit experiments described in Section 1.12. The prims is a thin quartz yarn with a thickness of one micron covered with a thin film of gold. We apply a positive potential to it and place two earthed flat plate electrodes on both sides of the yarn. The electron passes through the right or left side of the biprism, but

"Which side it goes through is random"

An electron shot from the electron gun may reach the biprism without being affected by the specimen and may also reach the biprism by hitting the specimen or passing by its side. Then, since the electrons undergo random motions, the entanglement of these different random motions is realized by the electron biprism.

In detail, one electron may pass to the left of the biprism without being influenced by the specimen and may pass to the right side. Or it may pass to the left of the biprism by hitting the specimen or passing by the side, sometimes going through the right side. **"Since which of these motions occurs is accidental, entanglements occur."**

We label motion through the biprism without being affected by the minute specimen (sample) as "random motion 1", and the motion through the biprism after hitting the specimen or passing by its side as "random motion 2."

Let the random motion 1 be described by

the complex evolution function 1: $\psi_1(t, x) = e^{R_1(t,x)+iS_1(t,x)}$,

and the random motion 2 be described by

the complex evolution function 2: $\psi_2(t, x) = e^{R_2(t,x)+iS_2(t,x)}$,

In fact, random motion 1 (resp., 2) corresponds to motion through the left (resp., right) side of the electron biprism, labeled by l (resp., r). Therefore, we must consider four random motions, determined by the generating functions

$$\{R_{1l}(t,x), S_{1l}(t,x)\}, \quad \{R_{1r}(t,x), S_{1r}(t,x)\},$$
$$\{R_{2l}(t,x), S_{2l}(t,x)\}, \quad \{R_{2r}(t,x), S_{2r}(t,x)\},$$

but here we will not use this to avoid complications.

The random motion obtained by entangling "random motion 1" and "random motion 2" is described by a linear combination of the complex evolution functions $\psi_1(t, x)$ and $\psi_2(t, x)$:

$$\psi(t, x) = \alpha_1\psi_1(t, x) + \alpha_2\psi_2(t, x), \tag{2.9.5}$$

where α_1 and α_2 are chosen appropriately, so that the normalization condition $\int |\psi(t, x)|^2 dx = 1$ is satisfied. Represent $\psi(t, x) = \alpha_1\psi_1(t, x) + \alpha_2\psi_2(t, x)$ in the exponential form

$$\psi(t, x) = e^{R_3(t,x)+iS_3(t,x)}. \tag{2.9.6}$$

The exponent $\{R_3(t, x), S_3(t, x)\}$ in (2.9.6) is the entanglement of the generating functions $\{R_1(t, x), S_1(t, x)\}$ and $\{R_2(t, x), S_2(t, x)\}$, and the function $\psi(t, x) = e^{R_3(t,x)+iS_3(t,x)}$ in (2.9.6) is the **entangled complex evolution function** of $\psi_1(t, x)$ and $\psi_2(t, x)$.

The stochastic process $\{X_t(\omega), \mathbf{Q_3}\}$ determined by the entangled complex evolution function $\psi(t, x)$ in (2.9.6) is the **entangled random motion** of the random motion 1 and random motion 2.

What is used in the particle theory of electron holography is the stochastic process $\{X_t(\omega), \mathbf{Q_3}\}$. (See Theorem 1.10.4 for the existence of entangled random motion.) It is described by the pair of evolution functions

$$\phi_3(t, x) = e^{R_3(t,x)+S_3(t,x)}, \quad \widehat{\phi}_3(t, x) = e^{R_3(t,x)-S_3(t,x)}, \qquad (2.9.7)$$

i.e., the entangled complex evolution function $\psi(t, x) = e^{R_3(t,x)+iS_3(t,x)}$.

Therefore, is a denotes the time at which the entanglement occurred,

$$X_t = X_a + \sigma B_{t-a} + \int_a^t \mathbf{a}_3(s, X_s)ds, \qquad (2.9.8)$$

where the distribution density of X_a is $\mu_a(x) = e^{2R_3(a,x)}$, determined by the generating function $R_3(a, x)$ at a, and $\mathbf{a}_3(s, x)$ is the entangled drift $\mathbf{a}_3(s, x) = \sigma^2 \nabla R_3(s, x) + \sigma^2 \nabla S_3(s, x)$. This is determined by the function $\tan(S_3(s, x))$ of the entangled drift potential $S_3(s, x)$.

As the above discussion makes plain, "entanglement (superposition)" is a concept concerning the random motion of **"one electron"**. It is therefore completely free from "wave interference".

Then, by Theorem 1.10.5, the distribution density $\mu_3(t, x)$ of the entangled random motion is given by the formula

$$\begin{aligned}
\mu_3(t, x) = {} & |\alpha_1|^2 e^{2R_1(t,x)} + |\alpha_2|^2 e^{2R_2(t,x)} \\
& + 2e^{R_1(t,x)+R_2(t,x)} \Re\left(\alpha\overline{\beta}\right) \cos(S_2(t, x) - S_1(t, x)) \qquad (2.9.9) \\
& + 2e^{R_1(t,x)+R_2(t,x)} \Im\left(\alpha\overline{\beta}\right) \sin(S_2(t, x) - S_1(t, x)).
\end{aligned}$$

The entanglement terms $\cos(S_2(t, x) - S_1(t, x))$ and $\sin(S_2(t, x) - S_1(t, x))$ constitute the mathematical core of the theory of electron holography.

Indeed, the distribution density given by formula (2.9.9) displays the difference $S_2(t, x) - S_1(t, x)$ of the drift potentials when the electrons hit specimens (minute magnets) or pass by.

If electrons are shot one after another from the electron gun, lines of magnetic force appear in the image. In the formula (2.9.9) for the distribution of electrons, the two terms $\cos(S_2(t, x) - S_1(t, x))$ and $\sin(S_2(t, x) - S_1(t, x))$ indicate the change in shade of the image depending on position. This shading represents lines of magnetic force in and around the specimen (minute magnet).

Tonomura and his collaborators showed by using electron holography that lines of magnetic force appear in the image. Figure 2.9 of magnetic field lines is taken from Tonomura (1990).

In Tonomura (1990) and Tonomura (1994) one can find many interesting applications of electron holography.

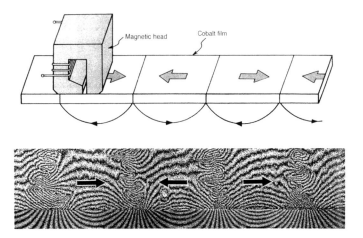

Figure 2.9

For example, assuming that $\alpha_1 = 1$ and $\alpha_2 = i$ in (2.9.9) (neglecting the normalization), the distribution density (2.9.9) of electrons, that is, the image (hologram), is given by

$$\mu_3(t,x) = e^{2R_1(t,x)} + e^{2R_2(t,x)} + 2e^{R_1(t,x)+R_2(t,x)} \sin(S_2(t,x) - S_1(t,x)).$$

The first term $e^{2R_1(t,x)}$ represents an image showing the distribution of electrons shot from the electron source and is irrelevant for the specimen. The second term $e^{2R_2(t,x)}$ is the image of the specimen (tiny magnet) seen by the electron microscope, previously given by formula (2.9.4). The third term

$$2e^{R_1(t,x)+R_2(t,x)} \sin(S_2(t,x) - S_1(t,x))$$

represents the lines of magnetic force in and around the specimen.

This technique allowing to see the distribution density $\mu_3(t,x)$ of the entangled random motion given by formula (2.9.9) is

"electron holography" or "phase contrast electron microscope".

In this section we applied the dynamics of random motion to entanglement and presented a "particle theory of electronic holography".

Incidentally, to actually visualize the image generated by the electrons appearing on the hologram, the image recorded on the hologram is enlarged and converted into an optical image by using a laser beam. Details on this can be found in specialized electronic holography manuals.

2.10 Escherichia coli and Meson models

The applications of the mechanics of random motion to quantum are not limited to quantum theory.

(i) Nagasawa (1981) deals with unique septation of mutant Escherichia coli using a one-dimensional mathematical model, in which E. coli is regarded as a group of particles formed by entangling a large number of molecules performing zigzag motions.

As molecular entanglements generate constant forces towards the center, we consider the Schrödinger equation

$$i\frac{\partial\psi}{\partial t} + \frac{1}{2}\sigma^2\frac{\partial^2\psi}{\partial x^2} - \kappa|x|\psi = 0. \tag{2.10.1}$$

In the model the stable distribution density $\mu = \phi^2$ obtained from a solution ϕ of the eigenvalue problem

$$\frac{1}{2}\sigma^2\frac{\partial^2\phi}{\partial x^2} + (\lambda - \kappa|x|)\phi = 0, \quad k > 0 \tag{2.10.2}$$

describes an E. coli corresponding to the eigenvalue λ. For details, we refer to Section 9.3 of Nagasawa (1993).

(ii) The dynamics of random motion was also applied to the problem of "division of the population of a biological group". Please refer to Nagasawa (1980, 1985, 1988).

(iii) The mass spectrum of mesons can be analyzed by using a mathematical model similar to that provided for E. coli in item (i). Specifically, we consider a model in which a meson is regarded as an aggregate of a large number of gluons which perform zigzag motions. Due to the entanglement of gluons, a constant force towards the center is generated. It is assumed that the solution ϕ of Schrödinger equation (2.10.2) describes the motion of the group of gluons. Then $\mu = \phi^2$ gives the distribution density of long gluon strings. Almost the complete mass spectra of mesons were reproduced by calculating the energy of the group of gluons, that is, the eigenvalues of the steady-state problem (2.10.2) above and adding the mass of the two quarks to it. The following table is taken from Nagasawa (1993) Section 9.4, page 236:

Case	Name	Mass(Mev)	computed	Model
Case 1:	$\pi^{\pm}$	139.5669	139.5669	(postulated)
Case 2:	$K^{\pm}$	493.67	495	(u or d, ϕ_3, 0, s)
Case 3:	η	549	545	(s, ϕ_3, 0, s)
Case 4:	ρ	769	756	(u or d, ϕ_6, 0, u or d)
Case 5:	ω	783	771	(u or d, ϕ_6, 1, u or d)
Case 6:	K^*	892	894	(u or d, ϕ_7, 1, s)
Case 7:	η'	958	944	(s, ϕ_7, 0, s)
Case 8:	$S^*(0)(975),$	$\delta(0)(980)$		not well established
Case 9:	$\phi(1)$	1020	1030	(s, ϕ_8, 0, s)
Case 10:	$H(1)$	1190	1178	(u or d, ϕ_{11}, 1, u or d)
Case 11:	$B(1)$	1233	1236	(u or d, ϕ_{12}, 0, u or d)
Case 13:	$Q_1(1)$	1270	1286	(s, ϕ_{12}, 0, u or d)
Case 14:	$f(2)$	1273	1266	(u or d, ϕ_{12}, 2, u or d)
Case 15:	$A_1(1)$	1275	1251	(u or d, ϕ_{12}, 1, u or d)
Case 17:	$D(1)$	1283	1278	(s, ϕ_{11}, 0, s)
Case 18:	$\epsilon(1)$	1300	1306	(u or d, ϕ_{13}, 0, u or d)
.....	....	....	....	......
Case 39:	D	1869	1863	(u or d, ϕ_{11}, 0, c)
.....	....	....	....	......
Case 42:	$F^{\pm}$	2021	(1986)	(s, ϕ_{12}, 0, c)
.....	....	....	....	......
Case 58:	J/Ψ	3097	3097	(c, ϕ_{19}, 0, c)
Case 59:	$\chi(0)$	3415	3390	(c, ϕ_{24}, 0, c)
Case 60:	$\chi(1)$	3510	3517	(c, ϕ_{26}, 0, c)
Case 61:	$\chi(2)$	3556	3571	(c, ϕ_{27}, 1, c)
Case 61:	η'_c	3590	3586	(c, ϕ_{27}, 2, c)
Case 63:	$\Psi(1)$	3686	3678	(c, ϕ_{29}, 1, c)
Case 64:	$\Psi(1)$	3770	3768	(c, ϕ_{31}, 0, c)
.....	....	....	....	......

2.11 High-Temperature Superconductivity

In 1911, Kamerlingh Onnes discovered the interesting phenomenon that when mercury (Hg) is brought to a cryogenic temperature, the electric resistance suddenly disappears at 4.2 K (Kelvin) and current flows continuously undisturbed. This phenomenon is called "superconductivity". After his discovery, researchers looked for metals with higher transition temperatures, but the highest one achieved was about 10 K. A breakthrough was made by Müller and Bednorz in 1986. They discovered that a ceramic La_2CuO_4 (precisely speaking, a percentage of La is replaced by Sr: $La_{2-x}Sr_xCuO_4$) exhibit superconductivity at 40 K. This was surprising, because the copper oxide La_2CuO_4 is a ceramic, which is essentially an insulator. The phenomenon discovered by Mueller and Bednorz is called "high temperature superconductivity". "High" is relative to the absolute temperature 4.2 K (Kelvin), but still in the cryogenic range compared to the normal temperature of 300 K. (By now, the highest transition temperature observed is 95 K for the Yttrium-based copper oxide high-temperature superconductor $YBa_2Cu_3O_{7-y}$. Superconductors with even higher transition temperatures have been discovered under high-pressure conditions.)

Let us take a look at the mathematical model of high-temperature superconductivity, considering the case of Müller and Bednorz's copper oxide (ceramic) La_2CuO_4. (Cf. Nagasawa (2014)) This ceramic consists of layers of copper oxide, which are conductive layers of electrons and will be called "yellow layers", and layers without copper oxide, which will be called "white layers", as illustrated in Figure 2.10.

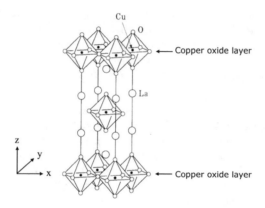

Figure 2.10

Mobile particles (electrons or positive electron holes) are confined to a yellow layer (which we model as $\mathbf{R}^2 \times [0, \ell]$, $\ell > 0$) and do not cross white layers.

Consider a yellow (copper oxide) layer. It contains a large number ($10^{20} \sim 10^{23}$) of mobile particles (electrons or holes). There are interactions among the particles, but as a vast particle system, it can be handled as an aggregate of particles that are moving independently, as a result of the propagation of chaos (Section 5.6).

Electrons are fermions, and do not obey the Bose-Einstein statistics, so we cannot immediately apply the theory of Bose-Einstein condensation to discuss the transition of an electron gas to superconductivity; therefore, additional ideas are needed.

In the widely accepted Bardeen-Cooper-Schrieffer theory (1957) of low-temperature superconductivity, two electrons are coupled and regarded as a boson, and the theory of Bose-Einstein condensation is then applied to these bosons. This is an interesting idea, and was successful in discussing low-temperature superconductors, but so far did not lead to a comprehensive understanding of high-temperature superconductivity, because it has been a difficult task to fit the crystal (layer) structure such as the plane where superconductivity occurs and the structure orthogonal to it for high-temperature superconductors in the Bardeen-Cooper-Schrieffer theory. The technique of "forming pairs" is rather a hindrance in their considerations. Since the discovery of high-temperature superconductivity, various theories have been proposed to explain it, but to the best of my knowledge, "a coherent theory of high-temperature superconductivity" is still missing.

Nagasawa (2014) formulated a "consistent theory of high-temperature superconductivity", based on a completely different, simple mathematical idea. We will explain it below.

To do that, because electrons are Fermi particles, some ingenuity is necessary. Namely, we consider that the layer structure of high-temperature superconductors plays a decisive role for the occurrence of superconductivity. Based on this, we look at "the components of the motions of mobile particles orthogonal to the copper-oxide layer" and regard them as "a set of one-dimensional Bose particles."

Consider a yellow (copper-oxide) layer $\mathbf{R}^2 \times [0, \ell]$ and take the direction orthogonal to the (x, y)-plane $\mathbf{R}^2$ as the z-axis. If we look at the crystal structure of the high-temperature superconductor La_2CuO_4 in this coordinate system, we find that the z-direction and (x, y)-directions play completely different roles. In fact, there are upper and lower boundaries in the z-direction, while there is no such restriction in the (x, y)-plane. Therefore, it is plausible to assume that the superconductive current occurs in the (x, y)-plane and the component along the z-axis concerns the transition to superconductivity.

This can be pictured as follows. Imagine a large number of needles standing on the (x, y)-plane (a forest of needles); the lengths of the needles are different and change with time. ("Needle" is just a visualization tool, and has no further meaning.) Each needle tip carries a mobile particle. If we watch two arbitrarily chosen needles in the forest, they are *allowed to have the same length and to perform the same up-down motion at the same time*, while the (x, y)-components of our particles may be different. Therefore, even though the mobile particles are fermions, their z-components may be treated as bosons.

We now collect all the needles and place them on the z-axis. (Then the needles' standing points on the (x, y)-plane no longer play a role.) We denote by $z(t)$ the value of a needle tip on z-axis, and look at the up-down motion of $z(t)$. What we observe is, therefore, the set {all "needle tips" }, that is, the set $\mathbf{Z} = \{z$-components of all mobile particles in a yellow layer $\mathbf{R}^2 \times [0, \ell]$ }. Because the lengths of needles may become the equal and the needles may perform simultaneously the same up-down motion, the set {*all "needle tips"*} is a system of one-dimensional Bose particles.

The model considered above can be formulated in terms of Schrödinger functions. We assume that the Schrödinger function

$$\psi_t((x, y, z)_1, (x, y, z)_2, (x, y, z)_3, \ldots)$$

that describes the motion of mobile particles (electrons or holes) in a yellow layer $\mathbf{R}^2 \times [0, \ell]$ can be decomposed as

$$\psi_t((x, y, z)_1, (x, y, z)_2, (x, y, z)_3, \ldots)$$
$$= \psi_t((x, y)_1, (x, y)_2, (x, y)_3, \ldots) \times \psi_t(z_1, z_2, z_3, \ldots),$$

where

$$\psi_t((x, y)_1, (x, y)_2, (x, y)_3, \ldots) \text{ is antisymmetric,}$$

while

$$\psi_t(z_1, z_2, z_3, \ldots) \text{ is symmetric.}$$

Then, under the assumptions of our mathematical model, we can state

Proposition 2.11.1. *The factor $\psi_t(z_1, z_2, z_3, \ldots)$ is the Schrödinger function for the z-component of the motion and describes an aggregate of one-dimensional Bose particles.*

Accordingly, we can apply the Bose-Einstein statistics to our one-dimensional boson gas on the z-axis, and discuss its Bose-Einstein condensation.

Remark. When the Schrödinger function for a set of particles is symmetric (i.e., it does not change under permutations), one has a set of Bose particles. This is equivalent to the definition of a Bose particle in the Bose-Einstein theory. By contrast, when the Schrödinger function for a set of particles is antisymmetric (i.e., it sign changes sign under odd permutations), we are dealing with a group of Fermi particles. For symmetric functions and antisymmetric functions, see, for example, Dirac (1930, 58) quantum mechanics, Tomonaga (1953) quantum mechanics II.

Therefore, under the preceding assumptions above, we can state that

The set $\mathbf{Z} = \{z\text{-components of all mobile particles (electrons or positive holes) in a yellow layer } \mathbf{R}^2 \times [0, \ell]\}$ is a boson gas and undergoes a Bose-Einstein condensation.

With this fact in mind, we now look at the motion of mobile particles in a yellow layer. Then the random motion orthogonal to the (x, y)-plane drops to the ground state at the time Bose-Einstein condensation occurs. What remains after that is therefore only the drift motion in the (x, y)-plane, which is actually *"the permanent electric current"* in the superconductor.

Remark. The random motion of the z-component acts as a perturbation to the electric current in the (x, y)-plane, i.e., as an electric resistance. This implies that if the z-component of the motion drops to the ground state, it produces a minimal unavoidable perturbation of that current. Therefore, there exists no better state for the electric current. This is the so-called superconductivity.

Consider the ceramic La_2CuO_4 studied by Müller and Bednorz. A mobile particle (electron or positive hole of an electron) is confined to a yellow layer $\mathbf{R}^2 \times [0, \ell]$, and does not move across white layers (here ℓ may be regarded as the thickness of a yellow layer, which is determined by the crystal structure of the specific high-temperature superconductor; this is a task for physicists.)

Because of the crystal structure, while the motion of each mobile particle along the z-axis is confined, the motion in the (x, y)-plane has no such restriction. We will show that, as indicated above, the z-component of the motion drops to the ground state and acts as a minimal electric resistance, while the (x, y)-component of the motion is a (permanent) electric current. Thus, in our superconducting crystal the z-axis and (x, y)-plane play completely different roles.

In other words, our mathematical model of superconductivity assumes that

"the set of z-components of the mobile particles represents a one-dimensional population of Bose particles",

or, equivalently,

"The Schrödinger function $\psi_t(z_1, z_2, z_3, \ldots)$ is symmetric".

This is the key point in our discussion here.

Then, to reiterate, in our model $\mathbf{Z} = \{z$-components of mobile particles in the yellow layer $\mathbf{R}^2 \times [0, \ell]$ } represents a population of one-dimensional Bose particles. Therefore, applying the Bose-Einstein statistics to the population $\mathbf{Z}$, we conclude that Bose-Einstein condensation occurs.

Once the random vertical motion in the direction orthogonal to the yellow layer (copper oxide layer) drops to the ground state at the time of condensation (that is, the needles are crushed and the length is minimized), only the flow on the (x, y)-plane remains. This flow is actually a superconducting current.

When the random motion on the z-axis is in the ground state, it no longer perturbs the electric current on the (x, y)-plane. This means that, in general the random motion on the z-axis is a perturbation for the motion in the (x, y)-plane (that is, current) and acts as an electrical resistance for that current. In the ground state, this random perturbation motion is minimized (and this minimal perturbation cannot be avoided). For the current there is no better choice, that is, no way of easy flow beyond this. This is superconductivity.

Back to the high-temperature superconductor La_2CuO_4 of Müller and Bednorz, recall that mobile particles (electrons or positive holes) stay in the yellow layer $\mathbf{R}^2 \times [0, \ell]$ and will not penetrate the white layer. The length ℓ is a parameter determined by the crystal structure. (The length ℓ is considered to be the effective thickness of the yellow layer, but the relationship between the actual crystal structure and ℓ is not discussed here.)

Applying our the mathematical model of superconductivity, let us first investigate the one-dimensional movement in z-direction. The mobile particles (positive holes or free electrons) move in the yellow layer and do not penetrate through the white layer. So, the one-dimensional random motion in the z-axial direction must be confined in a segment of length ℓ, or the white layers should be reflecting walls.

Since confinement is equivalent to reflection, we consider confinement. Let the center of the yellow layer $\mathbf{R}^2 \times [0, \ell]$ be at $z = \frac{\ell}{2}$, and consider the segment $[0, \ell]$ of the length ℓ. If the motion of a particle on the z-axis is confined to the interval $[0, \ell]$, then the eigenvalue problem governing the motion is

$$\frac{d^2\Phi}{dz^2} + \frac{2m}{\hbar^2}(\lambda - V(z))\Phi = 0,$$

where the potential function $V(z)$ is given by

$$V(z) = \begin{cases} 0 & \text{for } z \in [0, \ell], \\ \infty & \text{for } z \notin [0, \ell]. \end{cases}$$

In this case, the eigenfunctions are

$$\Phi = \sqrt{\frac{2}{\ell}} \sin\left(\frac{n\pi z}{\ell}\right), \quad n = 1, 2, 3, \dots,$$

with corresponding eigenvalues

$$\lambda = \frac{\hbar^2}{2m}\pi^2\frac{n^2}{\ell^2}, \quad n = 1, 2, 3, \dots, \tag{2.11.1}$$

which represent the energy values that the z-component of the motion can take.

Now consider the set $\mathbf{Z}$ introduced above. As explained earlier, in our mathematical model of superconductivity we can assume that $\mathbf{Z}$ represents a population of independent Bose particles. Therefore, the mean number $\langle n_\epsilon \rangle$ of particles with energy ϵ is given by the Bose-Einstein distribution (see Section 2.5.)

$$\langle n_\epsilon \rangle = \frac{1}{\exp((\epsilon - \mu)/k_\mathrm{B}T) - 1}, \tag{2.11.2}$$

where μ is the chemical potential, k_B is the Boltzmann constant and T is the temperature (cf. Einstein (1924, 1925)). (Since we use thermodynamics to obtain the distribution $\langle n_\epsilon \rangle$, this is not quantum theory, but a result of the classical theory.)

Therefore, the mean number of particles in excited states is

$$N_\mathrm{ex} = \sum_\epsilon \langle n_\epsilon \rangle.$$

Now let us substitute the value $\epsilon = \frac{\hbar^2}{2m}\pi^2\frac{n^2}{\ell^2}$ given in (2.11.1) in formula (2.11.2), and assume that $\mu = 0$ (although there is no positive justification). Then the mean number N_{ex} of particles in excited states is given by

$$N_{\text{ex}} = \sum_{n=1}^{\infty} \frac{1}{\exp\left(\frac{\hbar^2}{2m}\pi^2\frac{n^2}{\ell^2}/k_B T\right) - 1}.$$

We denote $a = \frac{\hbar^2}{2m}\pi^2\frac{1}{\ell^2}/k_B T \ll 1$ and replace summation by integration:

$$N_{\text{ex}} = \sum_{n=1}^{\infty} \frac{1}{\exp(an^2) - 1} = \int_1^{\infty} \frac{1}{\exp(ax^2) - 1}\, dx,$$

and then set $y = \exp(ax^2)$ to obtain

$$N_{\text{ex}} = \frac{1}{2\sqrt{a}}\int_{e^a}^{\infty} \frac{1}{y(y-1)}\frac{1}{\sqrt{\log y}}\, dy.$$

Then, since only values of y close to e^a contribute essentially to the integral, we have

$$N_{\text{ex}} \simeq \frac{1}{2a}\int_{e^a}^{\infty} \frac{1}{y(y-1)}\, dy.$$

As the primitive of $\frac{1}{y(y-1)} = -\frac{1}{y} + \frac{1}{y-1}$ is $\log\left|\frac{y-1}{y}\right|$, we further have

$$N_{\text{ex}} = \frac{1}{2a}\left[\log\left|\frac{y-1}{y}\right|\right]_{e^a}^{\infty} = \frac{1}{2a}\log\left|\frac{e^a}{e^a - 1}\right|.$$

Since $a \ll 1$, we conclude that

$$N_{\text{ex}} \simeq \frac{1}{2a}\log\frac{1}{a},$$

i.e.,

$$N_{\text{ex}} \simeq \frac{1}{2}\frac{k_B T}{\frac{\hbar^2}{2m}\pi^2\frac{1}{\ell^2}}\log\left(\frac{k_B T}{\frac{\hbar^2}{2m}\pi^2\frac{1}{\ell^2}}\right). \tag{2.11.3}$$

Now let N denote the number of particles per unit volume. If the temperature T is high enough and $N_{\text{ex}} > N$, then all particles are in excited states. If we now lower the temperature T, then N_{ex} decreases and N_{ex} becomes smaller than N.

Let us denote the critical temperature, at which $N_{\text{ex}} = N$, by T_c, Then

$$k_B T_c \simeq \frac{\hbar^2}{2m}\pi^2\frac{1}{\ell^2}\frac{2N}{\log 2N}, \tag{2.11.4}$$

that is, the critical temperature T_c is determined by a structure constant ℓ^2 of the crystal and the number N of particles in the electron gas.

According to Einstein (1924, 1925), the gas of particles is saturated at the temperature $T = T_c$. Since the number N_{ex} is less than N at $T < T_c$, $N_0 = N - N_{ex}$ particles must be condensed into the state of the lowest energy. This is the so-called Bose-Einstein condensation. In other words, the motion along the z-axis drops to the lowest energy state.

Definition 2.11.1. We will call the critical temperature T_c given by formula (2.11.4) *"to transition temperature to high-temperature superconductivity"*.

The transition temperature T_c of a high-temperature superconductor depends on $1/\ell^2$ and the number N, that is, on its crystal structure and the number of mobile particles per unit volume. Accordingly, the transition temperature T_c varies with the high-temperature superconductor.

Remark. Concerning the Bose-Einstein condensation in lower dimensions we refer to, e.g., Görlitz et al. (2001) and Section 2.6 of Pethick and Smith (2002).

When the temperature is lowered, the kinetic energy of particles decreases, but in our model the kinetic energy of the motion along the z-axis drops to the lowest energy at the transition temperature T_c, that is, the spatial symmetry of the motion of particles remarkably breaks down. In copper oxide La_2CuO_4, we believe that this is caused by the crystal structure of copper oxide.

Now assuming that the temperature T is below the transition temperature T_c, we investigate the motion of mobile particles in a yellow layer $\mathbf{R}^2 \times [0, \ell]$. According to Proposition 5.6.1, it is determined by a Schrödinger function (complex evolution function)

$$\psi_t(x, y, z) = e^{R(t,x,y,z)+iS(t,x,y,z)},$$

or, equivalently, by a pair of evolution functions

$$\phi_t(x, y, z) = e^{R(t,x,y,z)+S(t,x,y,z)}, \quad \widehat{\phi}_t(x, y, z) = e^{R(t,x,y,z)-S(t,x,y,z)}.$$

Specifically, when $\psi_t(x, y, z) = e^{R(t,x,y,z)+iS(t,x,y,z)}$ is known, the evolution drift $\mathbf{a}(t, x, y, z)$ is given by

$$\mathbf{a}(t, x, y, z) = \sigma^2 \nabla \left(R(t, x, y, z) + S(t, x, y, z) \right),$$

where $\sigma^2 = \hbar/m$. The time-reversed evolution drift $\widehat{\mathbf{a}}(t, x, y, z)$ is given by

$$\widehat{\mathbf{a}}(t, x, y, z) = \sigma^2 \nabla (R(t, x, y, z) - S(t, x, y, z)).$$

The paths $X_t = (x_t, y_t, z_t)$ of the motion of an electron are given by the stochastic differential equation

$$X_t = X_a + \sigma B_{t-a} + \int_a^t \mathbf{a}(s, X_s) ds,$$

where $B_t = (B_t^1, B_t^2, B_t^3)$ is the three-dimensional Brownian motion.

The evolution drift was introduced in Definition 1.4.3 for a single particle, but according to Proposition 5.6.1, since the Schrödinger function describes the motion of a population of electrons, we can considered that $\mathbf{a}(t, x, y, z)$ *is the drift vector for a population of particles.*

Due to Bose-Einstein condensation, the motion in the direction of the z-axis represents the lowest energy state

$$\Phi_0(z) = \sqrt{\frac{2}{\ell}} \, \sin\left(\frac{\pi z}{\ell}\right).$$

In this case the z-component of the drift is

$$\sigma^2 \frac{\partial \log \Phi_0(z)}{\partial z} = \sigma^2 \frac{\pi}{\ell} \cot\left(\frac{\pi z}{\ell}\right).$$

Therefore the z-component of the motion the z-axis is strongly repelled by the drift $\sigma^2 \frac{\pi}{\ell} \cot \frac{\pi z}{\ell}$ at the boundary points $z = 0$ and ℓ, and is attracted to $z = \frac{\ell}{2}$. Therefore, this motion is an oscillation about the point $z = \frac{\ell}{2}$, but stays mostly close to this point.

Moreover

$$\frac{\partial R(t, x, y, z)}{\partial z} + \frac{\partial S(t, x, y, z)}{\partial z} = \frac{\pi}{\ell} \cot\left(\frac{\pi z}{\ell}\right), \tag{2.11.5}$$

$$\frac{\partial R(t, x, y, z)}{\partial z} - \frac{\partial S(t, x, y, z)}{\partial z} = -\frac{\pi}{\ell} \cot\left(\frac{\pi z}{\ell}\right), \tag{2.11.6}$$

must hold, and consequently

$$\frac{\partial R(t, x, y, z)}{\partial z} = 0, \quad \frac{\partial S(t, x, y, z)}{\partial z} = \frac{\pi}{\ell} \cot\left(\frac{\pi z}{\ell}\right). \tag{2.11.7}$$

Thus, we can state

Lemma 2.11.1. *The components of the drift vector* $\mathbf{a}(t, x, y, z)$ *of a population of mobile particles (electrons and positive holes) are*

$$\left(\sigma^2 \frac{\partial (R(t, x, y) + S(t, x, y))}{\partial x}, \sigma^2 \frac{\partial (R(t, x, y) + S(t, x, y))}{\partial y}, \sigma^2 \frac{\pi}{\ell} \cot \frac{\pi z}{\ell}\right), \tag{2.11.8}$$

where $\sigma^2 = \hbar/m$.

For example, if we choose

$$\psi(t, x, y, z) = e^{i\frac{E}{\hbar} t} e^{R(x,y) + i(S(x,y) + \log(\sqrt{\frac{2}{\ell}} \sin \frac{\pi z}{\ell}))},$$

then on the (x, y)-plane one has the two-dimensional drift vector of mobile particles (electrons) with the components

$$\mathbf{a}(x, y) = \left(\sigma^2 \frac{\partial (R(x, y) + S(x, y))}{\partial x}, \sigma^2 \frac{\partial (R(x, y) + S(x, y))}{\partial y}\right). \tag{2.11.9}$$

To this drift is added the z-component $\sigma^2 \frac{\pi}{\ell} \cot \frac{\pi z}{\ell}$. But since this component is already in the lowest energy state, it does not perturb the drift $\mathbf{a}(x, y)$ on the (x, y)-plane. Therefore, $\mathbf{a}(x, y)$ represents the persistent current.

Remark. According to Theorem 3.1.2 stated later in Section 3.1, the kinetic energy is given by

$$K(t, x, y, z) = \frac{1}{4m} \left((m \mathbf{a}(t, x, y, z))^2 + (m \widehat{\mathbf{a}}(t, x, y, z))^2 \right)$$

$$= \frac{1}{2m} \hbar^2 \left((\nabla R(t, x, y, z))^2 + (\nabla S(t, x, y, z))^2 \right).$$

Therefore, if the kinetic energy is constant and the z-component of the drift vector increases, then the xy-component of the drift vector will naturally decrease. That is, "the z-component of the drift vector acts as an electrical resistance to the current $\mathbf{a}(x, y)$ in the (x, y)-plane given in equation (2.11.9)".

This justifies our choice to examine first the z-component of movement and discuss Bose-Einstein condensation.

Lemma 2.11.1 admits the following immediate consequence:

Proposition 2.11.2. *At the transition temperature T_c given by formula (2.11.4), the motion on the z-axis drops to the lowest energy state. Therefore, below the temperature T_c, the influence of the perturbation produced by the z-component on the population-drift $\mathbf{a}(x, y)$ of mobile particles (electrons or positive holes) is minimized. Since electrons (or positive holes) have electric charge, the drift $\mathbf{a}(x, y)$ is the persistent current.*

What happens when a magnetic field is applied to a superconductor (for example, copper oxide La_2CuO_4) in the superconductive state?

When the magnetic field is weak, the superconductor expulses the magnetic field. In other words, the magnetic field cannot enter the superconductor. This fact has been confirmed experimentally as the Meissner-Ochsenfeld effect. This is the most typical phenomenon characterizing the superconducting state. Why does this happen?

Our model provides an explanation. Suppose the magnetic field penetrates the superconductor. Then, a momentum component arises in the z-axis direction (see Section 2.2). As a result, a new momentum will be added to the z-axial motion, that is, the z-component of the motion leaves the ground state. In our model, "superconducting state" means that "the z-component of the motion is in the ground state". Therefore, this means that if the magnetic field penetrates the superconductor, the superconductive state will be broken. Hence, to maintain superconductivity, expulsion of the magnetic field is necessary.

This was confirmed experimentally. As the external magnetic field becomes stronger, it gradually penetrates the surface of the superconductor and the superconductive state eventually breaks down.

Low-temperature superconductivity

Our mathematical model can be applied to low-temperature superconductors. Let us consider the case when there are no white layers (i.e., low- temperature super-conductors). In this case we can take the x-, y- and z-axes arbitrarily. We assume that the motion in the direction of the z-axis n is attracted by a Hooke force and performs a random harmonic oscillation.

We regard the strength of Hooke's force as a characteristic physical quantity of each sample (for example, mercury Hg). (See Section 1.7 about Hooke's force.)

Set $\mu = \frac{1}{2}h\nu$ in equation (2.11.2), Then, since the energy of one-dimensional harmonic oscillations is $\epsilon = (n + \frac{1}{2})h\nu$, the average number of particles in the excited states is

$$N_{ex} = \sum_{n=1}^{\infty} \frac{1}{\exp(nh\nu/k_B T) - 1}. \tag{2.11.10}$$

(Hooke's force was discussed in Section 1.7, where the eigenvalue was $\lambda = (n + \frac{1}{2})\sigma k$. If we set $\sigma^2 = \hbar^2/m$, $k^2 = 4\pi^2 m\nu^2$ with the used notation in Section 1.7, then $\sigma k = h\nu$.)

Let us set $a = h\nu/k_B T$ in equation (2.11.10) and replace the sum by an integral to get

$$N_{ex} = \sum_{n=1}^{\infty} \frac{1}{\exp(an) - 1} = \int_1^{\infty} \frac{1}{\exp(ax) - 1}\, dx.$$

Changing the variables to $\exp(ax) = y$, we have

$$N_{ex} = \frac{1}{a} \int_{e^a}^{\infty} \frac{1}{y(y-1)}\, dy.$$

Since the indefinite integral of $\frac{1}{y(y-1)}$ is $\log\left|\frac{y-1}{y}\right|$, we further have

$$N_{ex} = \frac{1}{a}\left[\log\left|\frac{y-1}{y}\right| \right]_{e^a}^{\infty} = \frac{1}{a} \log\left|\frac{e^a}{e^a - 1}\right|$$

where $a = h\nu/k_B T \ll 1$, and so

$$N_{ex} \simeq \frac{1}{a} \log\left(\frac{1}{a}\right).$$

We conclude that

$$N_{ex} \simeq \frac{k_B T}{h\nu} \log\left(\frac{k_B T}{h\nu}\right) \tag{2.11.11}$$

We denote the critical temperature by T_c, at which $N_{ex} = N$. Then

$$k_B T_c \simeq h\nu \frac{N}{\log N}. \tag{2.11.12}$$

According to Einstein (1924, 1925), the gas of particles is saturated at the temperature $T = T_c$. Since the mean number N_{ex} is smaller than N at $T < T_c$, the remaining $N_0 = N - N_{\text{ex}}$ particles must be condensed into the state of the lowest energy, that is, Bose-Einstein condensation occurs. In other words, the motion along the z-axis drops to the lowest energy state.

Definition 2.11.2. The critical temperature T_c given at Equation (2.11.12) is called **transition temperature** of low-temperature superconductors.

Now, since the motion along the z-axis direction is at the lowest energy state because of the Bose-Einstein condensation, the z-component of flow is $-\sigma^2 h\nu z$. Then equations (2.11.5), (2.11.6) and (2.11.7), in which $\frac{\pi}{\ell}\cot\left(\frac{\pi z}{\ell}\right)$ is replaced by $-h\nu z$, hold. Therefore we have the following lemma:

Lemma 2.11.2. *The flow* $\mathbf{a}(t, x, y, z)$ *of a group of electrons in a low-temperature superconductor has the form*

$$\left(\sigma^2\frac{\partial(R(t,x,y) + S(t,x,y))}{\partial x}, \quad \sigma^2\frac{\partial(R(t,x,y) + S(t,x,y))}{\partial y}, \quad -\sigma^2 h\nu z\right), \quad (2.11.13)$$

where $\sigma^2 = \frac{\hbar}{m}$.

For example, if we choose

$$\psi(t, x, y, z) = e^{i\frac{E}{\hbar}t}e^{R(x,y)+i(S(x,y)-\frac{1}{2}h\nu z^2)},$$

then on the (x, y)-plane there is a two-dimensional flow of a group of electrons

$$\mathbf{a}(x, y) = \left(\sigma^2\frac{\partial(R(x,y) + S(x,y))}{\partial x}, \quad \sigma^2\frac{\partial(R(x,y) + S(x,y))}{\partial y}\right), \quad (2.11.14)$$

where (X, Y) denotes a two-dimensional vector, to which the component $-\sigma^2 h\nu z$ of z-axis direction is added. However, since this component is already in the lowest energy state, it does not disturb the flow $\mathbf{a}(x, y)$ on the xy-plane. Hence the flow $\mathbf{a}(x, y)$ represents the persistent current.

Lemma 2.11.2 immediately yields

Proposition 2.11.3. *For low-temperature superconductors, at the transition temperature T_c given by the formula (2.11.12), the z-component of the motion drops to the lowest energy state. Therefore, for temperatures $\leq T_c$, the perturbation produced by the z- component is minimal. Since electrons carry an electric charge, $\mathbf{a}(x, y)$ is the persistent current. That is, a superconductive state is realized.*

The transition temperature T_c to superconductivity given by formula (2.11.12) was obtained under the assumption that there are no white layers and the z-component of the motion is a one-dimensional random harmonic oscillation. Therefore, the factor ν contained in T_c is the main physical characteristic of a low-temperature superconductor.

We conclude the chapter with two remarks concerning our model.

Remark 1. In our high-temperature superconductivity model we considered that the motion of a one-dimensional **"Bose particle"**, which represents the component on the z-axis orthogonal to the (x, y)-plane of the yellow layer, produces a Bose-Einstein condensation. (In low-temperature superconductivity, the z-component of the motion has no influence, but we treated it in the same way.) This is a hypothesis. Note that we do not consider pairs of electrons (Fermi particles)in our model.

Remark 2. For a theory of low-temperature superconductivity, see, e.g., Feynman (2012), Chapter 10.

Chapter 3

Momentum, Kinetic Energy, Locality

In Chapter 1 we discussed the foundations of a new quantum theory and the paths of random motion under the action of external forces, and in Chapter 2 we gave some applications of our theory. In this chapter, we first define in Section 3.1 the momentum and kinetic energy, which are important physical quantities of random motion. Section 3.2 discusses momentum and kinetic energy in quantum mechanics.

Heisenberg claimed that "there is no path for the movement of an electron", but we will argue that his claim is wrong. In fact, Heisenberg's claim that

"the product of the standard deviation of position and standard deviation of momentum has a positive minimum $\hbar/2$"

will be proved to be erroneous.

Subsequently, we will explain de Broglie's concept of hidden variables and discuss Einstein's concept of locality. Bell argued that "local hidden variables cannot reproduce the correlation in quantum mechanics", and this was used as a basis for the claim that "quantum mechanics is the only theory". Bell's argument is based on a misunderstanding, which will be proved to be a simple mistake.

3.1 Momentum and Kinetic Energy

Let us define the momentum and kinetic energy of random motion of a particles of mass m.

Recall that in terms of the generating function $R(t, x)$ and the drift potential $S(t, x)$, the evolution function $\phi(t, x)$ and backward evolution function $\widehat{\phi}(t, x)$ are

M. Nagasawa, *Markov Processes and Quantum Theory*,
Monographs in Mathematics 109, https://doi.org/10.1007/978-3-030-62688-4_3

represented in exponential form as

$$\phi\left(t, x\right) = e^{R(t,x)+S(t,x)} \quad \text{and} \quad \widehat{\phi}\left(t, x\right) = e^{R(t,x)-S(t,x)}.$$

Then the evolution drift is given by

$$\mathbf{a}\left(t, x\right) = \sigma^2 \nabla \left(R\left(t, x\right) + S\left(t, x\right)\right),$$

where $\sigma^2 = \frac{\hbar}{m}$, and the backward evolution drift is given by

$$\widehat{\mathbf{a}}\left(t, x\right) = \sigma^2 \nabla \left(R\left(t, x\right) - S\left(t, x\right)\right).$$

Since the evolution drift $\mathbf{a}(t, x)$ is a generalization of the classical velocity to the case of random motion, it is natural to define the **momentum of random motion** of a particle of mass m as

$$m\mathbf{a}\left(t, x\right) = \hbar \nabla \left(R\left(t, x\right) + S\left(t, x\right)\right). \tag{3.1.1}$$

Moreover, using the time-reversed evolution drift $\widehat{\mathbf{a}}(t, x)$, we define the **time-reversed momentum** as

$$m\widehat{\mathbf{a}}\left(t, x\right) = \hbar \nabla \left(R\left(t, x\right) - S\left(t, x\right)\right), \tag{3.1.2}$$

The expectation $\overline{m\mathbf{a}}$ of the momentum $m\mathbf{a}$ of random motion is given by the formula

$$\begin{aligned}
\overline{m\mathbf{a}} &= \int m\mathbf{a}\left(t, x\right) \mu\left(t, x\right) dx \\
&= \int \hbar \nabla \left(R\left(t, x\right) + S\left(t, x\right)\right) e^{2R(t,x)} dx,
\end{aligned} \tag{3.1.3}$$

where the distribution density is given by

$$\mu\left(t, x\right) = \phi\left(t, x\right) \widehat{\phi}\left(t, x\right) = e^{2R(t,x)}. \tag{3.1.4}$$

Next, let us consider the variance of the momentum $m\mathbf{a}(t, x)$ of random motion, which will play an important role in subsequent sections.

According to the definition in statistics, the **variance** $\mathrm{Var}(m\mathbf{a})$ of the **momentum** $m\mathbf{a}(t, x)$ of random motion is

$$\mathrm{Var}(m\mathbf{a}) = \int \left(m\mathbf{a}\left(t, x\right) - \overline{m\mathbf{a}}\right)^2 \mu\left(t, x\right) dx. \tag{3.1.5}$$

Theorem 3.1.1. *The variance of the distribution of momentum is expressed in terms of the generating function $R(t, x)$ of distribution and the drift potential $S(t, x)$ as*

$$\mathrm{Var}(m\mathbf{a}) = \hbar^2 \int \left(\nabla(R + S)\right)^2 e^{2R} dx - \hbar^2 \left(\int (\nabla S) e^{2R} dx\right)^2. \tag{3.1.6}$$

Proof. Formula (3.1.6) is obtained by substituting the expression (3.1.1) in the definition (3.1.5) of the variance and then using (3.1.3) and the fact that

$$2\int (\nabla R)e^{2R}dx = \int \nabla(e^{2R})dx = 0.$$

The proof is complete. □

To define the kinetic energy of random motion, recall that the kinetic energy K of smooth classical motion is given in terms of the classical momentum $\mathbf{p}$ as

$$K = \frac{1}{2m}\mathbf{p}^2.$$

Since the time-reversed momentum of classical motion is $-\mathbf{p}$, we can write

$$K = \frac{1}{4m}\left(\mathbf{p}^2 + (-\mathbf{p})^2\right), \tag{3.1.7}$$

where the right-hand side is invariant under time reversal. Therefore, it is natural to define the **kinetic energy of random motion** $K(t,x)$ by

$$K(t,x) = \frac{1}{4m}\left((m\mathbf{a}(t,x))^2 + (m\widehat{\mathbf{a}}(t,x))^2\right). \tag{3.1.8}$$

This definition ensures that the kinetic energy of random motion does not depend on the direction of time.

Theorem 3.1.2. *The kinetic energy of random motion $K(t,x)$ is expressed in terms of the generating function $R(t,x)$ of distribution and the drift potential $S(t,x)$ as*

$$K(t,x) = \frac{1}{2m}\hbar^2\left((\nabla R(t,x))^2 + (\nabla S(t,x))^2\right). \tag{3.1.9}$$

Its expectation is given by

$$\int K(t,x)\mu(t,x)\,dx = \int \frac{1}{2m}\hbar^2\left((\nabla R)^2 + (\nabla S)^2\right)(t,x)\,e^{2R(t,x)}dx,$$

where the distribution density is $\mu(t,x) = \phi(t,x)\widehat{\phi}(t,x) = e^{2R(t,x)}$.

Proof. By (3.1.1) and (3.1.2),

$$\begin{aligned}
&\left(m\mathbf{a}(t,x)\right)^2 + (m\widehat{\mathbf{a}}(t,x))^2 \\
&= \left(\hbar\nabla\left(R(t,x) + S(t,x)\right)\right)^2 + \left(\hbar\nabla\left(R(t,x) - S(t,x)\right)\right)^2 \\
&= 2\hbar^2\left((\nabla R(t,x))^2 + (\nabla S(t,x))^2\right),
\end{aligned}$$

i.e., (3.1.9) holds. □

In Section 2.1, we discussed the random motion of particles for a given complex evolution equation (Schrödinger equation)

$$ i\hbar \frac{\partial \psi}{\partial t} + \frac{1}{2m}\hbar^2 \triangle \psi - V(x)\psi = 0 \tag{3.1.10} $$

Let us consider the case of stationary motion, and look for a solution in the form $\psi = e^{-i\frac{1}{\hbar}\lambda t}\varphi$. This yields the eigenvalue problem

$$ -\frac{1}{2m}\hbar^2 \triangle\varphi + V(x)\varphi = \lambda\varphi, \tag{3.1.11} $$

where the **eigenvalue λ represents energy.** We asserted this without proof in Section 2.1, where we used spherical coordinate (r, θ, η).

Let us prove that the eigenvalue λ represents the total energy of random motion of a particle.

Let $K(x)$ be the kinetic energy of random motion defined by equation (3.1.9). Since the motion is stationary, $K(x)$ does not depend on time. We will show that

$$ \lambda = \text{the expectation of the sum } K(X_t) + V(X_t), $$

where X_t is the random motion determined by the above eigenvalue problem. Therefore, λ represents the expectation of the total energy.

We write the eigenfunction as $\varphi(x) = e^{R(x)+iS(x)}$. Multiplying both sides of equation (3.1.11) by the complex conjugate $\overline{\varphi}$ of the function φ and integrating the result, we obtain

$$ -\frac{1}{2m}\hbar^2 \int \overline{\varphi} \triangle\varphi\, dx + \int \overline{\varphi} V(x)\varphi\, dx = \lambda \int \overline{\varphi}\varphi\, dx, \tag{3.1.12} $$

where φ is normalized: $\int \overline{\varphi}\varphi dx = 1$. The first integral on the left-hand side is

$$ -\int \overline{\varphi}(\triangle\varphi)\, dx = \int \nabla\overline{\varphi}\,\nabla\varphi\, dx $$

$$ = \int \overline{\varphi}(x)\big((\nabla R(x))^2 + (\nabla S(x))^2\big)\varphi(x)dx $$

$$ = \int \big((\nabla R(x))^2 + (\nabla S(x))^2\big)\mu(x)dx, $$

where $\mu(x) = \overline{\varphi}(x)\varphi(x) = e^{2R(x)}$. Therefore, equation (3.1.12) becomes

$$ \int \frac{1}{2m}\hbar^2\big((\nabla R(x))^2 + (\nabla S(x))^2\big)\mu(x)dx + \int V(x)\mu(x)dx = \lambda. \tag{3.1.13} $$

Since $\mu(x)$ is the distribution density of the random motion X_t, we have

$$ \lambda = \mathbf{P}\left[\frac{1}{2m}\hbar^2\big((\nabla R(X_t))^2 + (\nabla S(X_t))^2\big) + V(X_t)\right], \tag{3.1.14} $$

where, by formula (3.1.9), $K(x) = \dfrac{1}{2m}\hbar^2\big((\nabla R(x))^2 + (\nabla S(x))^2\big)$ is the kinetic energy of the random motion X_t. □

3.2 Matrix Mechanics

For the energy function, that is, the Hamiltonian

$$H(\mathbf{p}, \mathbf{q}) = \frac{1}{2m}\,\mathbf{p} \times \mathbf{p} + V(\mathbf{q})$$

Heisenberg (1925) discovered a new mathematical rule for computing the product $\mathbf{p} \times \mathbf{p}$.

Born analyzed Heisenberg's calculation rules and discovered that they represent a method of multiplying matrices, and co-authored a paper with Jordan (Born and Jordan (1925)). In that paper, they called the method for calculating the values of energy based on the matrix multiplication rules "matrix mechanics". In matrix mechanics, $\mathbf{p}$ and $\mathbf{q}$ are matrices, and the values of energy are obtained as the eigenvalues of the matrix $H(\mathbf{p}, \mathbf{q})$, that is, as the diagonal elements in the diagonalization of the matrix $H(\mathbf{p}, \mathbf{q})$.

Dirac (1926) generalized matrix mechanics as a non-commutative algebraic theory. He assumed that the commutation relations

$$p_j q_k - q_k p_j = \frac{\hbar}{i}\delta_{jk}, \quad \hbar = \frac{h}{2\pi},$$

hold for $\mathbf{q} = (q_1, q_2, q_3)$ and $\mathbf{p} = (p_1, p_2, p_3)$, where h is the Planck constant. Dirac developed his unique symbolic laws and formulated the general theory of quantum mechanics. See Dirac (1930, 58): *The Principles of Quantum Mechanics*, 1st, 4th ed., Oxford Univ. Press, New York.

When the Hamiltonian $H(\mathbf{p}, \mathbf{q})$ is given, we can calculate energy values by using matrix mechanics. However, there is no concept of "path of particle motion" in this theory, and there is no mathematical structure for analyzing it. So calling this theory "dynamics" means overestimating it a bit. Rather, it is easier to understand it as a "formalism of a method of calculating energy values by means of operators"

Einstein repeatedly asserted that "this theory is incomplete (semi-finished)".

Schrödinger developed wave mechanics based on considerations different from those matrix mechanics, but when one calculates energy values using Schrödinger's wave mechanics on one hand, and matrix mechanics on the other hand, one gets the same answer. Why?

Schrödinger published a paper (Schrödinger (1926)) on the relationship between the two theories and clarified the reason for this coincidence of results, which is stated next as Schrödinger's theorem.

Theorem 3.2.1. *Define the differential operator p_x by*

$$p_x = \frac{\hbar}{i} \frac{\partial}{\partial x}.$$

Then

$$\frac{\hbar}{i} \frac{\partial}{\partial x} x f - x \frac{\hbar}{i} \frac{\partial}{\partial x} f = \frac{\hbar}{i} f. \tag{3.2.1}$$

That is, if we set

$$\mathbf{q} = (x, y, z) \quad and \quad \mathbf{p} = (p_x, p_y, p_z) = \left(\frac{\hbar}{i} \frac{\partial}{\partial x}, \frac{\hbar}{i} \frac{\partial}{\partial y}, \frac{\hbar}{i} \frac{\partial}{\partial z} \right), \tag{3.2.2}$$

then $\mathbf{q}$ and $\mathbf{p}$ are operators defined on the function space $L^2\left(\mathbf{R}^3\right)$ and satisfy the commutation relation

$$p_x x - x p_x = \frac{\hbar}{i}.$$

(Analogous relations hold for y and p_y, and for z and p_z.)

Hence, if we substitute $\mathbf{q}$ and $\mathbf{p}$ given by (3.2.2) into the Hamiltonian $H(\mathbf{p}, \mathbf{q})$, we obtain the expression

$$H(\mathbf{p}, \mathbf{q}) = -\frac{\hbar^2}{2m} \left(\frac{\partial^2}{\partial x^2} + \frac{\partial^2}{\partial y^2} + \frac{\partial^2}{\partial z^2} \right) + V(x, y, z).$$

Therefore, the eigenvalue problem in wave mechanics can be recast as the eigenvalue problem for the operator $H(\mathbf{p}, \mathbf{q})$

$$H(\mathbf{p}, \mathbf{q})\psi - \lambda\psi = 0.$$

Born (1926) immediately acknowledged the importance of this fact. He analyzed a particle scattering problem and claimed that $|\psi_t(x, y, z)|^2$ is not the strength of a wave, as Schrödinger thought, but the distribution density of particles. After giving this "statistical interpretation" for $|\psi_t(x, y, z)|^2$, he brought Schrödinger's wave equation and wave function $\psi_t(x, y, z)$ into matrix mechanics.

The combination of matrix mechanics with the Schrödinger wave equation and wave function $\psi_t(x, y, z)$ as well as with the "statistical interpretation" (based on Schrödinger's theorem) will be called the "Born-Heisenberg theory (quantum mechanics)". For details on the "Born-Heisenberg theory (quantum mechanics)" see, e.g., Dirac (1994), Tomonaga (1952).

While Born interpreted $|\psi_t(x, y, z)|^2$ as the statistical distribution density of particles, he asserted nothing about what Schrödinger's function $\psi_t(x, y, z)$ itself represents. This became a decisive weak point of the Born-Heisenberg theory, causing confusion about its interpretation.

Comment. We repeat, to avoid misunderstandings. We know what $\psi_t(x, y, z)$ means: it is a complex evolution function. It leads to the path equation via a pair

$\{\phi(t, x, y, z), \widehat{\phi}(t, x, y, z)\}$ of evolution functions of random motion. "The decisive weak point of the Born-Heisenberg theory" is that in its framework the theoretical structure revealed by our mechanics of random motion is not understood.

Bohr claimed that the function $\psi_t(x, y, z)$ represents a quantum state, and called it state function. However, Bohr does not define what "state" exactly means.

Based on Born's idea, Dirac called $\psi_t(x, y, z)$ a probability amplitude. He just changed the name, but nothing new about $\psi_t(x, y, z)$ was added. Actually, $\psi_t(x, y, z)$ does not represent a probability or an amplitude.

Since Born and Heisenberg took the Hamiltonian as a starting point, when considering matrix mechanics, they must have considered that electrons are particles. However, since Schrödinger's wave mechanics (that is, wave equation and wave function) is incorporated into matrix mechanics, one does not know whether electrons are particles or waves.

Bohr argued that "the electron has a particle nature and a wave nature. It behaves like a particle in some cases, and in other cases behaves like a wave, that is, the particle nature and wave nature are complementary to each other." He called this "complementarity" and considered that the difficulty of the Born-Heisenberg's theory (quantum mechanics) is resolved by the complementarity principle.

In his later years, Schrödinger criticized Bohr's claim that the "wave and particle dilemma" had already been solved by the word "complementarity" (a word rather than a concept), and asserted that (p. 473 of Moore (1989))

"Complementarity is a 'thoughtless slogan'. If I were not thoroughly convinced that the man [Bohr] is honest and really believes in the relevance of his — I do not say theory but — sounding word, I should call it intellectually wicked."

A particle occupies a point in space-time and a wave spreads in space-time. Therefore, a physical object cannot be simultaneously a particle and a wave. Particles and waves belong to different categories. As mentioned in Chapter 1, in order to develop a mathematical theory for a physical phenomenon, we must first of all confirm a mathematical entity corresponding to the object to be handled. In Newtonian mechanics, the object to be handled is a particle, and the mathematical entity corresponding to "particle" is one point in time and space. Maxwell's electromagnetism deals with wave motion. The mathematical entity corresponding to "wave" is a function of the time and space variables (defined in space-time). However, in Born-Heisenberg's theory, the mathematical entity for handling quantum phenomena has not been determined.

In order to justify Bohr's concept of "having a particle nature and a wave nature", it is necessary to associate to it an appropriate mathematical entity. Although Bohr did not make a clear statement, it seems that he thought that such a mathematical entity is provided by a wave function $\psi_t(x, y, z)$. He affixed "particle" to the wave function. However, "a particle is a point in space-time" — this is the essence of the concept of particle. Bohr does not reveal the relationship between this fact and the wave function. Therefore, there is no basis to argue that "an object with both a particle nature and a wave nature is represented by a function $\psi_t(x, y, z)$". Bohr introduced the term "complementarity" instead of providing such a basis. But "complementary" is just a word, a dogma due to Bohr; this must be what Schrödinger wanted to assert.

In fact, quantum mechanics does not include mathematical entities corresponding to "object that a have both a particle nature and wave nature." Therefore the "particle–wave dilemma" cannot be solved by quantum mechanics.

As we have already demonstrated, what is important for quantum phenomena is not the "wave" concept, but the drift potential $S(t, x)$. The notions of drift potential and wave should not be confused. And there is a mathematical description for "particle motion with a drift potential": it is the random motion of a particle $\{X_t(\omega) ; \mathbf{Q}\}$ satisfying the equations of motion (1.4.1). "The random motion under external force $\{X_t(\omega) ; \mathbf{Q}\}$ evolves carrying the drift-potential $S(t, x)$". Moreover, the drift potential generates the evolution drift. The crucial point is that in the mechanics of random motion, the mathematical description is clearly separated into two concepts: "particle" and "(random) motion"; consequently, the "particle–wave dilemma" is eliminated. (The mathematical entities in Newtonian mechanics were "particle and smooth motion." Smooth motion is not associated with a drift potential.)

3.3   Function Representations of Operators

Before we start talking about "function representations of operators", which will be needed in the sequel, let us review basic notions and propositions on Hilbert spaces and linear operators on them.

Let H be a linear space endowed with an inner product $\langle f, g \rangle$, defined for elements $f, g \in H$. Define a norm $\|f\|$ on H by $\|f\| = \sqrt{\langle f, f \rangle}$, and define the distance of $f, g \in H$ by

$$\rho(f, g) = \|f - g\|.$$

If the linear space H is complete with respect to the distance $\rho(f, g)$, we call H a Hilbert space.

The space $L^2(\mathbf{R}^d)$ of square-integrable functions (i.e., functions f such that $\int_{\mathbf{R}^d} |f(x)|^2 dx < \infty$) is a typical Hilbert space. We equip the space $L^2(\mathbf{R}^d)$ with the inner product $\langle f, g \rangle$ defined by

$$\langle f, g \rangle = \int_{\mathbf{R}^d} \overline{f(x)} g(x) \, dx, \quad \text{for } f, g \in L^2(\mathbf{R}^d),$$

where $\overline{f(x)}$ is the complex conjugate of the function $f(x)$.

If A is a linear operator on the Hilbert space H, we define its norm by

$$\|A\| = \sup_{f \in H, f \neq 0} \frac{\|Af\|}{\|f\|}.$$

If $\|A\| < \infty$, the linear operator A is said to be bounded. When we treat only bounded linear operators, we call them simply "operators".

The operator A^* defined by the equality

$$\langle A^* f, g \rangle = \langle f, Ag \rangle \quad \text{for all } f, g \in H$$

is called the adjoint of the operator A, If $A = A^*$, then operator A is called self-adjoint.

If two operators A and B on a Hilbert space H satisfy

$$AB = BA,$$

then they are said to commute.

Let us first give the basic definition concerning function representations of operators (also known as "hidden variables").

Definition 3.3.1. Let $\mathbf{F}$ be a family of self-adjoint operators on a Hilbert space H. Take a vector $f \in H$ of norm $\|f\| = 1$. If there exists a probability space $\{\Omega, \mathcal{F}, \mathbf{P}\}$, and for each operator $A \in \mathbf{F}$ there exists a random variable $h_A(\omega)$ on $\{\Omega, \mathcal{F}, \mathbf{P}\}$ such that

$$\langle f, Af \rangle = \int h_A(\omega)\, d\mathbf{P}, \qquad (3.3.1)$$

then $h_A(\omega)$ is called a **function representation** of the self-adjoint operator $A \in \mathbf{F}$, and $\{(\Omega, \mathcal{F}, \mathbf{P}); h_A, A \in \mathbf{F}\}$ is called a **function representation** of the family $\{\mathbf{F}, f \in H\}$.

Remark. A function representation of $\{\mathbf{F}, f \in H\}$ is not uniquely determined. When handling a set of physical quantities, it is necessary to select an appropriate function representation. We will discuss Bell's inequality in Section 3.4, but in Bell (1964) and subsequent papers this point was neglected and the authors made a careless mistake.

Theorem 3.3.1. *Let $\mathbf{F}$ be a family of self-adjoint operators on a Hilbert space H such that any two operators $A, B \in \mathbf{F}$ commute. Let $f \in H$ be an arbitrary vector of norm $\|f\| = 1$. Then $\{\mathbf{F}, f \in H\}$ admits a function representation $\{(\Omega, \mathcal{F}, \mathbf{P}); h_A, A \in \mathbf{F}\}$.*

Proof. Let $\mathbf{B}$ be the smallest Banach algebra of bounded self-adjoint operators such that the family $\mathbf{F} \subset \mathbf{B}$.

Then there exist a compact space Ω and an isomorphism θ from the Banach algebra $\mathbf{B}$ to the function space $C(\Omega)$. Namely, for each $A \in \mathbf{B}$ there exists a function $h_A \in C(\Omega)$ such that $h_A = \theta A$ and

$$\|h_A\| = \|A\|,$$

where the norm of h_A is defined by $\|h_A\| = \sup_{\omega \in \Omega} |h_A(\omega)|$. (For a proof, see, e.g., Nagasawa (1959).) Define a linear functional on $C(\Omega)$ by

$$m(h_A) = \langle f, Af \rangle, \quad h_A \in C(\Omega). \tag{3.3.2}$$

Since $\|m\| = 1$, a theorem of Riesz, Markov and Kakutani (for a proof, Yosida (1965)) ensures that there exists a probability measure $\mathbf{P}(d\omega)$ on a measurable space $\{\Omega, \mathcal{F}\}$ such that

$$m(h_A) = \int h_A(\omega) \mathbf{P}(d\omega). \tag{3.3.3}$$

Combining (3.3.2) and (3.3.3), we have

$$\langle f, Af \rangle = \int h_A(\omega) \mathbf{P}(d\omega), \quad A \in \mathbf{B},$$

that is, (3.3.1) holds. Since $\mathbf{F}$ is a subset of the Banach algebra $\mathbf{B}$,

$$\{(\Omega, \mathcal{F}, \mathbf{P}); h_A, A \in \mathbf{F}\}$$

is a function representation of $\{\mathbf{F}, f \in H\}$. □

3.4   Expectation and Variance

(i) On expectation.

In probability theory, the expectation $\mathbf{P}[X]$ of a random variable $X(\omega)$ defined on a probability space $\{\Omega, \mathcal{F}, \mathbf{P}\}$ is defined as

$$\mathbf{P}[X] = \int X(\omega) \, d\mathbf{P},$$

i.e., it is the integral (mean value) of the random variable $X(\omega)$ with respect to the probability measure $\mathbf{P}$.

In quantum mechanics (Born-Heisenberg theory) the inner product

$$\langle \psi, A\psi \rangle$$

is interpreted as the expectation of the self-adjoint operator A, where $\psi \in H$ is a vector with $\|\psi\| = 1$.

If one deals with just one operator A, then, by Theorem 3.3.1, there exist a random variable $h_A(\omega)$ and a probability measure $\mu = \mathbf{P}(d\omega)$ such that

$$\langle \psi, A\psi \rangle = \int h_A(\omega)\, d\mu.$$

Since the right-hand side is the expectation of a function representation $h_A(\omega)$ of the operator A, the above interpretation in quantum mechanics is reasonable. However, when handling multiple operators simultaneously, things are not so simple.

Looking back at Definition 3.3.1, if we consider a Hilbert space H and a unit-norm vector $\psi \in H$ and if $\{\mu; h_A, A \in \mathbf{F}\}$ is a function representation of a set $\mathbf{F}$ of self-adjoint operators, then

$$\langle \psi, A\psi \rangle = \int h_A(\omega)d\mu, \quad A \in \mathbf{F}.$$

The right-hand side is the expectation of the function representation (random variable). So, if a function representation exists, the above interpretation is valid for a set of self-adjoint operators. The function representation is, however, not unique.

In the following we adopt a function representation $\{\mu; h_A, A \in \mathbf{F}\}$. This is physically meaningful in the framework of our mechanics of random motion.

(ii) On variance.

In probability theory, the variance of a random variable h_A is defined, following the definition in statistics, as

$$\mathrm{Var}(h_A) = \int \left(h_A - \int h_A\, d\mu \right)^2 d\mu.$$

In quantum mechanics, Robertson (1929) defined the variance of a self-adjoint operator A as

$$\langle \psi, (A - \langle \psi, A\psi \rangle)^2 \psi \rangle, \quad \psi \in H, \tag{3.4.1}$$

where $\|\psi\| = 1$; this definition was adopted in many textbooks.

However, as it is readily see, the equality

$$\langle \psi, (A - \langle \psi, A\psi \rangle)^2 \psi \rangle = \int \left(h_A - \int h_A\, d\mu \right)^2 d\mu$$

holds only in very special cases. In fact, if we think carefully, there is no reason for assuming that "the inner product (3.4.1) is the variance", because we do not know what kind of physical quantity the operator $(A - \langle \psi, A\psi \rangle)^2$ represents.

To clarify this point, we consider

(iii) The case of the Schrödinger operator.

To treat a concrete situation, consider the Schrödinger function (complex evolution function)

$$\psi_t(x) = e^{R(t,x)+iS(t,x)},$$

and the Schrödinger operator (momentum operator)

$$\mathbf{p} = \frac{\hbar}{i}\nabla,$$

and then consider the set of operators $\mathbf{F} = \{x, \mathbf{p}, \mathbf{p}^2\}$.

We first take the operator $\mathbf{p} = \frac{\hbar}{i}\nabla$. Then

$$
\begin{aligned}
\langle \psi_t, \mathbf{p}\psi_t \rangle &= \hbar \int \overline{\psi}_t(x) \frac{1}{i}\nabla\psi_t(x)\, dx \\
&= \hbar \int \overline{\psi}_t(x) \left(\frac{1}{i}\nabla R(t,x) + \nabla S(t,x) \right) \psi_t(x)\, dx \qquad (3.4.2) \\
&= \int \hbar \left(\frac{1}{i}\nabla R(t,x) + \nabla S(t,x) \right) \mu_t(x)\, dx,
\end{aligned}
$$

where the probability distribution density is $\mu_t(x) = \overline{\psi}_t(x)\psi_t(x) = e^{2R(t,x)}$.

Since the operator $\mathbf{p}$ is self-adjoint, $\langle \psi_t, \mathbf{p}\psi_t \rangle$ is a real number. Therefore, the imaginary part of the right-hand side vanishes:

$$\int \hbar\big(\nabla R(t,x)\big)\mu_t(x)dx = 0. \qquad (3.4.3)$$

Hence we can write the right-hand side of (3.4.2) without i, as

$$\langle \psi_t, \mathbf{p}\psi_t \rangle = \int \hbar\big(\nabla R(t,x) + \nabla S(t,x)\big)\mu_t(x)dx. \qquad (3.4.4)$$

This shows that

$$\hbar\big(\nabla R(t,x) + \nabla S(t,x)\big)$$

is a function representation of the momentum operator $\mathbf{p} = \frac{\hbar}{i}\nabla$.

As shown in Section 3.1 (see (3.1.1)), the momentum of the random motion $\{X_t(\omega); \mathbf{Q}\}$ is

$$m\mathbf{a}(t,x) = \hbar\nabla\big(R(t,x) + S(t,x)\big),$$

where $R(t,x)$ is the generating function of distribution and $S(t,x)$ is the drift potential.

Therefore, the above function representation of the momentum operator $\mathbf{p} = \frac{\hbar}{i}\nabla$ coincides with the momentum of the random motion. As already noted, the function representation is not unique, but we adopt

$$h_{\mathbf{p}}(t,x) = \hbar\nabla\big(R(t,x) + S(t,x)\big), \qquad (3.4.5)$$

as the function representation of the momentum operator $\mathbf{p}$, i.e.,

$$\langle \psi_t, \mathbf{p}\psi_t \rangle = \int h_{\mathbf{p}}(t, x)\mu_t(x)\, dx. \tag{3.4.6}$$

Incidentally, since the expectation of $\hbar \nabla R(t, x)$ with respect to the measure $\mu_t(x)\, dx$ vanishes by (3.4.3), $\hbar \nabla S(t, x)$ is also a function representation of the operator $\mathbf{p}$. However, we cannot consider (and adopt) this representation as the momentum, because $\hbar \nabla S(t, x)$ alone does not determine the paths of the random motion.

As already noted, function representations of operators are not uniquely determined. Therefore, as a function representation it is necessary to select something physically meaningful. (Fényes (1952) adopted $\hbar \nabla S(t, x)$, but he did not reach definitive conclusions about the uncertainty principle.)

Next, we take the operator $\mathbf{p}^2 = \left(\frac{\hbar}{i}\nabla\right)^2$. Then

$$\int \overline{\psi}_t(x)\left(\mathbf{p}^2\psi_t(x)\right) dx = \int \hbar^2\, \nabla\overline{\psi}_t(x)\, \nabla\psi_t(x)\, dx$$

$$= \int \hbar^2\, \overline{\psi}_t(x)\left((\nabla R(t, x))^2 + (\nabla S(t, x))^2\right)\psi_t(x)\, dx$$

$$= \int \hbar^2\left((\nabla R(t, x))^2 + (\nabla S(t, x))^2\right)\mu_t(x)\, dx,$$

where $\mu_t = \overline{\psi}_t\psi_t$.

Thus, the following proposition holds true:

Proposition 3.4.1. *Let* $\mathbf{p} = \frac{\hbar}{i}\nabla$. *Then*

$$h_{\mathbf{p}^2} = \hbar^2\left((\nabla R)^2 + (\nabla S)^2\right) \tag{3.4.7}$$

is a function representation of the operator $\mathbf{p}^2$. *That is,*

$$\langle \psi_t, \mathbf{p}^2\psi_t \rangle = \int h_{\mathbf{p}^2}(t, x)\mu_t(x)\, dx.$$

We carefully formulate the following theorem to avoid possible misunderstandings.

Theorem 3.4.1. *Let* h_p *be the function representation* h_p *of the momentum operator* $\mathbf{p}$ *given by formula* (3.4.5) *and* $h_{\mathbf{p}^2}$ *be the function representation of the operator* $\mathbf{p}^2$ *given by formula* (3.4.7). *Then:*

(i) *If* $\int \nabla R \nabla S \mu_t dx \neq 0$, *i.e., there is a correlation between* ∇R *and* ∇S, *then*

$$\int (h_{\mathbf{p}})^2(t, x)\mu_t(x)\, dx \neq \int h_{\mathbf{p}^2}(t, x)\mu_t(x)\, dx, \tag{3.4.8}$$

where the probability distribution density is $\mu_t(x) = e^{2R(t, x)}$.

(ii) *Moreover, if* $\nabla R \nabla S \neq 0$, *then*

$$(h_{\mathbf{p}})^2 \neq h_{\mathbf{p}^2}. \tag{3.4.9}$$

Proof. Since the function representation of the momentum operator $\mathbf{p}$ is $h_{\mathbf{p}}(t, x) = \hbar \nabla \big(R(t, x) + S(t, x) \big)$, we have

$$(h_{\mathbf{p}})^2 = \hbar^2 \left((\nabla R)^2 + (\nabla S)^2 \right) + 2\hbar^2 \nabla R \nabla S. \tag{3.4.10}$$

Integrating both sides of equality (3.4.10) with respect to $\mu_t(x)\,dx$ and looking at equation (3.4.7), we complete the proof. $\square$

To re-emphasize:

The function representation $h_{\mathbf{p}^2}$ of the square $\mathbf{p}^2$ of the momentum operator $\mathbf{p}$ is not equal to the square $(h_{\mathbf{p}})^2$ of the function representation h_p of the operator $\mathbf{p}$.

This will prove to be an important point when we will discuss Heisenberg's uncertainty principle.

Theorem 3.4.2 (Nagasawa (2012)). *Let $\psi(t, x) = e^{R(t,x)+iS(t,x)}$ satisfy the complex evolution equation*

$$i\frac{\partial \psi}{\partial t} + \frac{1}{2}\sigma^2 \triangle \psi - V(t, x)\,\psi = 0. \tag{3.4.11}$$

Then the equality

$$\int (h_{\mathbf{p}})^2(t, x)\mu_t(x)\,dx = \int h_{\mathbf{p}^2}(t, x)\,\mu_t(x)\,dx$$

holds only when

$$\int \left\{ \frac{\partial R}{\partial t} + \frac{1}{2}\sigma^2 \triangle S \right\} \mu_t\,dx = 0. \tag{3.4.12}$$

Proof. Substituting $\psi(t, x) = e^{R(t,x)+iS(t,x)}$ in equation (3.4.11), dividing by $\psi(t, x)$ and separating the imaginary part, we see that

$$\frac{\partial R}{\partial t} + \frac{1}{2}\sigma^2 \triangle S + \sigma^2 \nabla S \cdot \nabla R = 0. \tag{3.4.13}$$

By Theorem 3.4.1, this concludes the proof. $\square$

Remark. The equality $(h_{\mathbf{p}})^2 = h_{\mathbf{p}^2}$ holds only when:

$$\frac{\partial R}{\partial t} + \frac{1}{2}\sigma^2 \triangle S \equiv 0. \tag{3.4.14}$$

When the distribution generation function $R(t, x)$ depends on time, equation (3.4.12) or equation (3.4.14) holds. It may be well the case that both equations can hold only if $R(t, x)$ does not depend on time.

Now, let $h_A(t, x)$ be a function representation of a physical quantity (self-adjoint operator) A. It is natural, following Nagasawa (2009), to define the variance $\mathrm{Var}(A)$ of the physical quantity A as the variance $\mathrm{Var}(h_A)$, i.e., to set

$$\mathrm{Var}(A) = \mathrm{Var}(h_A) = \int (h_A)^2 (t, x)\, \mu_t(x)\, dx - \left(\int h_A(t, x)\, \mu_t(x)\, dx \right)^2.$$

Then we have

Proposition 3.4.2. *The variance* $\mathrm{Var}(\mathbf{p})$ *of the momentum operator* $\mathbf{p} = \frac{\hbar}{i} \nabla$ *is*

$$\mathrm{Var}(\mathbf{p}) = \int (h_{\mathbf{p}})^2 (t, x) \mu_t(x)\, dx - \left(\int h_{\mathbf{p}}(t, x)\, \mu_t(x)\, dx \right)^2, \qquad (3.4.15)$$

where $(h_{\mathbf{p}})^2 = \hbar^2 \left((\nabla R)^2 + (\nabla S)^2 \right) + 2\hbar^2 \nabla R\, \nabla S.$

When one deals with the variance of the momentum operator $\mathbf{p}$, the following result must be kept in mind.

Theorem 3.4.3. *Let* $\psi_t(x) = e^{R(t,x)+iS(t,x)}$. *The quantity*

$$\langle \psi_t, (\mathbf{p} - \langle \psi_t, \mathbf{p}\psi_t \rangle)^2 \psi_t \rangle,$$

which is used as the variance of the momentum operator $\mathbf{p}$ *in quantum mechanics,* **is not** *the variance of the operator* $\mathbf{p}$ *in general. It coincides with the variance only when there is no correlation between* ∇R *and* ∇S, *that is, only when* $\int \nabla R \nabla S \mu_t dx = 0$. *In particular, if* $\psi_t(x)$ *is a solution of equation (3.4.11), this happens only when (3.4.12) holds.*

Proof. We have

$$\langle \psi_t, (\mathbf{p} - \langle \psi_t, \mathbf{p}\psi_t \rangle)^2 \psi_t \rangle = \langle \psi_t, \mathbf{p}^2 \psi_t \rangle - (\langle \psi_t, \mathbf{p}\psi_t \rangle)^2$$

$$= \int h_{\mathbf{p}^2}(t, x)\, \mu_t(x)\, dx - \left(\int h_{\mathbf{p}}(t, x)\, \mu_t(x)\, dx \right)^2.$$

By Theorem 3.4.1, if $\int \nabla R \nabla S \mu_t dx \neq 0$, the right-hand side is not equal to

$$\int (h_{\mathbf{p}})^2 (t, x)\mu_t(x)\, dx - \left(\int h_{\mathbf{p}}(t, x)\, \mu_t(x)\, dx \right)^2.$$

Therefore, by formula (3.4.15) of Proposition 3.4.2,

$$\langle \psi_t, (\mathbf{p} - \langle \psi_t, \mathbf{p}\psi_t \rangle)^2 \psi_t \langle \neq \mathrm{Var}(\mathbf{p})$$

in general. That is, $\langle \psi_t, (\mathbf{p} - \langle \psi_t, \mathbf{p}\psi_t \rangle)^2 \psi_t \rangle$ does not coincide with the variance $\mathrm{Var}(\mathbf{p})$ of the momentum operator $\mathbf{p}$. $\qquad \square$

3.5   The Heisenberg Uncertainty Principle

The Heisenberg uncertainty principle (Heisenberg (1927)) has been said to touch the essence of the Born-Heisenberg theory (quantum mechanics). (See Kennard (1927), Chapter 3 of Jammer (1974).)

However, one can ask whether Heisenberg's uncertainty principle really holds true. Looking at history, Einstein expressed his doubt concerning the principle, and intense controversy surrounded it over the years. In this section we will give a counterexample in which the uncertainty principle fails. In other words, Heisenberg's uncertainty principle is not correct. We explain this next.

Robertson (1929) had the idea to substitute the vectors

$$X = (x - \langle \psi_t, x\psi_t \rangle)\psi_t, \quad Y = (p_x - \langle \psi_t, p_x\psi_t \rangle)\psi_t$$

in the Cauchy-Schwarz inequality

$$\sqrt{\langle X, X \rangle}\sqrt{\langle Y, Y \rangle} \geq \frac{1}{2}\left|\langle X, Y \rangle - \langle Y, X \rangle\right|, \tag{3.5.1}$$

and obtained the inequality

$$\sqrt{\langle \psi_t, (x - \langle \psi_t, x\psi_t \rangle)^2\psi_t \rangle}\ \sqrt{\langle \psi_t, (p_x - \langle \psi_t, p_x\psi_t \rangle)^2\psi_t \rangle} \geq \frac{1}{2}\hbar. \tag{3.5.2}$$

We call this the Robertson inequality.

Then Robertson claimed that his inequality implies the so-called Heisenberg-Kennard inequality, which asserts that

The product of the standard deviation of the position and the standard deviation of the momentum cannot be smaller than $\frac{1}{2}\hbar$.

This assertion is called the **Heisenberg uncertainty principle**. (Heisenberg (1927), Kennard (1927))

However, Robertson's claim is incorrect, because the term

$$\sqrt{\langle \psi_t, (p_x - \langle \psi_t, p_x\psi_t \rangle)^2\psi_t \rangle}$$

in the left-hand side of his inequality is not the standard deviation of the momentum, except in the very special case when ψ_t satisfies the condition shown in Theorem 3.4.3.

Therefore, there is no proof for the "Heisenberg-Kennard inequality", i.e., for the "Heisenberg uncertainty principle."

Exceptions are not allowed in "Heisenberg's uncertainty principle". In other words, if there is just one example that does not satisfy the "Heisenberg-Kennard inequality", it invalidates the principle.

Let us prove the following theorem.

Theorem 3.5.1 (Nagasawa (2009)). *The product*

$$\sqrt{\mathrm{Var}\,(x)}\sqrt{\mathrm{Var}\,(p_x)}$$

of the standard deviation $\sqrt{\mathrm{Var}\,(x)}$ of the x-component of position and the standard deviation $\sqrt{\mathrm{Var}\,(p_x)}$ of the x-component of momentum p_x has no positive lower bound. It can be arbitrarily small, that is, the Heisenberg-Kennard inequality does not hold.

Proof. We compute the standard deviation $\sqrt{\mathrm{Var}\,(p_x)}$ of the x-component p_x of the momentum. To this end we use the function representation of the x-component of the operator **p** given by

$$h_{p_x} = \hbar\frac{\partial R}{\partial x} + \hbar\frac{\partial S}{\partial x}.$$

The variance $\mathrm{Var}\,(p_x)$ of p_x is given by the variance $\mathrm{Var}\,(h_{p_x})$ of h_{p_x}. By the definition (3.4.15),

$$\mathrm{Var}\,(p_x) = \mathrm{Var}\,(h_{p_x})$$
$$= \int \hbar^2 \left(\frac{\partial R}{\partial x} + \frac{\partial S}{\partial x}\right)^2 \mu_t\,dx - \left(\int \hbar\frac{\partial S}{\partial x}\mu_t\,dx\right)^2, \qquad (3.5.3)$$

where $\mu_t = \overline{\psi}_t\psi_t = e^{2R}$ is the distribution density. We fix this (i.e., we fix the generating function of distribution R).

Since the drift potential S can be chosen arbitrarily, we take, for example, $S = -\gamma R$. The Schrödinger function $\psi_t(x) = e^{R(t,x)+iS(t,x)}$ is a solution of the Schrödinger equation with a suitable potential function. Then the second term on the right-hand side of formula (3.5.3) is

$$\left(\int \hbar\frac{\partial S}{\partial x}\mu_t\,dx\right)^2 = \left(\frac{\gamma\hbar}{2}\int \frac{\partial}{\partial x}e^{2R}\,dx\right)^2 = 0,$$

because the distribution density e^{2R} vanishes at infinity.

Hence, the right-side of equation (3.5.3) is equal to

$$(1-\gamma)^2 \int e^{2R}\,dx\hbar^2 \left(\frac{\partial R}{\partial x}\right)^2.$$

This can be made arbitrarily small by choosing γ.

Since $\mu_t = \overline{\psi}_t\psi_t = e^{2R}$ is fixed, the variance of the x-component of position does not change. Therefore, $\sqrt{\mathrm{Var}\,(x)}\sqrt{\mathrm{Var}\,(p_x)}$ can be made arbitrarily small by choosing γ suitably. This completes the proof. $\qquad\square$

Remark. For convenience, we give here a proof of the Cauchy-Schwarz inequality (3.5.1).

Let μ be a probability measure, and let $X, Y \in L^2(\mu)$ be complex-valued random variables. Normalizing, we set

$$\overline{X} = \frac{X}{\sqrt{\langle X, X \rangle}} \quad \text{and} \quad \overline{Y} = \frac{Y}{\sqrt{\langle Y, Y \rangle}}.$$

Then

$$\langle i\overline{X} + \overline{Y}, i\overline{X} + \overline{Y} \rangle = \|\overline{X}\|^2 + \|\overline{Y}\|^2 - i\left(\langle \overline{X}, \overline{Y} \rangle - \langle \overline{Y}, \overline{X} \rangle\right) \geq 0,$$

where $\|\overline{X}\|^2 = 1$ and $\|\overline{Y}\|^2 = 1$. Hence

$$2 \geq |\langle \overline{X}, \overline{Y} \rangle - \langle \overline{Y}, \overline{X} \rangle|.$$

Returning to the original vectors X and Y, we obtain the Schwarz inequality

$$\sqrt{\langle X, X \rangle} \sqrt{\langle Y, Y \rangle} \geq \frac{1}{2} |\langle X, Y \rangle - \langle Y, X \rangle|.$$

To avoid misunderstandings, we remark that Robertson's inequality (3.5.2) is correct, but "in general it does not lead to a proposition about the deviation of momentum". Consider the complex evolution function $\psi_t(x) = e^{R(t,x)+iS(t,x)}$. When $\int \nabla R \nabla S \mu_t \, dx = 0$, i.e., when there is no correlation between ∇R and ∇S, Robertson's inequality becomes a proposition about the standard deviation of momentum, and it reads:

"If the complex evolution function $\psi_t(x) = e^{R(t,x)+iS(t,x)}$ satisfies the condition $\int \nabla R \nabla S \mu_t \, dx = 0$, then the product of the standard deviation of position x and the standard deviation of momentum p_x can not be smaller than $\frac{1}{2} \hbar$."

Examples "demonstrating that the uncertainty principle holds" provided in many quantum mechanics books satisfy the condition $\int \nabla R \nabla S \mu \, dx = 0$. (Cf. Theorem 3.4.3.)

Consider, for instance, the following example given in Section 53 of Pauling and Wilson (1935):

$$\psi = A \exp\left(-\frac{(x - x_0)^2}{4\delta^2} + i \frac{p_0(x - x_0)}{\hbar}\right),$$

where x_0, p_0, and δ are constants. Since the normalization constant plays no role, we neglect it. Then

$$R = -\frac{(x - x_0)^2}{4(\delta)^2} \quad \text{and} \quad S = \frac{p_0(x - x_0)}{\hbar},$$

so

$$\int \frac{\partial R}{\partial x} \frac{\partial S}{\partial x} \mu \, dx = 0.$$

In this example the distribution density is

$$\mu = \psi\overline{\psi} = A \exp\left(-\frac{(x-x_0)^2}{2\delta^2}\right),$$

i.e., the normal distribution with variance δ^2. Therefore,

$$\Delta x = \sqrt{\text{Var}(x)} = \delta$$

and

$$h_p = \hbar\left(\frac{\partial R}{\partial x} + \frac{\partial S}{\partial x}\right) = -\frac{\hbar}{2\delta^2}(x-x_0) + p_0.$$

Let $\tilde{p} = p - \langle\psi, p\psi\rangle$. Since in this example $h_{(\tilde{p})^2} = (h_{\tilde{p}})^2$, by Theorem 3.4.1,

$$\langle\psi, (\tilde{p})^2\psi\rangle = \int (h_{\tilde{p}})^2 \psi\overline{\psi}\, dx = \frac{\hbar^2}{4\delta^4}\delta^2 = \frac{\hbar^2}{4\delta^2}.$$

Therefore,

$$\Delta p = \sqrt{\langle\psi, (\tilde{p})^2\psi\rangle} = \frac{\hbar}{2\delta},$$

and so, in this example,

$$\Delta x\, \Delta p = \frac{\hbar}{2}.$$

We claim that it cannot be argued that the "Heisenberg's uncertainty principle holds" just because there is such an example (or Heisenberg's example of microscope error).

Remarks.

(i) What Heisenberg (1927) discussed was not the "Heisenberg-Kennard inequality," but optical microscope errors (scattering experiment of photons and electrons).

(ii) "Inequality" is one of the ways to discuss the problem of Heisenberg's problem, but it obscures what the real question was, namely, whether the "paths of motion of electrons "exist or not. He wanted to argue that "there is no path for the motion of electrons"; the inequality was the means for that. We showed that in fact "there are paths of motion of electrons and they can be calculated". So the uncertainty principle is erroneous.

(iii) Heisenberg insisted that the uncertainty principle reveals the essence of quantum mechanics, and in the 1930's a fierce controversy on the uncertainty principle took place between Bohr and Einstein. Einstein expressed his doubts about the uncertainty principle and devised several thought experiments attempting to show that it does not hold. (Cf. Chapter 5 of Jammer (1974).) In retrospect, to insist that the uncertainty principle does not hold, it should have indicated, while discussing thought experiments. that it is a mistake of mathematical origin.

(iv) Robertson's inequality (3.5.2) is correctly derived from the Cauchy-Schwarz inequality (3.5.1), i.e., it allows no exceptions. It is a pity that Robertson misinterpreted $\sigma(p) = \sqrt{\langle \psi_t, (p_x - \langle \psi_t, p_x \psi_t \rangle)^2 \psi_t \rangle}$ to be the standard deviation of the momentum, after which all textbooks repeatedly stated this with no criticism. It was Theorem 3.4.3 that showed for the first time that in general the above $\sigma(p)$ is not the deviation: in order for it to be a deviation, it is necessary that $\int \nabla R \nabla S \mu_t \, dx = 0$. If this additional condition is satisfied, then the left-hand side of Robertson's inequality coincides with the quantity $\sqrt{\mathrm{Var}(x)} \sqrt{\mathrm{Var}(p_x)}$ of Theorem 3.5.1. However, as we showed in Theorem 3.5.1, this quantity can be made arbitrarily small. In the example of Theorem 3.5.1, $\int \nabla R \nabla S \mu_t \, dx \neq 0$, and Robertson's $\sigma(p)$ is not the standard deviation of the momentum That is, Theorem 3.5.1 exhibits an example in which Robertson's inequality gives erroneous results. The fact that Robertson did not notice this additional condition became a blind spot in the subsequent discussions of the so-called uncertainty principle.

3.6 Kinetic Energy and Variance of Position

We have seen above that if we use the Cauchy-Schwarz inequality to discuss the variance of momentum we run into a difficulty. However, we will show that using the Cauchy-Schwarz inequality (3.5.1) we can derive a new inequality for the product of the expectation of the kinetic energy and the variance of position. (See Nagasawa (2012))

Theorem 3.6.2. *Let* $\mathbf{x} = (x, y, z)$ *and* $\mathbf{p} = (p_x, p_y, p_z)$. *Then*

$$\langle \psi_t, (x - \langle \psi_t, x\psi_t \rangle)^2 \psi_t \rangle \, \langle \psi_t, \frac{1}{2m} \mathbf{p}^2 \psi_t \rangle \geq \frac{1}{4} \frac{1}{2m} \, \hbar^2. \qquad (3.6.1)$$

(This holds also for y and z.) In other words,

"The product of the variance of the coordinate x (also y and z) and the expected value of the kinetic energy is larger than or equal to $\frac{1}{4} \frac{1}{2m} \hbar^2$."

Proof. Setting $X = (x - \langle \psi_t, x\psi_t \rangle)\psi_t$ and $Y = p_x \psi_t$ in (3.5.1), we obtain

$$\langle \psi_t, \tilde{x}^2 \psi_t \rangle \, \langle \psi_t, p_x^2 \psi_t \rangle \geq \frac{1}{4} |\langle \psi_t, (xp_x - p_x x) \psi_t \rangle|^2,$$

where $\tilde{x} = x - \langle \psi_t, x\psi_t \rangle$. Since

$$(p_x x - x p_x) \psi_t = \frac{\hbar}{i} \psi_t$$

holds for any ψ_t, we see that

$$\langle \psi_t, (x - \langle \psi_t, x\psi_t \rangle)^2 \psi_t \rangle \langle \psi_t, p_x^2 \psi_t \rangle \geq \frac{1}{4} \hbar^2.$$

Moreover, we obviously have

$$\langle \psi_t, (x - \langle \psi_t, x\psi_t \rangle)^2 \psi_t \rangle \langle \psi_t, p_y^2 \psi_t \rangle \geq 0$$

and

$$\langle \psi_t, (x - \langle \psi_t, x\psi_t \rangle)^2 \psi_t \rangle \langle \psi_t, p_z^2 \psi_t \rangle \geq 0.$$

Adding these inequalities, we obtain the inequality (3.6.1). ☐

By Proposition 3.4.1, the kinetic energy $\frac{1}{2m}\mathbf{p}^2$ admits the function representation

$$K(x,t) = \frac{1}{2m}\hbar^2 \left((\nabla R(t,x))^2 + (\nabla S(t,x))^2 \right),$$

that is,

$$\langle \psi_t, \frac{1}{2m}\mathbf{p}^2\psi_t \rangle = \int \frac{1}{2m}\hbar^2 \left((\nabla R(t,x))^2 + (\nabla S(t,x))^2 \right) \mu_t(x)\, dx.$$

Let us point out here that inequality (3.6.1) has a significant physical meaning: if the kinetic energy is increased, the bound on the dispersion of position becomes smaller. Conversely, as the kinetic energy becomes smaller, the dispersion of the position increases accordingly. Examples demonstrating the importance of this result are provided below.

Example 1. In the double-slit experiments carried out by Tonomura, the dispersion of position is reduced by increasing the kinetic energy of electrons, and not by increasing the dispersion of the momentum. For larger values of the dispersion of the momentum their experiments would not work.

Example 2. Conversely, when the kinetic energy is small, for example, at the ground state, inequality (3.6.1) shows that the variance of the position cannot be made arbitrarily small. This, for example, implies that there is a limit to Bose-Einstein condensation. The variance of position must be large at the ground state energy, and then it is expected that the fluidity of the population of particles will occur. This fact is used to explain theoretically the superfluidity at very low temperatures. This is what was pointed out in the discussion of Bose-Einstein condensation in Section 2.5.

Example 3. In the mathematical model of high-temperature superconductivity presented in Section 2.11, superfluidity was achieved. Inequality (3.6.1) may be used in the discussion of superfluidity.

3.7 Theory of Hidden Variables

If we look at history, the problem of function representation of operators goes back to the theory of hidden variables in quantum theory. Let us look at the rough story of the theory of hidden variables discussed in the 1930s.

To consider an example, suppose the self-adjoint operator Λ has the eigenvalues $\lambda_1, \lambda_2, \ldots$, with associated eigenfunctions $\psi_1, \psi_2, \ldots$. In the Born-Heisenberg theory, the operator Λ represents a physical quantity, and under the eigenfunction ψ_j, the physical quantity Λ takes the value λ_j, for $j = 1, 2, \ldots$.

However, for

$$\psi = \sqrt{\alpha_1}\psi_1 + \sqrt{\alpha_2}\psi_2 + \cdots, \quad \alpha_j > 0 \sum_{j=1}^{\infty} \alpha_i = 1,$$

physical quantity Λ does not take a determined value. In this case, one considers that the value of the physical quantity (that is, the operator) Λ is one of the eigenvalues $\lambda_1, \lambda_2, \ldots$, and the probability that the value is λ_j is α_j, for $j = 1, 2, \ldots$.

In this way, since in general the value of the physical quantity can not be determined by $\{\Lambda, \psi\}$, one introduces an additional, new variable (parameter) ω and one considers $\{\Lambda, \psi, \omega\}$, thereby obtaining a physical quantity that takes a determined value. This was the idea of de Broglie, resulting in the "theory of hidden variable".

Actually, this leads to a function representation of $\{\Lambda, \psi\}$. In fact, if we let $\Omega = \{\omega_1, \omega_2, \ldots\}$, and define a function representation $h_\Lambda(\omega)$ of Λ by

$$h_\Lambda(\omega_j) = \lambda_j, \qquad j = 1, 2, \ldots,$$

then $h_\Lambda(\omega_1)$ can be understood as a physical quantity taking a fixed value. In this case the corresponding probability measure is given by

$$\mathbf{P}\left[\{\omega_j\}\right] = \alpha_j, \quad j = 1, 2, \ldots.$$

However, "the theory of hidden variables" implies that the Born-Heisenberg theory is incomplete; hence, people who believed the Born-Heisenberg theory is complete discussed theories that deny the hidden variable theory. (Actually, "hidden variable" is a slightly naughty and rude term that Pauli used to attack de Broglie.)

In his book *The Mathematical Foundations of Quantum Mechanics*, published in 1932, von Neumann "proved" that hidden variables were impossible as a matter of principle.

However, it has been pointed out that his "proof" is problematic. (Cf. Chapter 7 of Jammer (1974).)

Nevertheless, von Neumann's theorem has been accepted for a long time by supporters of the Born-Heisenberg theory as a theorem prohibiting the existence of hidden variables.

By the way, von Neumann gives a definition of "the hidden variable theory" in his book in which he assumes that the following property holds: the hidden variable h_{A+B} is the sum of the hidden variable h_A of operator A and the hidden variable h_B of the operator B. That is,

$$h_{A+B} = h_A + h_B. \tag{3.7.1}$$

However, Selleri (1983) noted that relation (3.7.1) does not hold in the case of spin operators, and pointed out that consequently von Neumann's discussion on hidden variables has no generality. He does not assert that there are mistakes in von Neumann's proof, but pointed out that there is an error in the premise of discussion, so von Neumann's theorem has no meaning.

Let us explain this. Let

$$\sigma_z = \begin{pmatrix} 1 & 0 \\ 0 & -1 \end{pmatrix}, \quad \sigma_y = \begin{pmatrix} 0 & -i \\ i & 0 \end{pmatrix}$$

and

$$u^+ = \begin{pmatrix} 1 \\ 0 \end{pmatrix}.$$

Then

$$\langle u^+, \sigma_z u^+ \rangle = 1, \quad \langle u^+, \sigma_y u^+ \rangle = 0$$

so

$$\langle u^+, (\sigma_z + \sigma_y) u^+ \rangle = 1.$$

Define a probability measure on the open interval $\Omega = \left(-\frac{\pi}{2}, \frac{\pi}{2} \right)$ by

$$\mathbf{P}[d\omega] = \frac{1}{2} \cos \omega \, d\omega$$

and consider the functions

$$h_{\sigma_z}(\omega) = 1, \text{ for } \omega \in \Omega, \quad h_{\sigma_y}(\omega) = \begin{cases} -1, & \text{for } -\frac{\pi}{2} \leq \omega < 0, \\ 1, & \text{for } 0 \leq \omega < \frac{\pi}{2}, \end{cases}$$

and

$$h_{\sigma_z + \sigma_y}(\omega) = \begin{cases} -\sqrt{2}, & \text{for } -\frac{\pi}{2} < \omega < -\frac{\pi}{4}, \\ \sqrt{2}, & \text{for } -\frac{\pi}{4} \leq \omega < \frac{\pi}{2}. \end{cases}$$

Then,

$$\int h_{\sigma_z}(\omega) \mathbf{P}[d\omega] = 1, \quad \int h_{\sigma_y}(\omega) \mathbf{P}[d\omega] = 0,$$

and

$$\int h_{\sigma_z + \sigma_y}(\omega) \mathbf{P}[d\omega] = -\int_{-\frac{\pi}{2}}^{-\frac{\pi}{4}} \frac{\sqrt{2}}{2} \cos \omega \, d\omega + \int_{-\frac{\pi}{4}}^{\frac{\pi}{2}} \frac{\sqrt{2}}{2} \cos \omega \, d\omega = 1.$$

Therefore $\{\Omega, \mathbf{P}; h_{\sigma_z}, h_{\sigma_y}, h_{\sigma_z + \sigma_y}\}$ gives a function representation of the set of operators $\{\sigma_z, \sigma_y, \sigma_z + \sigma_y, u^+\}$.

However, since

$$h_{\sigma_z}(\omega) + h_{\sigma_y}(\omega) = \begin{cases} 0, & \text{for } -\frac{\pi}{2} < \omega < 0, \\ 2, & \text{for } 0 \leq \omega < \frac{\pi}{2}, \end{cases}$$

relation (3.7.1) does not hold:

$$h_{\sigma_z + \sigma_y}(\omega) \neq h_{\sigma_z}(\omega) + h_{\sigma_y}(\omega). \tag{3.7.2}$$

This completes the proof of Selleri's claim.

Remark. In von Neumann's book it is implicitly assumed that the probability measure **P** for computing the expected value does not change for each set of operators to be handled, but is determined universally: **P** does not depend on which set of operators is considered and satisfies

$$\langle \psi, A\psi \rangle = \int h_A(\omega) \, d\mathbf{P},$$

for all operators A. This is too demanding: in general, such a universal probability measure does not exist.

By the way, to show that von Neumann claim is wrong, it is enough to exhibit one hidden variable theory, that is, to give a counterexample to the non-existence claim.

Such a counterexample example exists. We have relation (3.7.1) for a set of commuting operators, and, as shown in Theorem 3.3.1, every set of commuting operators admits a function representation (hidden variable theory).

In addition, although the mechanics of random motion is not a theory of function representation of operators, it can be interpreted as a viable theory of hidden variables, and it exists.

By the way, although we have discussed hidden variables, is the existence of such variables (i.e., function representations of operators) significant for the Born-Heisenberg theory (quantum mechanics)?

To argue that hidden variables have serious meanings, it is necessary to show that function representation (i.e., random variables) have solid physical meanings, and that they are indispensable. This point was insufficiently addressed in the conventional discussions of hidden variables.

Whether hidden variables exist or not was thought to be mostly unimportant to Born-Heisenberg's theory (quantum mechanics); in the Born-Heisenberg operator theory, it was considered sufficient to interpret the inner product $\langle \psi, A\psi \rangle$ as the expected value of the operator (physical quantity) A.

We explained in Section 3.4 that this idea is erroneous. Concretely speaking, if we follow this idea, the variance of the physical quantity cannot be defined.

Re-examining the issue, we now recognize that adding hidden variables in the Born-Heisenberg theory is meaningful, and moreover that the existence of hidden variables implicitly indicates that

"separately from the Born-Heisenberg theory, there should exist a consistent theory in which hidden variables (random variables) are physical quantities."

This is certainly true. Mechanics of random motion is exactly such a theory. In the mechanics of random motion, physical quantities are given as random variables, and these random variable (that is, hidden variables) have a physical meaning.

In addition, as clear from the definition (3.3.1) of a function representation, function representations are not uniquely determined, so to determine physically meaningful function representations, we need a consistent theory. We can regard our mechanics of random motion as one of such theories.

Einstein co-authored a paper with Podolsky and Rosen in 1935 (Einstein-Podolsky-Rosen (1935)), in which they put forward the following alternative: the Born-Heisenberg theory is complete, or the physical reality is local. They insisted that "the theory (quantum mechanics) of Born-Heisenberg is incomplete", because "physical phenomena (laws) must be local".

We interpret function representations as mathematical entities corresponding to the "physical reality" of Einstein-Podolsky-Rosen. Then, if the Born-Heisenberg theory is complete, the function representation is generally nonlocal. But that is not true. So

"the Born-Heisenberg theory cannot be complete, i.e., it is incomplete."

This is now a obvious fact, but at that time it was not considered to be so obvious. Bohr regarded the paper by Einstein, Podolsky and Rosen as a serious attack against Born-Heisenberg's theory and expressed his opinion, denying it immediately. After that, a complicated controversy on this issue continued, finally to be settled by Nagasawa (1997), Nagasawa and Schröder (1997), and Nagasawa (2012).

As a matter of fact, Bohr's argument that the Born-Heisenberg's theory was complete was widely accepted until the issue was finally settled. (Even now many of people think so.) Bell's paper (Bell (1964)) supported Bohr's opinion. Specifically, Bell insisted that

"local hidden variables (function representations) cannot reproduce the correlations in quantum mechanics."

If Bell's assertion is correct, quantum mechanics is complete.

The fact that the physical phenomena (laws) are nonlocal is a serious thing, and Einstein strongly denied it, saying if so it is "telepathy". Those who understood that "the world is nonlocal" was shown in Bell's paper developed various philosophical arguments. As a matter of fact, when dealing with concrete physical phenomena, non-locality does not have a serious impact in the meantime, and Bell's "proof" is extremely easy (see Proposition 3.9.1) and seems there is no mistake. So, many physicists seem to admit it dubiously. (Jammer expressed doubts in Section 7.7 of Jammer (1974).) Bell's claim is based on so-called Bell's inequality. However, **Bell's inequality has nothing to do with locality**. Therefore, Bell's argument is a mistake. We will explain this in the following.

3.8   Einstein's Locality

Let us consider an example. Suppose a pair of particles with spin are generated at the origin and assume that the particles go to points A and B, which lie at a large distance from one another. Stern-Gerlach magnets are installed at A and B, and we observe the spin of the two particles. At the time of observation, the orientations of the Stern-Gerlach magnets at the points A and B are **a** and **b**, respectively.

Consider the three so-called spin 2×2 matrices on $\mathbf{R}^2$

$$\sigma_x = \begin{pmatrix} 0 & 1 \\ 1 & 0 \end{pmatrix}, \quad \sigma_y = \begin{pmatrix} 0 & -i \\ i & 0 \end{pmatrix}, \quad \sigma_z = \begin{pmatrix} 1 & 0 \\ 0 & -1 \end{pmatrix} \qquad (3.8.1)$$

and set

$$\boldsymbol{\sigma} = (\sigma_x, \sigma_y, \sigma_z).$$

Let the norm of the vectors $\mathbf{a} = (a_1, a_2, a_3)$ and $\mathbf{b} = (b_1, b_2, b_3)$ be 1, and set

$$S(\mathbf{a}, \mathbf{1}) = \boldsymbol{\sigma}\mathbf{a} \otimes \mathbf{1}, \quad S(\mathbf{1}, \mathbf{b}) = \mathbf{1} \otimes \boldsymbol{\sigma}\mathbf{b}, \qquad (3.8.2)$$

where

$$\boldsymbol{\sigma}\mathbf{a} = \begin{pmatrix} a_3 & a_1 - a_2 i \\ a_1 + a_2 i & -a_3 \end{pmatrix}.$$

Then $S(\mathbf{a}, \mathbf{1}) = \boldsymbol{\sigma}\mathbf{a} \otimes \mathbf{1}$ and $S(\mathbf{1}, \mathbf{b}) = \mathbf{1} \otimes \boldsymbol{\sigma}\mathbf{b}$ are commuting operators on $\mathbf{R}^2 \times \mathbf{R}^2$. Next, consider the vectors

$$u^+ = \begin{pmatrix} 1 \\ 0 \end{pmatrix}, \quad u^- = \begin{pmatrix} 0 \\ 1 \end{pmatrix} \qquad (3.8.3)$$

and define

$$\psi = \frac{u^+ \otimes u^- - u^- \otimes u^+}{\sqrt{2}}. \qquad (3.8.4)$$

Let us explain the physical interpretation of the operators $S(\mathbf{a}, \mathbf{1}) = \boldsymbol{\sigma}\mathbf{a} \otimes \mathbf{1}$ and $S(\mathbf{1}, \mathbf{b}) = \mathbf{1} \otimes \boldsymbol{\sigma}\mathbf{b}$.

Once the two particles are generated at the origin, their spins remain unchanged. That is, it is assumed that they do not depend on time and, by definition (3.8.4), the spin is either up or down; ψ does not depend on time or position.

The particles head towards the points A and B. The operator $S(\mathbf{a}, \mathbf{1}) = \boldsymbol{\sigma}\mathbf{a} \otimes \mathbf{1}$ (resp., $S(\mathbf{1}, \mathbf{b}) = \mathbf{1} \otimes \boldsymbol{\sigma}\mathbf{b}$) is interpreted as representing the spin of the particle detected by the Stern-Gerlach magnet at the point A (resp., B).

Let us compute the expectations $\langle \psi, \boldsymbol{\sigma}\mathbf{a} \otimes \mathbf{1}\psi \rangle$ and $\langle \psi, \mathbf{1} \otimes \boldsymbol{\sigma}\mathbf{b}\psi \rangle$. First, since

$$u^+ \boldsymbol{\sigma}\mathbf{a}u^+ = (1,0) \begin{pmatrix} a_3 & a_1 - a_2 i \\ a_1 + a_2 i & -a_3 \end{pmatrix} \begin{pmatrix} 1 \\ 0 \end{pmatrix} = a_3,$$

$$u^- \boldsymbol{\sigma}\mathbf{a}u^- = (0,1) \begin{pmatrix} a_3 & a_1 - a_2 i \\ a_1 + a_2 i & -a_3 \end{pmatrix} \begin{pmatrix} 0 \\ 1 \end{pmatrix} = -a_3,$$

$$u^+ \boldsymbol{\sigma}\mathbf{a}u^- = (1,0) \begin{pmatrix} a_3 & a_1 - a_2 i \\ a_1 + a_2 i & -a_3 \end{pmatrix} \begin{pmatrix} 0 \\ 1 \end{pmatrix} = a_1 - a_2 i,$$

$$u^- \boldsymbol{\sigma}\mathbf{a}u^+ = (0,1) \begin{pmatrix} a_3 & a_1 - a_2 i \\ a_1 + a_2 i & -a_3 \end{pmatrix} \begin{pmatrix} 1 \\ 0 \end{pmatrix} = a_1 + a_2 i,$$

we have

$$\langle \psi, \boldsymbol{\sigma}\mathbf{a} \otimes \mathbf{1}\psi \rangle = 0, \quad \langle \psi, \mathbf{1} \otimes \boldsymbol{\sigma}\mathbf{b}\psi \rangle = 0, \tag{3.8.5}$$

which indicates that if we observe the spins, their expectations are zero. Moreover

$$\langle \psi, (\boldsymbol{\sigma}\mathbf{a} \otimes \mathbf{1}) (\mathbf{1} \otimes \boldsymbol{\sigma}\mathbf{b}) \psi \rangle = -\mathbf{ab}, \tag{3.8.6}$$

where $\mathbf{ab} = a_1 b_1 + a_2 b_2 + a_3 b_3$. This indicates that when a pair of particles are generated at the origin there is a "correlation" between the operators $S(\mathbf{a}, 1) = \boldsymbol{\sigma}\mathbf{a} \otimes 1$ and $S(1, \mathbf{b}) = \mathbf{1} \otimes \boldsymbol{\sigma}\mathbf{b}$.

We consider four spin-correlation experiments with different orientations of the Stern-Gerlach magnets at the points A and B, as follows:

Experiment 1. Magnet orientations are $\{\mathbf{a}, \mathbf{b}\}$,

Experiment 2. Magnet orientations are $\{\mathbf{a}, \mathbf{c}\}$,

Experiment 3. Magnet orientations are $\{\mathbf{b}, \mathbf{c}\}$,

Experiment 4. Magnet orientations are $\{\mathbf{b}, \mathbf{b}\}$,

where, for example, $\{\mathbf{a}, \mathbf{b}\}$ indicates that the direction of the Stern-Gerlach magnet at A is $\mathbf{a}$, and the direction of Stern-Gerlach magnet at B is $\mathbf{b}$. We assume that these experiments are carried out independently and do not interact in any way.

To discuss these experiments, we take the set of operators

$$\begin{aligned}
\mathbf{F} = \Big\{ & S^{\{1\}}(\mathbf{a}, 1), \ S^{\{1\}}(1, \mathbf{b}), \ S^{\{1\}}(\mathbf{a}, 1)S^{\{1\}}(1, \mathbf{b}); \\
& S^{\{2\}}(\mathbf{a}, 1), \ S^{\{2\}}(1, \mathbf{c}), \ S^{\{2\}}(\mathbf{a}, 1)S^{\{2\}}(1, \mathbf{c}); \\
& S^{\{3\}}(\mathbf{b}, 1), \ S^{\{3\}}(1, \mathbf{c}), \ S^{\{3\}}(\mathbf{b}, 1)S^{\{3\}}(1, \mathbf{c}); \\
& S^{\{4\}}(\mathbf{b}, 1), \ S^{\{4\}}(1, \mathbf{b}), \ S^{\{4\}}(\mathbf{b}, 1)S^{\{4\}}(1, \mathbf{b}) \Big\}
\end{aligned} \tag{3.8.7}$$

and the ψ given by formula (3.8.4), and form the pair $\{\mathbf{F}, \psi\}$. Let

$$\{\Omega, \mathbf{P}; h_A, A \in \mathbf{F}\} \tag{3.8.8}$$

be a function representation of the pair $\{\mathbf{F}, \psi\}$.

As already noted, the function representation is not uniquely determined, so we have to set it up carefully. We will introduce a function representation that discriminates among the four experiments, so that the non-uniqueness has no effect. Its specific form is provided in Section 3.10. From now on we follow Nagasawa (2012).

Now let us introduce function representations for the individual experiments.

Experiment 1: Let the function representations of the operators $S^{\{1\}}(\mathbf{a}, 1)$ and $S^{\{1\}}(1, \mathbf{b})$ be

$$h_{\mathbf{a},1}^{\{1\}}(\omega) \quad \text{and} \quad h_{1,\mathbf{b}}^{\{1\}}(\omega).$$

Suppose both take the value ± 1. Then, by (3.8.5), we have

$$\mathbf{P}[h_{\mathbf{a},1}^{\{1\}}] = 0 \quad \text{and} \quad \mathbf{P}[h_{1,\mathbf{b}}^{\{1\}}] = 0.$$

Moreover, by (3.8.6),

$$\mathbf{P}[h_{\mathbf{a},1}^{\{1\}} h_{1,\mathbf{b}}^{\{1\}}] = -\mathbf{ab}.$$

Experiment 2: Let the function representations of the operators $S^{\{2\}}(\mathbf{a}, 1)$ and $S^{\{2\}}(1, \mathbf{c})$ be

$$h_{\mathbf{a},1}^{\{2\}}(\omega) \quad \text{and} \quad h_{1,\mathbf{c}}^{\{2\}}(\omega).$$

Suppose both take the value ± 1. Then, again by (3.8.5), we have

$$\mathbf{P}[h_{\mathbf{a},1}^{\{2\}}] = 0 \quad \text{and} \quad \mathbf{P}[h_{1,\mathbf{c}}^{\{2\}}] = 0.$$

Moreover, by (3.8.6),

$$\mathbf{P}[h_{\mathbf{a},1}^{\{2\}} h_{1,\mathbf{c}}^{\{2\}}] = -\mathbf{ac}.$$

Note that Experiments 1 and 2 are independent, so the random variables $h_{\mathbf{a},1}^{\{1\}}$ and $h_{\mathbf{a},1}^{\{2\}}$ are independent and of course different functions:

$$h_{\mathbf{a},1}^{\{1\}} \neq h_{\mathbf{a},1}^{\{2\}}. \tag{3.8.9}$$

Experiment 3: Let the function representations of the operators $S^{\{3\}}(\mathbf{b}, 1)$ and $S^{\{3\}}(1, \mathbf{c})$ be

$$h_{\mathbf{b},1}^{\{3\}}(\omega)) \quad \text{and} \quad h_{1,\mathbf{c}}^{\{3\}}(\omega).$$

Suppose both take the value ± 1. Then, by (3.8.5), we have

$$\mathbf{P}[h_{\mathbf{b},1}^{\{3\}}] = 0 \quad \text{and} \quad \mathbf{P}[h_{1,\mathbf{c}}^{\{3\}}] = 0.$$

Moreover, by (3.8.6),

$$\mathbf{P}[h_{\mathbf{b},1}^{\{3\}} h_{1,\mathbf{c}}^{\{3\}}] = -\mathbf{bc}.$$

Again, Experiments 2 and 3 are independent, so the random variables $h_{1,\mathbf{c}}^{\{2\}}$ and $h_{1,\mathbf{c}}^{\{3\}}$ are independent and of course different functions:

$$h_{1,\mathbf{c}}^{\{2\}} \neq h_{1,\mathbf{c}}^{\{3\}}. \tag{3.8.10}$$

Experiment 4: Let the function representations of the operators $S^{\{4\}}(\mathbf{b}, 1)$ and $S^{\{4\}}(1, \mathbf{b})$ be

$$h_{\mathbf{b},1}^{\{4\}}(\omega)) \quad \text{and} \quad h_{1,\mathbf{b}}^{\{4\}}(\omega).$$

Suppose both take the value ± 1, and

$$h_{\mathbf{b},1}^{\{4\}}(\omega) h_{1,\mathbf{b}}^{\{4\}}(\omega) = -1.$$

Then, by (3.8.5),

$$\mathbf{P}[h_{\mathbf{b},1}^{\{4\}}] = 0 \quad \text{and} \quad \mathbf{P}[h_{1,\mathbf{b}}^{\{4\}}] = 0.$$

Moreover, by (3.8.6),

$$\mathbf{P}[h_{\mathbf{b},1}^{\{4\}} h_{1,\mathbf{b}}^{\{4\}}] = -1.$$

Again, note that Experiments 3 and 4 are independent, so the random variables $h_{\mathbf{b},1}^{\{3\}}$ and $h_{\mathbf{b},1}^{\{4\}}$ are independent, and of course different functions. Furthermore, since Experiments 4 and 1 are independent, also the random variables $h_{1,\mathbf{b}}^{\{4\}}$ and $h_{1,\mathbf{b}}^{\{1\}}$ are independent and different. That is,

$$h_{\mathbf{b},1}^{\{3\}} \neq h_{\mathbf{b},1}^{\{4\}} \quad \text{and} \quad h_{1,\mathbf{b}}^{\{4\}} \neq h_{1,\mathbf{b}}^{\{1\}}. \tag{3.8.11}$$

What is important in our discussion is the "locality of hidden variables (function representations)." Einstein, Podolsky and Rosen (1935) where the first who took up and discussed the "locality of hidden variables." The concept of "locality" used by Bell (1964) is adopted from the paper of Einstein, Podolsky and Rosen (1935).

Let us define the "locality of hidden variables (function representations)" following Einstein, Podolsky and Rosen (1935).

Definition 3.8.1. Consider the set of operators

$$\mathbf{F} = \{S(\mathbf{a}, 1), \ S(1, \mathbf{b}), \ S(\mathbf{a}, 1)S(1, \mathbf{b})\}, \tag{3.8.12}$$

and the ψ given by (3.8.4). Then the function representation of $\{\mathbf{F}, \psi\}$ is said to be **local** if

$$\{\Omega, \mathbf{P}; \ h_{\mathbf{a},1}, \ h_{1,\mathbf{b}}, \ h_{\mathbf{a},1}h_{1,\mathbf{b}}\} \tag{3.8.13}$$

satisfies the following **locality condition**:

(L) The random variable $h_{\mathbf{a},1}$ does not depend on vector $\mathbf{b}$ at location B, and the random variable $h_{1,\mathbf{b}}$ does not depend on vector $\mathbf{a}$ at location A.

An important point that should be emphasized here is that

"Locality" as defined concerns "a single experiment," with a fixed orientation of magnets.

When several different locality experiments are discussed simultaneously, we require that locality holds for each individual experiment under consideration.

Therefore, we assume that the hidden variables for Experiments 1, 2, 3, 4,

$$\{h_{a,1}^{\{1\}},\, h_{1,b}^{\{1\}}\}, \quad \{h_{a,1}^{\{2\}},\, h_{1,c}^{\{2\}}\}, \quad \{h_{b,1}^{\{3\}},\, h_{1,c}^{\{3\}}\}, \quad \text{and} \quad \{h_{b,1}^{\{4\}},\, h_{1,b}^{\{4\}}\}, \qquad (3.8.14)$$

are local.

Furthermore, when several experiments that are carried out independently are discussed, it is assumed that the function representations of the individual experiments are independent of one another.

This is the case with our Experiments 1, 2, 3 and 4, so we assume that the corresponding function representations are independent of one another.

What we have discussed above is a general theory of function representations for the four spin-correlation experiments and the locality of the function representation for each individual experiment.

3.9   Bell's Inequality

In this section we will argue that Bell's claim that "local function representation can not reproduce correlation in quantum mechanics" is erroneous. His argument is based on the so-called Bell's inequality. However, Bell's inequality has nothing to do with locality, as we will explain below. Bell's error comes from a misunderstanding caused by the imperfection of mathematical symbols he used.

Let us analyze Bell's paper (Bell (1964)) following Nagasawa (2012).

Bell (1964) implicitly introduced additional conditions between the hidden variables used in the discussion of spin correlation in Experiments 1, 2, 3, 4. Namely, he assumed that

$$h_{a,1}^{\{2\}} = h_{a,1}^{\{1\}}, \quad h_{1,c}^{\{2\}} = h_{1,c}^{\{3\}}, \quad h_{b,1}^{\{4\}} = h_{b,1}^{\{3\}}, \quad \text{and} \quad h_{1,b}^{\{4\}} = h_{1,b}^{\{1\}}. \qquad (3.9.1)$$

Note that Bell's relations (3.9.1) reject all the relations (3.8.9), (3.8.10) and (3.8.11) expressing the mutual independence of the experiments. In other words, Bell's relations (3.9.1) constitute a "condition that denies the mutual independence of the four experiments", but has *no relation to locality*.

We will call (3.9.1) "Bell's additional condition" or "Bell's dependence."

In fact, Bell ignored and did not append the upper indices {1}, {2}, {3}, and {4} of hidden variables in (3.9.1), which clearly distinguish among the four experiments 1, 2, 3, 4. This was a mistake of Bell in the symbolic law, leading to failure. Moreover, he mistook "Bell's dependence (3.9.1)" for "locality" — simply a misunderstanding.

Bell's dependence (3.9.1) expresses the mutual relationships of the four experiments with changed orientation of the magnets, and it is completely irrelevant for the notion of locality (L) introduced in Definition 3.8.1 for a single experiment with fixed orientations of the two magnets.

Next, let us examine "Bell's inequality." In discussing the spin correlation of hidden variables for the Experiments 1, 2, 3, and 4, Bell used **Bell's hidden variables that satisfy his additional condition (3.9.1)**. Using the symbols introduced above, we can state

Proposition 3.9.1. *Under Bell's additional condition* (3.9.1), *the Bell inequality*

$$\left| \mathbf{P}[h_{a,1}^{\{1\}} h_{1,b}^{\{1\}}] - \mathbf{P}[h_{a,1}^{\{2\}} h_{1,c}^{\{2\}}] \right| \le 1 + \mathbf{P}[h_{b,1}^{\{3\}} h_{1,c}^{\{3\}}] \tag{3.9.2}$$

holds

Proof. Since $h_{b,1}^{\{4\}} h_{1,b}^{\{4\}} = -1$, we have

$$h_{a,1}^{\{1\}} h_{1,b}^{\{1\}} - h_{a,1}^{\{2\}} h_{1,c}^{\{2\}} = h_{a,1}^{\{1\}} h_{1,b}^{\{1\}} + h_{a,1}^{\{2\}} h_{1,c}^{\{2\}} h_{b,1}^{\{4\}} h_{1,b}^{\{4\}}. \tag{3.9.3}$$

By Bell's dependence (3.9.1), equation (3.9.3) can be recast as

$$h_{a,1}^{\{1\}} h_{1,b}^{\{1\}} - h_{a,1}^{\{2\}} h_{1,c}^{\{2\}} = h_{a,1}^{\{1\}} h_{1,b}^{\{1\}} + h_{a,1}^{\{1\}} h_{1,c}^{\{3\}} h_{b,1}^{\{3\}} h_{1,b}^{\{1\}}$$

$$= h_{a,1}^{\{1\}} h_{1,b}^{\{1\}} \left(1 + h_{b,1}^{\{3\}} h_{1,c}^{\{3\}} \right).$$

Since $h_{a,1}^{\{1\}} h_{1,b}^{\{1\}} \le 1$,

$$h_{a,1}^{\{1\}} h_{1,b}^{\{1\}} - h_{a,1}^{\{2\}} h_{1,c}^{\{2\}} \le 1 + h_{b,1}^{\{3\}} h_{1,c}^{\{3\}}.$$

Taking the expectation of both sides, we obtain

$$\left| \mathbf{P}[h_{a,1}^{\{1\}} h_{1,b}^{\{1\}}] - \mathbf{P}[h_{a,1}^{\{2\}} h_{1,c}^{\{2\}}] \right| \le 1 + \mathbf{P}[h_{b,1}^{\{3\}} h_{1,c}^{\{3\}}].$$

The proof is complete. □

The proof of the proposition makes no use whatsoever of the locality of hidden variables: only Bell's additional condition (3.9.1) was needed. Therefore, **Bell's inequality (3.9.2) has nothing to do with locality.**

We claim that Bell's additional condition (3.9.1) is not only "totally unrelated to locality", it is also an "unreasonable condition" incompatible with the function representations of the spin correlation Experiments 1, 2, 3, and 4. Let us show this.

Proposition 3.9.2. *Let the random variables*

$$\{h_{a,1}^{\{1\}}, h_{1,b}^{\{1\}}, h_{a,1}^{\{2\}}, h_{1,c}^{\{2\}}, \ldots\}$$

listed in (3.8.14) *be function representations of the spin-correlation Experiments* 1, 2, 3, 4. *Then Bell's additional condition* (3.9.1) *cannot be satisfied. That is, Bell's additional condition* (3.9.1) *is incompatible with the function representations of the Experiments* 1, 2, 3, 4.

Proof. Suppose, by contradiction, that the random variables

$$\{h_{a,1}^{\{1\}}, h_{1,b}^{\{1\}}, h_{a,1}^{\{2\}}, h_{1,c}^{\{2\}}, \ldots\}$$

listed in (3.8.14) are function representations of spin-correlation Experiments 1, 2, 3, 4 and Bell's additional condition (3.9.1) is satisfied. Then, by Proposition 3.9.1, Bell's inequality holds. We will show that this leads to a contradiction.

Indeed, suppose the vectors $\mathbf{b} = (b_1, b_2, b_3)$ and $\mathbf{c} = (c_1, c_2, c_3)$ are orthogonal to one another. Then, since $\{h_{b,1}^{\{3\}}, h_{1,c}^{\{3\}}\}$ is a function representation of Experiment 3,

$$\mathbf{P}\left[h_{b,1}^{\{3\}} h_{1,c}^{\{3\}}\right] = -\mathbf{bc} = 0.$$

Therefore, the right-hand side of Bell's inequality (3.9.2) is equal to 1. Moreover, since $\{h_{a,1}^{\{1\}}, h_{1,b}^{\{1\}}\}$ and $\{h_{a,1}^{\{2\}}, h_{1,c}^{\{2\}}\}$ are function representations of Experiment 1 and Experiment 2, respectively,

$$\mathbf{P}\left[h_{a,1}^{\{1\}} h_{1,b}^{\{1\}}\right] = -\mathbf{ab}, \quad \mathbf{P}\left[h_{a,1}^{\{2\}} h_{1,c}^{\{2\}}\right] = -\mathbf{ac}.$$

Therefore, if we set $\mathbf{a} = \dfrac{\mathbf{b} - \mathbf{c}}{\|\mathbf{b} - \mathbf{c}\|}$, then

$$\mathbf{P}\left[h_{a,1}^{\{1\}} h_{1,b}^{\{1\}}\right] - \mathbf{P}\left[h_{a,1}^{\{2\}} h_{1,c}^{\{2\}}\right] = -\mathbf{a}(\mathbf{b} - \mathbf{c}) = -\frac{\|\mathbf{b} - \mathbf{c}\|^2}{\|\mathbf{b} - \mathbf{c}\|}$$

$$= -\|\mathbf{b} - \mathbf{c}\| = -\sqrt{2}.$$

Thus, (3.9.2) leads to a contradiction: $\sqrt{2} \leq 1$, as claimed. $\square$

Remark. In his paper, Bell insisted that "if hidden variables are local, we must meet Bell's inequality (3.9.2). However, since the inequality (3.9.2) leads to contradiction, the local hidden variables cannot reproduce correlations in quantum mechanics."

Bell's claim is, of course, erroneous. Indeed, as shown in Proposition 3.9.1, Bell's inequality (3.9.2) has nothing to do with the locality, because it is derived from Bell's additional condition of (3.9.1), but not from the locality. As shown in Proposition 3.9.2, the additional condition (3.9.1) itself is an irrational condition.

In addition, in an attempt to verify Bell's assertion, Aspect et al. (1982) performed experiments and showed that Bell's inequality was not actually satisfied. Then people who agreed with Bell interpreted this experimental result a verification of non-locality regarded this as an epoch-making contribution to science. However, this was a misunderstanding, because Bell's inequality is completely unrelated to locality, as demonstrated by Proposition 3.9.1. If we look back with the present knowledge, we see that the experiments of Aspect et al. (1982) actually support the correctness of our Proposition 3.9.2.

3.10 Local Spin Correlation Model

In the previous section we have argued that Bell's claim is erroneous. But Einstein-Podolsky-Rosen's statement that "physical reality is local" has not been proven. We assume that the mathematical entity corresponding to "physical reality" is a function representation. Therefore, in this section, we show that a local function representation exists for the spin correlation experiment, thereby establishing the validity of Einstein-Podolsky-Rosen's claim that "physical reality is local."

In what follows we will examine the experiment by modifying in various ways the orientations $\{\mathbf{a}, \mathbf{b}\}$ of the Stern-Gerlach magnets at the points A and B. Specifically

Experiment 1. Orientations are $\{\mathbf{a}, \mathbf{b}\}$;

Experiment 2. Orientations are $\{\mathbf{a}', \mathbf{b}'\}$;

Experiment 3. Orientations are $\{\mathbf{a}'', \mathbf{b}''\}$;

and so on. It is supposed that this can be done not only three times, but also many times, changing the orientation of the magnets. Furthermore, the experiments are carried out independently and they do not influence one another. They may be carried out in parallel, regardless of the context and ordering.

To treat these experiments simultaneously, we introduce the set of operators

$$\mathbf{F} = \Big\{ S(\mathbf{a}, \mathbf{1}), \ S(\mathbf{1}, \mathbf{b}), \ S(\mathbf{a}, \mathbf{1})S(\mathbf{1}, \mathbf{b}); \ \text{pair } \{\mathbf{a}, \mathbf{b}\} \text{ is arbitrary} \Big\}, \qquad (3.10.1)$$

where the operator $S(\mathbf{a}, \mathbf{1}) = \boldsymbol{\sigma}\mathbf{a} \otimes \mathbf{1}$ (resp., $S(\mathbf{1}, \mathbf{b}) = \mathbf{1} \otimes \boldsymbol{\sigma}\mathbf{b}$) represents the spin of a particle detected at location A (resp., B). Consider the pair $\{\mathbf{F}, \psi\}$, where ψ is given by formula (3.8.4).

Let

$$\{\Omega, \mathbf{P} \ ; \ h_{\mathbf{a},\mathbf{1}}^{\{\mathbf{a},\mathbf{b}\}}, \ h_{\mathbf{1},\mathbf{b}}^{\{\mathbf{a},\mathbf{b}\}}, \ h_{\mathbf{a},\mathbf{1}}^{\{\mathbf{a},\mathbf{b}\}} h_{\mathbf{1},\mathbf{b}}^{\{\mathbf{a},\mathbf{b}\}}; \ \text{pair } \{\mathbf{a}, \mathbf{b}\} \text{ is arbitrary}\}, \qquad (3.10.2)$$

be a function representation of the pair $\{\mathbf{F}, \psi\}$, where the superscript $\{\mathbf{a}, \mathbf{b}\}$ in $h_{\mathbf{a},\mathbf{1}}^{\{\mathbf{a},\mathbf{b}\}}$ and $h_{\mathbf{1},\mathbf{b}}^{\{\mathbf{a},\mathbf{b}\}}$ indicates that we are dealing with an experiment with Stern-Gerlach magnets with orientation $\mathbf{a}$ and $\mathbf{b}$ at the locations A and B.

An important question here is whether the function representation of the set of operators $\mathbf{F}$ is local or non-local. (Cf. Einstein-Podolsky-Rosen (1935), Bell (1964))

We adopt the definition of locality given earlier.

Definition 3.10.1. Let $\mathbf{F}$ be the set of operators (3.10.1) and let ψ be given by formula (3.8.4). The function representation (3.10.2) of $\{\mathbf{F}, \psi\}$ is said to be **local** if

(L) the random variable $h_{\mathbf{a},1}^{\{\mathbf{a},\mathbf{b}\}}$ does not depend on the vector $\mathbf{b}$ at the location B, and the random variable $h_{1,\mathbf{b}}^{\{\mathbf{a},\mathbf{b}\}}$ does not depend on the vector $\mathbf{a}$ at the location A.

In this case we call $h_{\mathbf{a},1}^{\{\mathbf{a},\mathbf{b}\}}$ and $h_{1,\mathbf{b}}^{\{\mathbf{a},\mathbf{b}\}}$ "**local hidden variables.**"

Then we can state the following result:

Theorem 3.10.1 (Nagasawa (2012)). *Let $\mathbf{F}$ be the set of operators in (3.10.1) and let ψ be given by formula (3.8.4). Then the pair $\{\mathbf{F}, \psi\}$ admits the local function representation (3.10.2).*

Proof. Let $W = \{0,1\} \times \{0,1\}$, and define functions $h_1(i,j)$ and $h_2(i,j)$ on W by

$$h_1(0,j) = 1, \quad h_1(1,j) = -1, \quad j = 0,1, \tag{3.10.3}$$

and

$$h_2(i,0) = -1, \quad h_2(i,1) = 1, \quad i = 0,1. \tag{3.10.4}$$

The function $h_1(i,j)$ does not depend on the second variable j, and the function $h_2(i,j)$ does not depend on the first variable i. That is, both functions are defined locally.

Let $\Omega_{\{\mathbf{a},\mathbf{b}\}}$ denote a copy of the set $W = \{0,1\} \times \{0,1\}$ for each pair $\{\mathbf{a}, \mathbf{b}\}$ and consider the direct product space

$$\Omega = \prod_{\{\mathbf{a},\mathbf{b}\}} \Omega_{\{\mathbf{a},\mathbf{b}\}},$$

whose elements are $\omega = \left(\dots, \omega_{\mathbf{a},\mathbf{b}}, \dots, \omega_{\mathbf{a}',\mathbf{b}'}, \dots\right)$, where $\dots, \omega_{\mathbf{a},\mathbf{b}} \in \Omega_{\{\mathbf{a},\mathbf{b}\}}, \dots,$ $\omega_{\mathbf{a}',\mathbf{b}'} \in \Omega_{\{\mathbf{a}',\mathbf{b}'\}}, \dots.$

Using the functions h_1 and h_2, we define the random variables $h_{\mathbf{a},1}^{\{\mathbf{a},\mathbf{b}\}}(\omega)$ and $h_{1,\mathbf{b}}^{\{\mathbf{a},\mathbf{b}\}}(\omega)$ on Ω by

$$\begin{aligned} h_{\mathbf{a},1}^{\{\mathbf{a},\mathbf{b}\}}(\omega) &= h_1(\omega_{\mathbf{a},\mathbf{b}}), \\ h_{1,\mathbf{b}}^{\{\mathbf{a},\mathbf{b}\}}(\omega) &= h_2(\omega_{\mathbf{a},\mathbf{b}}), \end{aligned} \qquad \text{for } \omega = \left(\dots, \omega_{\mathbf{a},\mathbf{b}}, \dots, \omega_{\mathbf{a}',\mathbf{b}'}, \dots\right). \tag{3.10.5}$$

Again, the superscript $\{\mathbf{a}, \mathbf{b}\}$ in $h_{1,\mathbf{b}}^{\{\mathbf{a},\mathbf{b}\}}(\omega)$ is a parameter indicating that the orientations of the Stern-Gerlach magnets in the experiment under consideration are $\mathbf{a}$ and $\mathbf{b}$.

The definition (3.10.5) of the function $h_{\mathbf{a},1}^{\{\mathbf{a},\mathbf{b}\}}(\omega)$ (resp., $h_{1,\mathbf{b}}^{\{\mathbf{a},\mathbf{b}\}}(\omega)$) shows that it does not depend on the vector $\mathbf{b}$ (resp., $\mathbf{a}$).

Therefore, the functions $h_{\mathbf{a},1}^{\{\mathbf{a},\mathbf{b}\}}(\omega)$ and $h_{1,\mathbf{b}}^{\{\mathbf{a},\mathbf{b}\}}(\omega)$ satisfy the locality condition (L).

Next, for each pair $\{\mathbf{a}, \mathbf{b}\}$ we define a probability measure $\mathbf{P}_{\{\mathbf{a},\mathbf{b}\}} = p = \{p_{ij}; i, j = 0, 1\}$ on the space $W = \{0, 1\} \times \{0, 1\}$ by

$$p_{00} = p_{11} = \frac{1 + \mathbf{ab}}{4}, \quad p_{01} = p_{10} = \frac{1 - \mathbf{ab}}{4}, \quad (3.10.6)$$

and then consider the direct product of the measures $\mathbf{P}^{\{\mathbf{a},\mathbf{b}\}}$ on the space W,

$$\mathbf{P} = \prod_{\{\mathbf{a},\mathbf{b}\}} \mathbf{P}_{\{\mathbf{a},\mathbf{b}\}}. \quad (3.10.7)$$

(Cf. p. 152 (2) of Halmos (1950) for direct product measures.) Direct product measures are frequently used in the theory of hidden variables.

The direct product probability measure $\mathbf{P}$ is common for all pairs $\{\mathbf{a}, \mathbf{b}\}$, i.e., it is a universal measure.

Since $\mathbf{P}$ is a direct product measure, if we change the orientations of Stern-Gerlach magnets, i.e., take a pair $\{\mathbf{a}', \mathbf{b}'\}$ different from $\{\mathbf{a}, \mathbf{b}\}$, the sets of random variables $\{h_{\mathbf{a},1}^{\{\mathbf{a},\mathbf{b}\}}, h_{1,\mathbf{b}}^{\{\mathbf{a},\mathbf{b}\}}, h_{\mathbf{a},1}^{\{\mathbf{a},\mathbf{b}\}} h_{1,\mathbf{b}}^{\{\mathbf{a},\mathbf{b}\}}\}$ and $\{h_{\mathbf{a}',1}^{\{\mathbf{a}',\mathbf{b}'\}}, h_{1,\mathbf{b}'}^{\{\mathbf{a}',\mathbf{b}'\}}, h_{\mathbf{a}',1}^{\{\mathbf{a}',\mathbf{b}'\}} h_{1,\mathbf{b}'}^{\{\mathbf{a}',\mathbf{b}'\}}\}$ are mutually independent.

The expected values of the random variables $h_{\mathbf{a},1}^{\{\mathbf{a},\mathbf{b}\}}(\omega)$ and $h_{1,\mathbf{b}}^{\{\mathbf{a},\mathbf{b}\}}(\omega)$ are

$$\int h_{\mathbf{a},1}^{\{\mathbf{a},\mathbf{b}\}}(\omega) d\mathbf{P}[d\omega] = 0 \quad \text{and} \quad \int h_{1,\mathbf{b}}^{\{\mathbf{a},\mathbf{b}\}}(\omega) \mathbf{P}[d\omega] = 0,$$

and their correlation is

$$\int h_{\mathbf{a},1}^{\{\mathbf{a},\mathbf{b}\}}(\omega) h_{1,\mathbf{b}}^{\{\mathbf{a},\mathbf{b}\}}(\omega) \mathbf{P}[d\omega] = -\mathbf{ab}.$$

Comparing these relations with (3.8.5) and (3.8.6), we see that, for any pair $\{\mathbf{a}, \mathbf{b}\}$,

$$\langle \psi, \boldsymbol{\sigma}\mathbf{a} \otimes \mathbf{1}\psi \rangle = \int h_{\mathbf{a},1}^{\{\mathbf{a},\mathbf{b}\}}(\omega) \mathbf{P}[d\omega],$$

$$\langle \psi, \mathbf{1} \otimes \boldsymbol{\sigma}\mathbf{b}\,\psi \rangle = \int h_{1,\mathbf{b}}^{\{\mathbf{a},\mathbf{b}\}}(\omega) \mathbf{P}[d\omega]$$

and

$$\langle \psi, (\boldsymbol{\sigma}\mathbf{a} \otimes \mathbf{1})\,(\mathbf{1} \otimes \boldsymbol{\sigma}\mathbf{b})\,\psi \rangle = \int h_{\mathbf{a},1}^{\{\mathbf{a},\mathbf{b}\}}(\omega) h_{1,\mathbf{b}}^{\{\mathbf{a},\mathbf{b}\}}(\omega) \mathbf{P}[d\omega]. \qquad (3.10.8)$$

Hence,

$$\{\Omega, \mathbf{P}\,;\, h_{\mathbf{a},1}^{\{\mathbf{a},\mathbf{b}\}},\ h_{1,\mathbf{b}}^{\{\mathbf{a},\mathbf{b}\}},\ h_{\mathbf{a},1}^{\{\mathbf{a},\mathbf{b}\}} h_{1,\mathbf{b}}^{\{\mathbf{a},\mathbf{b}\}};\ \text{pair } \{\mathbf{a},\mathbf{b}\} \text{ is arbitrary}\}$$

is a function representation of $\{\mathbf{F}, \psi\}$. We have already noted that the random variables $h_{\mathbf{a},1}^{\{\mathbf{a},\mathbf{b}\}}(\omega)$ and $h_{1,\mathbf{b}}^{\{\mathbf{a},\mathbf{b}\}}(\omega)$ satisfy the locality condition (L). Therefore, the above function representations are local. The probability measure $\mathbf{P}$ does not depend on the vectors $\{\mathbf{a},\mathbf{b}\}$, it is a universal measure. The proof is complete. □

Hence, for the Experiments 1, 2, 3, 4 discussed in the preceding section, we established the following theorem:

Theorem 3.10.2. *For the Experiments* $1, 2, 3, 4$, *take the set of operators* $\mathbf{F}$ *and the element* ψ *given by formula* (3.8.4). *Then the pair* $\{\mathbf{F}, \psi\}$ *admits the local function representation*

$$\begin{aligned}
\{\Omega, \mathbf{P}\,;\, & h_{\mathbf{a},1}^{\{1\}}, h_{1,\mathbf{b}}^{\{1\}},\ h_{\mathbf{a},1}^{\{1\}} h_{1,\mathbf{b}}^{\{1\}};\ h_{\mathbf{a},1}^{\{2\}}, h_{1,\mathbf{c}}^{\{2\}},\ h_{\mathbf{a},1}^{\{2\}} h_{1,\mathbf{c}}^{\{2\}}; \\
& h_{\mathbf{b},1}^{\{3\}}, h_{1,\mathbf{c}}^{\{3\}},\ h_{\mathbf{b},1}^{\{3\}} h_{1,\mathbf{c}}^{\{3\}};\ h_{\mathbf{b},1}^{\{4\}}, h_{1,\mathbf{b}}^{\{4\}},\ h_{\mathbf{b},1}^{\{4\}} h_{1,\mathbf{b}}^{\{4\}}\}.
\end{aligned} \qquad (3.10.9)$$

The pairs $\{h_{\mathbf{a},1}^{\{1\}}, h_{1,\mathbf{b}}^{\{1\}}\}, \{h_{\mathbf{a},1}^{\{2\}}, h_{1,\mathbf{c}}^{\{2\}}\}, \{h_{\mathbf{b},1}^{\{3\}}, h_{1,\mathbf{c}}^{\{3\}}\}, \{h_{\mathbf{b},1}^{\{4\}}, h_{1,\mathbf{b}}^{\{4\}}\}$ *are local hidden variables.*

Remark. Equality (3.10.8) says that if we compute the left- and right-hand sides independently, then the results coincide. However, the right-hand side can be derived from the left-hand side. Indeed, the left-hand side of (3.10.8) is

$$\begin{aligned}
\langle \psi, (\boldsymbol{\sigma}\mathbf{a} \otimes \mathbf{1})\,(\mathbf{1} \otimes \boldsymbol{\sigma}\mathbf{b})\,\psi \rangle \\
= \frac{1}{2}\Big\{ & (u^{+}\boldsymbol{\sigma}\mathbf{a}u^{+})(u^{-}\boldsymbol{\sigma}\mathbf{b}u^{-}) + (u^{-}\boldsymbol{\sigma}\mathbf{a}u^{-})(u^{+}\boldsymbol{\sigma}\mathbf{b}u^{+}) \\
& - (u^{+}\boldsymbol{\sigma}\mathbf{a}u^{-})(u^{-}\boldsymbol{\sigma}\mathbf{b}u^{+}) - (u^{-}\boldsymbol{\sigma}\mathbf{a}u^{+})(u^{+}\boldsymbol{\sigma}\mathbf{b}u^{-}) \Big\} \\
= -\mathbf{ab},
\end{aligned}$$

which we decompose as

$$-\frac{1+\mathbf{ab}}{4} + \frac{1-\mathbf{ab}}{4} + \frac{1-\mathbf{ab}}{4} - \frac{1+\mathbf{ab}}{4}.$$

In terms of the functions $h_1(i,j)$ and $h_2(i,j)$ defined in (3.10.3) and (3.10.4), the last expression is

$$\begin{aligned}
h_1(0,0)h_2(0,0)\frac{1+\mathbf{ab}}{4} + h_1(0,1)h_2(0,1)\frac{1-\mathbf{ab}}{4} \\
+ h_1(1,0)h_2(1,0)\frac{1-\mathbf{ab}}{4} + h_1(1,1)h_2(1,1)\frac{1+\mathbf{ab}}{4},
\end{aligned}$$

which in view of the definition (3.10.6) of the probability measure $p = \{p_{ij}; i, j = 0, 1\}$ is further equal to

$$h_1(0,0)h_2(0,0)p_{00} + h_1(0,1)h_2(0,1)p_{01}$$
$$+ h_1(1,0)h_2(1,0)p_{10} + h_1(1,1)h_2(1,1)p_{11}.$$

Finally, by (3.10.5) and (3.10.7), the last expression is equal to

$$\int h_{a,1}^{\{a,b\}}(\omega)h_{1,b}^{\{a,b\}}(\omega)\mathbf{P}[d\omega],$$

i.e., with the right-hand side of equality (3.10.8).

3.11 Long-Lasting Controversy and Random Motion

The matrix mechanics developed by Born, Heisenberg and Jordan was a mathematical method to compute energy values. Heisenberg noticed that there is no need to think about the path of the motion of an electron when computing energy values with matrix mechanics, and he insisted that the "path of an electron's motion (trajectory) can not be observed and so it should not be included in the theory." He claimed his uncertainty principle in order to reinforce his argument. Einstein criticized Heisenberg's claim, asserting that "theory decides what will be observed." Matrix mechanics (quantum mechanics) does not have a mathematical structure that would allow it to discuss paths of motion (trajectories).

Confusion of recognition occurs, because Heisenberg discusses "path of motion" using matrix mechanics (quantum mechanics) nonetheless. For example, since the double-slit problem raised by Einstein is related to the path of electron's motion, one can not understand it with matrix dynamics (quantum mechanics). So in double-slit problem "mysterious inexplicable" discussion was done. The double-slit problem for electrons (and photons) was solved by the mechanics of random motion. Experiments using electrons (photons) are explained in Section 1.12 (resp., 1.13).

Well, in the history of quantum theory there was a different way from the matrix mechanics developed by Born, Heisenberg and Jordan. Let us look back on it.

When Bohr created his model of the hydrogen atom, he determined the radius of the circular orbit that an electron traces around the nucleus from the hydrogen spectrum. De Broglie tried to give a logic to it. Suppose the circular orbit lies in the (x, y)-plane. We suppose that electrons move up and down in the z-axis direction and are moving at a constant speed on the circular orbit of the xy-plane. Then, in order that the motion of an electron becomes a stationary motion, when the electron returns to the starting point, the z-axis direction must come to the same place as the departure point. The radius of the circular orbit of the electron is determined from this fact. This is the so-called de Broglie's "wave theory of electron". It is a misunderstanding that "de Broglie insisted that an electron is a wave". He was thinking that "an electron is a particle and make wave motion" (cf. de Broglie (1943)). De Broglie was considering the stationary motion of electrons. Schrödinger extended it and thought about "what kind of wave equation should be followed if the electron performed stationary wave motion". De Broglie did not discuss the wave equation. Schrödinger expressed it as an eigenvalue problem of the wave equation, solved it, and showed that eigenvalue gave the hydrogen spectrum. Thus Bohr's problem of hydrogen atom model was solved.

Schrödinger further advanced, giving not only the stationary equation but a time dependent equation that electron should follow. That is the so-called "Schrödinger equation". Schrödinger advanced one step more and thought that an electron was a wave given by the solution of the Schrödinger equation. This is the so-called "wave theory of electrons" of Schrödinger. (Electromagnetic wave is a "wave" described by Maxwell's wave equation. In the same way, he thought that electron was a "wave" described by Schrödinger wave equation.)

This was a bit over-kill. If so, Schrödinger's electron is in conflict with the fact that "an electron has the basic electric charge e, and this electric charge e is observed always at one point as a solid, and it is never observed spatially spread out."

This is a fatal weak point of the wave theory of electron. Schrödinger noticed its weak point and thought of the theory of random motion of particles to restore the "particle theory" of electrons. That was in 1931 (Schrödinger (1931)), at that time the idea of random motion was taken into quantum theory for the first time. However, he could not complete the theory of random motion, and could not develop his idea further on. That was impossible with mathematics at that time. (The mathematical method that enables it, i.e., theory of stochastic processes was developed after 1950.) And he stopped there. Therefore it was not possible to realize that

"The Schrödinger equation is not a wave equation, but a complex evolution equation, and the Schrödinger function $\psi_t(x) = e^{R(t,x)+iS(t,x)}$ is a complex evolution function, where $R(t,x)$ is the generating function of distribution and $S(t,x)$ is drift-potential."

De Broglie asserted that the "motion of an electron is accompanied by a wave." Let us replace "motion of an electron" by "random motion of an electron" and replace "wave" by "drift potential." Then one recovers a basic statement of the mechanics of random motion

"The random motion $X_t(\omega)$ of an electron carries the drift potential $S(t,x)$."

So the physical picture that de Broglie draws about the motion of an electron is very close to the physical picture in mechanics of random motion. The basic statement above is meant to emphasize the difference between the random motion of electrons and classical Brownian motion: the classical Brownian motion does not possess a drift potential $S(t, x)$.

In matrix mechanics (quantum mechanics), people gave up thinking about the concept of "motion" of an electron, but physicists usually study "motions." Schrödinger considered the motion of waves and assumed that the corresponding wave functions $\psi(t, x)$ are described by his wave equation. Then he thought that the quantity $|\psi(t, x)|^2$ is observable because it represents the intensity of the wave $\psi(t, x)$. Considering the scattering motion of particles, Born claimed that $|\psi(t, x)|^2$ represents the statistical distribution density of particles. At that time, he did not provide an interpretation of the function $\psi(t, x)$ itself. Bohr claimed that the function $\psi(t, x)$ represents a state. (He uses the term "state" vaguely, without providing a precise definition.) Schrödinger and Einstein argued heatedly about this, and controversy continued.

One example is provided by the famous "Schrödinger's cat" experiment. The function $\psi(t, x)$ can be mixed. He refuted Bohr's argument that "function $\psi(t, x)$ represents a state", asserting that there is no mixture of two states where the cat is dead and the cat is alive.

Einstein, Podolsky and Rosen argued that "whether quantum mechanics is an incomplete theory or whether the locality of physical reality holds" is an alternative. They thought that locality holds. Bohr regarded Einstein's argument as an attack against quantum mechanics and immediately rebutted. Einstein ignored Bohr's refutation. We claim that Bohr's argument was wrong, because we have shown in Section 3.10 that locality is an established fact. In other words, Einstein's assertion that physical reality is local and quantum mechanics is incomplete was correct. But I do not know what Einstein's "perfect theory" was. Although he was thinking about a theory like general relativity as the basis of quantum theory, he talked frequently about ensembles (see Nagasawa (2003) pp. 53–56) in describing his opinion. As it is, he may have anticipated the possibility of statistical and probabilistic theory (Brownian motion?)

On the other hand, there is no evidence that Born, Heisenberg and Bohr thought of notions such as the random motion (Brownian motion) of electrons.

Edward Condon visited Born in 1926 and stayed in Göttingen. He published in Vol.15 of Physics Today (1962) a paper entitled "60 years of quantum physics", in which he wrote (partial quote):

"I remember that David Hilbert was lecturing on quantum theory that fall, although he was in very poor health at that time. . . . But that is not the point of my story. What I was going to say is that Hilbert was having a great laugh on Born and Heisenberg and the Göttingen theoretical physicists because when they first discovered matrix mechanics they were having, of course, the same kind of trouble that everybody else had in trying to solve problems and to manipulate and really do things with matrices. So they went to Hilbert for help, and Hilbert said the only times that he had ever had anything to do with matrices was when they came up as a sort of by-product of the eigenvalues of the boundary-value problem of a differential equation. So if you look for the differential equation which has these matrices you can probably do more with that. They had thought it was a goofy idea and that Hilbert did not know what he was talking about, so he was having a lot of fun pointing out to them that they could have discovered Schrödinger's wave mechanics six months earlier if they had paid a little more attention to him."

So Hilbert asserted that matrices arise in dealing with the eigenvalue problem for a differential equation and represent just a mathematical technique.

Schrödinger found his wave equation because he thought that, according to de Broglie's wave theory of matter, the physical entity of an electron was a wave. Since Born and Heisenberg must have thought that electrons are particles, if Born and Heisenberg would have taken Hilbert's advice seriously, and arrived at the idea that small particles could not avoid random motion, then they might have found a wave equation. Born and Heisenberg lacked the way of understanding the problem such that random motion is inevitable in the motion of tiny particles to be handled in quantum theory, but rather it is an inherent property.

We already mentioned that there was an intense controversy between Bohr and Einstein in the 1930s, centered around the uncertainty principle. In that controversy, it has been said that Einstein's assertion was broken by Bohr who adopted the general theory of relativity assuming that the uncertainty principle held. (Cf. Jammer: Chapter 5 of *The Philosophy of Quantum Mechanics.*) This story must be corrected. Bohr's argument was wrong, and Einstein who expressed doubts about uncertainty relations was right, because in Section 3.6 we have proved that the uncertainty principle does not hold. Furthermore, electrons have path equations. So the uncertainty principle is meaningless.

We commented on the controversy regarding the Born-Heisenberg's theory (quantum mechanics) and pointed out some issues to which one should pay attention. Let us add a little more historical explanations.

Remark 1. Born-Heisenberg's theory (quantum mechanics) is an algebraic theory of operators based on the matrices and, as Hilbert asserted, a mathematical tool for calculating eigenvalues (energy). We will explain this:

In classical mechanics we get answers to problems by solving the equation of motion. However, depending on the problem, the answer may be immediately obtained from the conservation law of total energy $\frac{1}{2m}\mathbf{p}^2 + V = E$, where $\frac{1}{2m}\mathbf{p}^2$ is the kinetic energy. For example, you can calculate how much destructive power a falling stone has if you know from what height it starts, so in this case there is no need to solve the equation of motion.

In quantum theory, the calculation of total energy, and especially of kinetic energy was a problem. In Bohr's quantum theory, the method of calculation of the kinetic energy $\frac{1}{2m}\mathbf{p}^2$ was different from that in classical mechanics. It was Heisenberg who noticed this point. Born recognized that Heisenberg's method was a matrix calculation method, namely:

(i) Physical quantities of classical mechanics are replaced by matrices (or self-adjoint operators).

(ii) These matrices (or self-adjoint operators) satisfy commutation relations.

This method was called "quantization." It provides an effective, practical means to derive quantum mechanics from classical mechanics. To do this, one gives the Hamiltonian, i.e., the classical energy of the physical object under study, and one applies the above "quantization" to it.

This is the Born-Heisenberg quantum mechanics. In other words, it was a method based on the law of conservation of energy. Therefore, just as in classical mechanics Newton's equation of motion was not needed in the law of conservation of energy, in quantum mechanics "what kind of equation of motion governed the motion of particles" did not matter.

Unknown at that time, of course, was that the success of this approach was due to the fact that the motion of a particle is governed by a pair of solutions $\phi(t,x) = e^{R(t,x)+S(t,x)}$, $\widehat{\phi}(t,x) = e^{R(t,x)-S(t,x)}$ of the equations of motion (1.4.1) and that the complex evolution function satisfies the Schrödinger equation (1.9.3) (and, needless to say, that these equations are equivalent (Theorem 1.9.1)).

Incidentally, what physicists wanted to know in quantum theory was not only how to calculate energy values. So they added various interpretations to "quantization," enriching it. To give a basic idea:

(a) Matrices (self-adjoint operators) represent physical quantities.

(b) Physical quantities represented by non-commuting operators cannot be observed simultaneously.

(c) Uncertainty principle, complementarity principle.

(d) The Schrödinger function expresses a state.

Einstein disagreed with the idea that "operators are physical quantities", and there was controversy on the subject, but it was not settled. All the items (a), (b), (c) and (d) are now rejected by our theory: (a) is invalid, while (b), (c) and (d) are errors.

Let us note one more thing. In this theory (quantum mechanics), it is not known what the nature of the physical object being handled is. We started with the Hamiltonian of a classical particle, but since the momentum was replaced by an operator (matrix), it was not sure whether the physical object being handled is a particle or not. Furthermore, since the Schrödinger's wave equation was accepted in the Born-Heisenberg quantum mechanics, it became increasingly unclear whether an electron is a particle or a wave.

In summary, sticking to the interpretation of physical quantities as operators, the Born-Heisenberg theory (quantum mechanics) did not reach the understanding that microscopic particles perform a Brownian motion. In other words, quantum mechanics was a theory which improved the "method of calculation of energy values" in the old quantum theory (see Tomonaga (1952)). That prevented people to carry out a discussion about how particles move (dynamics).

Feynman (1948) was an exception. He followed the analogy with Brownian motion and tried to construct a complex-valued measure on the space of paths. He did not succeed, but his method was effective for a perturbative calculus.

Remark 2. On the other hand, Schrödinger considered that his equation described waves. That was the wave theory of electrons. When Born adopted the Schrödinger equation in matrix mechanics, the idea that the equation described a wave was (implicitly) adopted. This created a conceptual confusion of "compatibility between particle and wave behavior (Bohr's complementarity)." In fact, the equation derived by Schrödinger was not a wave equation, but a complex evolution equation, which described the evolution of random motion. We clarified this point, thereby divorcing the concept of "wave" from the Schrödinger function $\psi_t(x) = e^{R(t,x)+iS(t,x)}$.

Remark 3. Commenting on the controversial aspects of quantum theory, we briefly talked about quantum mechanics in *Remark* 1 and wave mechanics in *Remark* 2, and explained that the controversies in both theories were resolved by the mechanics of random motion. According to the mechanics of random motion, one can specify an "entrance function" and an "exit function" for the motion of an electron (quantum particle). This was impossible with classical Markov processes or conventional quantum theory. We explained this in the first chapter. In Sections 1.12 and 1.13, we showed that the double-slit problem can be resolved theoretically by the mechanics of random motion, something that was not possible with quantum mechanics. (Note that being able to specify the "exit function" is the strong feature of the mechanics of random motion.) The application of the mechanics of random motion was explained in Chapter 2. We demonstrated that by analyzing the paths of motion (trajectories) using the path equation we can predict new phenomena. For instance, in Section 2.6, we analyzed the path of random motion in a space-time domain and showed that Einstein's classical relativity theory does not hold (particles may leave the light cone). This is not predictable in the Born-Heisenberg theory (quantum mechanics).

To summarize briefly, the mechanics of random motion is a theory different from quantum mechanics. It is based on the random hypothesis, and is more encompassing than quantum mechanics. It made it possible to analyze the path of motion of particles, and it dispelled many of the mysteries affecting quantum mechanics.

Appendix 1. Quantum theory contained two tenets that are incompatible with the mechanics of random motion. One is the claim that "there is no path of motion of an electron". Another is the claim that "quantum mechanics is "the only" theory." The former is based on Heisenberg's inequality. The latter is based on Bell's inequality. Let us comment on them a bit.

We begin with the first claim.

Let us assume that there is something that can be called the path of an electron (trajectory). Then, in order to describe it, we must be able to determine, at a space-time point, the position and the velocity or momentum of the electron.

"However, the Heisenberg-Kennard inequality, that is, the uncertainty principle, holds. Then the position and the momentum cannot be determined simultaneously. Consequently, the path of motion (trajectory) cannot be defined, that is, it does not exist."

This is Heisenberg's claim, based on the uncertainty principle: "there is no path for the motion of electrons."

Certainly, quantum mechanics cannot discuss the concept of the path of motion of an electron. Another theory is necessary, and the mechanics of random motion is such a theory: in it "the path of the motion" can be defined.

For simplicity, assuming that the vector potential $\mathbf{b}(s, x) = 0$ and starting at time $t = 0$, the path is described by the (path) equation

$$X_t = X_0 + \sigma B_t + \int_0^t \mathbf{a}(s, X_s)ds.$$

This equation involves the position x at each time s and the drift vector $\mathbf{a}(s, x) = \sigma^2 \nabla \left(R(s, x) + S(s, x) \right)$, or, alternatively, the position x at each time s and the momentum $m\mathbf{a}(s, x) = \hbar \nabla \left(R(s, x) + S(s, x) \right)$.

This fact has nothing to do with whether the Heisenberg-Kennard's inequality (equation of the uncertainty principle) holds or not.

For instance, in the Pauling-Wilson example given in Section 3.5, we considered the function

$$\psi = A \exp \left(-\frac{(x - x_0)^2}{4\delta^2} + i \frac{p_0 (x - x_0)}{\hbar} \right).$$

In this example, the Heisenberg-Kennard inequality holds, and since

$$h_p = \hbar \left(\frac{\partial R}{\partial x} + \frac{\partial S}{\partial x} \right) = -\frac{\hbar}{2\delta^2} (x - x_0) + p_0,$$

the path equation is

$$X_t = X_0 + \sigma B_t + \int_0^t \left(-\frac{\hbar}{2m\delta^2} (X_s - x_0) + \frac{1}{m}p_0 \right) ds.$$

That is, even though the Heisenberg-Kennard inequality (expressing the uncertainty principle) holds, the path of motion (trajectory) exists. Therefore, Heisenberg's claim that "there is no path of the motion of electrons because of the uncertainty principle" is incorrect.

Now, on the second tenet, that "quantum mechanics is the only theory." Let us assume that there is a valid theory other than quantum mechanics. That theory should not conflict at least with the conclusions reached by means of quantum mechanics. As candidate of such a theory one should keep in mind the theory of hidden variables.

According to Bell's argument, "local hidden variables cannot reproduce correlations in quantum mechanics."

If so, since quantum mechanics is a correct theory, "the theory of hidden variables", which cannot reproduce the conclusions in quantum mechanics, is not a valid theory; that is, "quantum mechanics is the only theory."

As we have shown already in Sections 3.9 and 3.10, Bell's assertion which is the premise of this claim is wrong.

Appendix 2. Here we make a remark on the so-called stochastic mechanics of Nelson (1966), although it is a different story from that we have discussed so far.

Using conditional expectations, Nelson considers a stochastic process $X(t)$, introduced the quantities

$$DX(t) = \lim_{h \downarrow 0} \mathbf{P} \left[\frac{X(t+h) - X(t)}{h} \mid X(t) \right] \tag{3.11.1}$$

$$\widehat{D}X(t) = \lim_{h \downarrow 0} \mathbf{P} \left[\frac{X(t) - X(t-h)}{h} \mid X(t) \right], \tag{3.11.2}$$

and adopted

$$\frac{1}{2}(D\widehat{D}X(t) + \widehat{D}DX(t)) = -\frac{1}{m}\nabla V(X(t)) \tag{3.11.3}$$

as the equation of motion of the process $X(t)$.

Then, he claimed that the quantity $DX(t)$ in (3.11.3) can be regarded as the acceleration of the process $X(t)$ and that equation (3.11.3) is the stochastic Newton equation of motion for $X(t)$. He showed that the Schrödinger equation can be derived from that equation.

A brief examination makes it clear that Nelson's equation of motion (3.11.3) is quite different from the equations of motion (1.4.1) we adopted. His equation (3.11.3) is nonlinear, and it is not easy to solve and find the stochastic process $X(t)$; concerning this, see Carlen (2006) and Faris (2006). Since Nelson's equation of motion plays no role in our mechanics of random motion, we will merely remark that our equations of motion (1.4.1) involve the notions of entrance and exit functions (the importance of which was demonstrated in the text), whereas Nelson's equation of motion (3.11.3) does not.

The equation of motions (1.4.1) can be extended to relativistic motion. In collaboration with Hirosi Tanaka, we have shown that this generates a purely jump process that jumps randomly over the path of continuous random motion; for details, refer to Chapters 7 through 10 of Nagasawa (2000).

Chapter 4

Markov Processes

In the mechanics of random motion, Markov processes are responsible for the kinematics. In Section 4.1, we first introduce the Markov processes, which are a powerful tool in the study of the mechanics of random motion, and show that Markov processes determine semigroups of operators. In Section 4.2, we describe the transformation of a probability measure by a multiplicative functional, and we prove the Kac and Maruyama-Girsanov formulas. In Section 4.3, we show that changing the time scale alters the severity of randomness. In Section 4.4 we discuss the duality and time reversal of Markov processes. In Section 4.5 we define the "terminal time" and explain time reversal from the terminal time. Finally, in Section 4.6, we discuss the time reversal of the random motion governed by the equations of motion.

4.1 Time-Homogeneous Markov Proces-ses

Consider a particle that moves in a certain space S, called a state space. For example, S can be the d-dimensional Euclidean space $\mathbf{R}^d$. Calling S "state space" is just a custom in the theory of Markov processes; the term "state" has no particular meaning. A Markov process evolving in the state space S is denoted by $\{X_t, t \geq 0; \mathcal{F}_t, \mathbf{P}_x, x \in S\}$.

If the transition probability $P(s, x, t, B)$ of the Markov process depends only on the difference of the two times t and s, that is, if there exists a function $P(t, x, B)$ of three variable such that

$$P(s, x, t, B) = P(t - s, x, B), \quad s \leq t,$$

then the Markov process $\{X_t, t \geq 0; \mathcal{F}_t, \mathbf{P}_x, x \in S\}$ is said to be **time-homogeneous**.

$P(t, x, B)$ is the probability that a particle which left from a point $x \in S$ reaches a region B after time t. We will call $P(t, x, B)$ **"time-homogeneous transition probability"**. To emphasize that the transition probability is a probability

© The Author(s), under exclusive license to Springer Nature Switzerland AG 2021
M. Nagasawa, *Markov Processes and Quantum Theory*,
Monographs in Mathematics 109, https://doi.org/10.1007/978-3-030-62688-4_4

measure, we write
$$P(t, x, dy) = \mathbf{P}_x[X_t \in dy]; \tag{4.1.1}$$
this is the probability that a particle that leaves from the point x lies in a small region dy after time t.

In this chapter, unlike the previous one, we deal mainly with time-homogeneous Markov processes, but we will explain in Section 4.6 that this is not a limitation.

Since this section deals only with time-homogeneous Markov processes, we simply call $\{X_t, t \geq 0; \mathcal{F}_t, \mathbf{P}_x\}$ a Markov process, and call $P(t, x, B)$ its transition probability.

By definition, the Markov property means that
$$\mathbf{P}_x \left[Gf\left(X_{t+s}\right) \right] = \mathbf{P}_x \left[G\mathbf{P}_{X_s} \left[f\left(X_t\right) \right] \right], \tag{4.1.2}$$

for any bounded $\mathcal{F}_s$-measurable function G. As a special case of relation (4.1.2), we have
$$\mathbf{P}_x \left[g\left(X_s\right) f\left(X_{t+s}\right) \right] = \mathbf{P}_x \left[g\left(X_s\right) \mathbf{P}_{X_s} \left[f\left(X_t\right) \right] \right]. \tag{4.1.3}$$
The choice $g \equiv 1$ in (4.1.3) gives
$$\mathbf{P}_x \left[f\left(X_{t+s}\right) \right] = \mathbf{P}_x \left[\mathbf{P}_{X_s} \left[f\left(X_t\right) \right] \right].$$

Moreover, if we take $f(x) = \mathbf{1}_B(x)$, then by (4.1.1) the left-hand side and right-hand side become
$$\mathbf{P}_x \left[X_{t+s} \in B \right] = P(s + t, x, B),$$
and
$$\mathbf{P}_x \left[\mathbf{P}_{X_s}[X_t \in B] \right] = \int \mathbf{P}_x[X_s \in dy] P(t, y, B)$$
$$= \int P\left(s, x, dy\right) P\left(t, y, B\right),$$

respectively. We arrived at the Chapman-Kolmogorov equation
$$P\left(s + t, x, B\right) = \int P\left(s, x, dy\right) P\left(t, y, B\right). \tag{4.1.4}$$

Thus, the Markov property (4.1.2) implies the Chapman- Kolmogorov equation (4.1.4).

In our discussion of Markov processes we will often apply methods and results of functional analysis.

Let the state space S be locally compact, and let $C_0(S)$ denote the space of continuous functions on S that vanish at infinity. We define an operator P_t on $C_0(S)$ by
$$P_t f(x) = \mathbf{P}_x \left[f\left(X_t\right) \right],$$

and assume that the following continuity condition is satisfied:

$$\lim_{t \to 0} P_t f(x) = f(x), \quad f \in C_0(S).$$

In terms of the transition probability $P(t, x, dy) = \mathbf{P}_x[X_t \in dy]$,

$$P_t f(x) = \int P(t, x, dy) f(y).$$

Consequently, the Chapman-Kolmogorov equation (4.1.4) can be recast as

$$P_{t+s} f(x) = P_s P_t f(x), \quad s, \ t \geq 0. \tag{4.1.5}$$

Therefore, P_t is a semigroup of operators on the function space $C_0(S)$, and so we can assert that the semigroup property (4.1.5) and the Chapman-Kolmogorov equation (4.1.4) are equivalent.

So, the Markov property of a stochastic process is expressed by the semi-group property of the family operators P_t, which is a functional analysis concept. This enables us to apply functional analysis in the study of Markov processes. The effectiveness of this approach was first demonstrated in 1954 in an paper by Feller, and subsequently the theory of Markov processes developed rapidly. See, for instance, Itô and McKean (1965).

Remarks.

(i) Semigroups of operators on Banach spaces were studied, e.g., by Yosida (1948). For example, the space $C_0(S)$ of continuous functions considered in this chapter is a Banach space with the norm

$$\|f\| = \sup_{x \in S} |f(x)|, \quad f \in C_0(S).$$

(ii) Definition of Banach spaces: A linear space B equipped with a norm $\|f\|$, $f \in B$, is called a Banach space if B is complete with respect to the distance $\rho(f - g) = \|f - g\|$, $f, g \in B$ (i.e., every Cauchy sequence converges).

(iii) *A semigroup of operators* $\{P_t, t \geq 0\}$ *on* $C_0(S)$ *such that* $P_t f \geq 0$ *for all* $f \geq 0$ *and* $\|P_t\| = 1$ *determines a time-homogeneous Markov process on* S.

Proof. Suppose the semigroup P_t on $C_0(S)$ is such that $P_t f \geq 0$ for all $f \geq 0$ and $\|P_t\| = 1$. Set

$$\ell_{t,x}(f) = P_t f(x), \quad \text{for } f \in C_0(S).$$

Then $\ell_{t,x}$ is a bounded linear functional on the Banach space $C_0(S)$, and by the Riesz-Markov-Kakutani Theorem (cf. Yosida (1965)) there exists a probability measure $P(t, x, dy)$ such that

$$\ell_{t,x}(f) = \int P(t, x, dy) f(y).$$

Now $P(t, x, dy)$ is the transition probability of a Markov process. Therefore, the semigroup of operators $\{P_t, t \geq 0\}$ on $C_0(S)$ determines a time-homogeneous Markov process on S, as claimed.   □

For time-homogeneous Markov processes, we define the shift operator θ_s acting on paths $\omega(t)$ by

$$\theta_s \omega(t) = \omega(s + t).$$

The shifted path $\theta_s \omega(t)$ is the view of the path $\omega(t)$ from the time s on. Then, since

$$X_{t+s}(\omega) = \omega(t + s) = \theta_s \omega(t) = X_t(\theta_s \omega),$$

the Markov property (4.1.2) can be written as

$$\mathbf{P}_x \left[Gf\left(X_t\left(\theta_s \omega \right) \right) \right] = \mathbf{P}_x \left[G \mathbf{P}_{X_s} \left[f\left(X_t \right) \right] \right], \tag{4.1.6}$$

where on the left $\theta_s \omega$ is the path from the time s, and the position at the time t of the path $\theta_s \omega$ is $X_t(\theta_s \omega)$.

This means that $f(X_t(\theta_s \omega))$ under the condition G until time s is equal to the expected value $\mathbf{P}_{X_s}[f(X_t)]$ when the process started from X_s. This is the meaning of the Markov property (4.1.6).

Henceforth, we will use this form (4.1.6) in calculations involving the Markov property.

In the study of semigroups of operators, a useful role is played by resolvent operators. If $\{P_t\}$ is a semigroup of operators on $C_0(S)$, its resolvent operator G_α is defined by the formula

$$G_\alpha f(x) = \int_0^\infty e^{-\alpha t} P_t f(x)\, dt = \mathbf{P}_x \left[\int_0^\infty e^{-\alpha t} f(X_t)\, dt \right]. \tag{4.1.7}$$

The meaning of the resolvent operator will be explained in Proposition 4.1.2.

Proposition 4.1.1. *The resolvent operator satisfies the resolvent equation*

$$G_\alpha f(x) - G_\beta f(x) + (\alpha - \beta) G_\alpha G_\beta f(x) = 0. \tag{4.1.8}$$

Proof. By the semigroup property,

$$(\alpha - \beta) G_\alpha G_\beta f(x)$$

$$= (\alpha - \beta) \int_0^\infty e^{-\alpha s} ds \int_0^\infty e^{-\beta t} P_s P_t f(x)\, dt$$

$$= (\alpha - \beta) \int_0^\infty e^{-\alpha s} ds \int_0^\infty e^{-\beta t} P_{s+t} f(x) dt$$

$$= (\alpha - \beta) \int_0^\infty e^{-\alpha s} ds \int_s^\infty e^{-\beta(r-s)} P_r f(x) dr$$

$$= (\alpha - \beta) \int_0^\infty e^{-(\alpha - \beta)s} ds \left(\int_0^\infty e^{-\beta r} P_r f(x) dr - \int_0^s e^{-\beta r} P_r f(x) dr \right)$$

$$= G_\beta f(x) - (\alpha - \beta) \int_0^\infty e^{-(\alpha - \beta)s} ds \int_0^s e^{-\beta r} P_r f(x) dr,$$

which upon exchanging the order of integration in the second term is equal to

$$G_\beta f(x) - G_\alpha f(x),$$

as claimed. □

The generator A of a semigroup of operators P_t is defined for $f \in C_0(S)$ by

$$Af(x) = \lim_{t \downarrow 0} \frac{1}{t} (P_t f(x) - f(x)) \tag{4.1.9}$$

whenever the limit on the right-hand side exists. The set of all functions $f \in C_0(S)$ for which the limit exists is called the domain of the generator A and is denoted by $\mathcal{D}(A)$.

Example. The semigroup corresponding to the one-dimensional Brownian motion is

$$P_t f(x) = \int_{\mathbf{R}} \frac{1}{\sqrt{2\pi t}} \exp\left(-\frac{(y-x)^2}{2t} \right) f(y) dy.$$

Its the generator A acts as

$$Af(x) = \lim_{t \downarrow 0} \frac{1}{t} (P_t f(x) - f(x)) = \frac{1}{2} \frac{\partial^2}{\partial x^2} f.$$

That is, $A = \frac{1}{2} \frac{\partial^2}{\partial x^2}$, and the domain $\mathcal{D}(A)$ is the set of all twice continuously differentiable functions $f \in C_0(S)$. For other examples, see Section 2.8.

Proposition 4.1.2. *The generator A and the resolvent operator G_α are connected by the relation*

$$G_\alpha^{-1} = \alpha - A, \tag{4.1.10}$$

with

$$\mathcal{D}(A) = \{G_\alpha f(x); f \in C_0(S)\}. \tag{4.1.11}$$

The domain $\mathcal{D}(A)$ is dense in the space $C_0(S)$, and $u = G_\alpha f$ gives a solution of the equation

$$(\alpha - A)u = f. \tag{4.1.12}$$

Proof. First we show that $\mathcal{D}(A) \supset \{G_\alpha f(x); f \in C_0(S)\}$.

Let $g = G_\alpha f(x)$. Since

$$P_t g = P_t G_\alpha f(x) = P_t \int_0^\infty e^{-\alpha s} P_s f(x) ds = \int_0^\infty e^{-\alpha s} P_t P_s f(x) ds$$

$$= \int_0^\infty e^{-\alpha s} P_{t+s} f(x) ds = \int_t^\infty e^{-\alpha(r-t)} P_r f(x) dr,$$

we have

$$P_t G_\alpha f(x) - G_\alpha(x) = \left(e^{\alpha t} - 1\right) G_\alpha f(x) - e^{\alpha t} \int_0^t e^{-\alpha r} P_r f(x) dr.$$

Therefore,

$$\frac{P_t G_\alpha f(x) - G_\alpha f(x)}{t} = \frac{e^{\alpha t} - 1}{t} G_\alpha f(x) - e^{\alpha t} \frac{1}{t} \int_0^t e^{-\alpha r} P_r f(x) dr.$$

Let $t \downarrow 0$. Then, since the second term on the right-hand side converges to $-f(x)$, we get

$$AG_\alpha f(x) = \alpha G_\alpha f(x) - f(x).$$

This shows that $g = G_\alpha f \in \mathcal{D}(A)$ and $\alpha g - Ag = f$.

Conversely, take any $g \in \mathcal{D}(A)$ and set $f = \alpha g - Ag$. Then, as shown above, $g_0 = G_\alpha f$ satisfies $f = \alpha g_0 - Ag_0$. Since the operator $\alpha - A$ is one-to-one, $g = g_0$. Thus, we proved that (4.1.11) holds. Since $f = G_\alpha^{-1} G_\alpha f$, the right-hand side of the above equation is $(\alpha - G_\alpha^{-1}) G_\alpha f(x)$. Hence

$$AG_\alpha f(x) = (\alpha - G_\alpha^{-1}) G_\alpha f(x),$$

i.e., (4.1.10) holds. Since $\lim_{\alpha \to \infty} \alpha G_\alpha f(x) = f(x)$, we see that the domain $\mathcal{D}(A)$ in (4.1.11) is dense in the space $C_0(S)$. Moreover, $u = G_\alpha f$ is a solution of equation $(\alpha - A)u = f$. $\square$

Remark. Since the semi-group P_t satisfies

$$\frac{d}{dt} P_t f(x) = AP_t f(x),$$

one often writes $P_t f(x) = e^{At} f(x)$.

Comments. The theory of semigroups of operators of K. Yosida and the theory of the stochastic differential equation of K. Itô played an important role in the study of Markov process since 1950. See, e.g., K. For the theory of operator semigroups, see, e.g., Yosida (1948).

4.2 Transformations by M-Functionals

In this section, we consider applications to random motions governed by the equations of motions, so we treat the case of time-inhomogeneous processes.

We will discuss the following two problems.

Problem 1. For simplicity, we assume that σ^2 is a constant and the vector potential $\mathbf{b} = 0$. Consider the first of the equations of motion (1.4.1):

$$\frac{\partial \phi}{\partial t} + \frac{1}{2} \sigma^2 \triangle \phi + c(x)\phi = 0.$$

This is the equation of Brownian motion

$$\frac{\partial \phi}{\partial t} + \frac{1}{2}\sigma^2 \triangle \phi = 0$$

with an added potential function $c(x)$. Therefore, it can be obtained through a transformation of the probability measure $\mathbf{P}_{(s,x)}$ of the Brownian motion. This was an idea of Kac (1951), inspired by the work of Feynman (1948) on the Schrödinger equation.

Remark. Feynman (1948) discovered a method to obtain a solution of the Schrödinger equation with a potential function $c(x)$,

$$i\frac{\partial \psi}{\partial t} + \frac{1}{2}\sigma^2 \triangle \psi + c(x)\psi = 0,$$

from the Schrödinger equation without the potential function,

$$i\frac{\partial \psi}{\partial t} + \frac{1}{2}\sigma^2 \triangle \psi = 0.$$

This provided in a way a new type of formula, not previously known in the theory of partial differential equations.

Apparently, Kac thought about the above problem after hearing an inaugural lecture that Feynman delivered at Cornell University.

Problem 2. The path equation

$$X_t = X_a + \sigma B_{t-a} + \int_a^t \mathbf{a}\left(s, X_s\right) ds$$

is an equation with a drift term $\mathbf{a}(s, x)$ added to the Brownian motion

$$X_t = X_a + \sigma B_{t-a}.$$

Therefore, the probability measure $\mathbf{Q}_{(s,x)}$ of the motion with drift can be obtained from the probability measure $\mathbf{P}_{(s,x)}$ of the Brownian motion through a transformation. This problem was considered by Maruyama (1954) and Girsanov (1960).

The two problems formulated above are solved by transforming the Brownian motion by means of a so-called "multiplicative functional", as explained below.

Let $\{X_t; \mathbf{P}_{(s,x)}, (s,x) \in [a,b] \times \mathbf{R}^d\}$ be a Markov process, and let $\mathcal{F}_s^t$ be a σ-algebra. Take

$$r \le r_1 \le r_2 \le \cdots \le s \le s_1 \le s_2 \le \cdots \le t \le b.$$

Then generalizing the proof of the Markov property (1.2.14), we have

$$\mathbf{P}_{(r,x)}[g(X_{r_1},\ldots,X_s)f(X_{s_1},\ldots,X_t)]$$
$$= \mathbf{P}_{(r,x)}\left[g(X_{r_1},\ldots,X_s)\mathbf{P}_{(s,X_s)}[f(X_{s_1},\ldots,X_t)]\right],$$

which implies "the generalized Markov property"

$$\mathbf{P}_{(r,x)}[GF] = \mathbf{P}_{(r,x)}\left[G\,\mathbf{P}_{(s,X_s)}[F]\right], \tag{4.2.1}$$

where G is $\mathcal{F}_r^s$-measurable and F is $\mathcal{F}_s^t$-measurable.

Definition 4.2.1. Let $\mathcal{F}_s^t$ be a σ-algebra which is increasing in t. A system of random variables $\{M_s^t(\omega)\}$ such that

(i) $M_s^t(\omega)$ is $\mathcal{F}_s^t$-measurable;

(ii) $M_s^t(\omega) = M_s^r(\omega)\,M_r^t(\omega), \quad s \le r \le t,$

will be called a **"multiplicative functional"**. If, in addition,

(iii) $\mathbf{P}_{(s,x)}[M_s^t] = 1,$

then the multiplicative functional$\{M_s^t(\omega)\}$ is said to be **"normalized"**.

Let $\{M_s^t(\omega)\}$ be a normalized multiplicative functional. We define a new probability measure $\mathbf{Q}_{(s,x)}$ by

$$\mathbf{Q}_{(s,x)}[F] = \mathbf{P}_{(s,x)}\left[M_s^b F\right], \tag{4.2.2}$$

for any bounded $\mathcal{F}_s^b$-measurable function F. Thus,

$$\frac{d\mathbf{Q}_{(s,x)}}{d\mathbf{P}_{(s,x)}} = M_s^b,$$

which we will often write in the form $\mathbf{Q}_{(s,x)} = \mathbf{P}_{(s,x)}M_s^b$.

Theorem 4.2.1. *If the probability measure $\mathbf{P}_{(s,x)}$ of a Markov process $\{X_t; \mathbf{P}_{(s,x)}\}$ is replaced by the probability measure $\mathbf{Q}_{(s,x)} = \mathbf{P}_{(s,x)}M_s^b$ defined by equation (4.2.2), then one obtains a new stochastic process $\{X_t; \mathbf{Q}_{(s,x)}\}$. The transformed stochastic process $\{X_t; \mathbf{Q}_{(s,x)}\}$ is a Markov process.*

Proof. Let $a \le s \le r \le t \le b$ and let F be a bounded $\mathcal{F}_s^t$-measurable function. Since M_t^b is $\mathcal{F}_t^b$-measurable, the generalized Markov property (4.2.1) of X_t implies that

$$\mathbf{P}_{(s,x)}\left[M_s^b F\right] = \mathbf{P}_{(s,x)}\left[M_s^t M_t^b F\right] = \mathbf{P}_{(s,x)}\left[M_s^t F\mathbf{P}_{(t,X_t)}\left[M_t^b\right]\right],$$

which because of the normalization condition $\mathbf{P}_{(t,X_t)}\left[M_t^b\right] = 1$ is equal to $\mathbf{P}_{(s,x)}\left[M_s^t F\right]$. That is, if F is $\mathcal{F}_s^t$-measurable, then

$$\mathbf{Q}_{(s,x)}[F] = \mathbf{P}_{(s,x)}\left[M_s^b F\right] = \mathbf{P}_{(s,x)}\left[M_s^t F\right].$$

Next, for any bounded $\mathcal{F}_s^r$-measurable function G,

$$\mathbf{Q}_{(s,x)}[Gf(t,X_t)] = \mathbf{P}_{(s,x)}\left[M_s^t Gf(t,X_t)\right] = \mathbf{P}_{(s,x)}\left[M_s^r GM_r^t f(t,X_t)\right],$$

which in view of the fact that $M_r^t f(t, X_t)$ is $\mathcal{F}_r^t$-measurable and the generalized Markov property (4.2.1) of $\{X_t; \mathbf{P}_{(s,x)}\}$, is equal to

$$\mathbf{P}_{(s,x)} \left[M_s^r G \mathbf{P}_{(r,X_r)} \left[M_r^t f(t, X_t) \right] \right],$$

and hence, by the definition of the probability measure $\mathbf{Q}_{(s,x)}$, to

$$\mathbf{Q}_{(s,x)} \left[G \mathbf{Q}_{(r,X_r)} \left[f(t, X_t) \right] \right].$$

Thus,

$$\mathbf{Q}_{(s,x)}[G f(t, X_t)] = \mathbf{Q}_{(s,x)} \left[G \mathbf{Q}_{(r,X_r)} \left[f(t, X_t) \right] \right],$$

i.e., $\{X_t; \mathbf{Q}_{(s,x)}, (s,x) \in [a, b] \times \mathbf{R}^d\}$ is a Markov process, as claimed. $\qquad\square$

Example 1. The Kac functional.

Definition 4.2.2. For a bounded scalar function $c(s, x)$, the Kac functional is defined by

$$K_s^t = \exp \left(\int_s^t c(r, X_r) \, dr \right).$$

Remark. The scalar function $c(s, x)$ does not need to be bounded: it suffices that $c^+(s, x) = c(s, x) \vee 0$ be bounded. Therefore, the Coulomb potential is allowed. (For more general cases, see Chapter 6 of Nagasawa (1993).)

Now, even though we can define a measure $\mathbf{P}_{(s,x)}^K$ by $\mathbf{P}_{(s,x)}^K = \mathbf{P}_{(s,x)} K_s^b$, this is not a probability measure because $\mathbf{P}_{(s,x)}^K[1] \neq 1$. Therefore, we cannot directly associate a stochastic process with the measure $\mathbf{P}_{(s,x)}^K$. We will explain what should be done in this case later on.

In the following, σ will be a constant. We assume that

$$X_t = x + \sigma B_{t-a} + \int_a^t \sigma^2 \mathbf{b}(s, X_s) \, ds$$

determines a stochastic process $\{X_t; \mathbf{P}_{(s,x)}, (s,x) \in [a, b] \times \mathbf{R}^d\}$. In other words, the stochastic process X_t is determined by the equation

$$\frac{\partial u}{\partial s} + \frac{1}{2} \sigma^2 \Delta u + \sigma^2 \mathbf{b}(s, x) \cdot \nabla u = 0.$$

(In the following story, the drift $\mathbf{b}(s, x)$ does not play an essential role. So you may simply write $X_t = x + \sigma B_{t-a}$.)

Define an operator $P_{s,t}$ by

$$P_{s,t} f(x) = \mathbf{P}_{(s,x)} \left[K_s^t f(X_t) \right].$$

Then for $s < r < t$ we have

$$P_{s,t}f(x) = \mathbf{P}_{(s,x)}\left[K_s^r K_r^t f(X_t)\right] = \mathbf{P}_{(s,x)}\left[K_s^r \mathbf{P}_{(r,X_r)}\left[K_r^t f(X_t)\right]\right]$$
$$= \mathbf{P}_{(s,x)}\left[K_s^r P_{r,t}f(X_r)\right] = P_{s,r}P_{r,t}f(x).$$

That is,

$$P_{s,t} = P_{s,r}P_{r,t}.$$

Next, let us find what equation the function

$$u(s,x) = \mathbf{P}_{(s,x)}\left[K_s^b f(X_b)\right]$$

satisfies. To this end we use Itô's formula (1.3.10) to compute

$$d\left(K_s^t f(X_t)\right) = d\left(K_s^t\right) f(X_t) + K_s^t df(X_t) + d\left(K_s^t\right) df(X_t).$$

Since

$$d\left(K_s^t\right) = K_s^t c(t, X_t)\, dt,$$

and

$$df(X_t) = \nabla f(X_t) \cdot \left(\sigma dB_t + \sigma^2 \mathbf{b}(t, X_t)\, dt\right) + \frac{1}{2}\sigma^2 \triangle f(X_t)\, dt,$$

we have

$$d\left(K_s^t f(X_t)\right) = K_s^t \left(\frac{1}{2}\sigma^2 \triangle + \sigma^2 \mathbf{b}(t, X_t) \cdot \nabla + c(t, X_t)\right) f(X_t)\, dt$$
$$+ K_s^t \nabla f(X_t) \cdot \sigma dB_t.$$

Let us integrate both sides of this equality. Then, since the expectation of the stochastic integral vanishes, we have

$$\mathbf{P}_{(s,x)}\left[K_s^t f(X_t)\right] - f(x) = \mathbf{P}_{(s,x)}\left[\int_s^t K_s^r A(r) f(X_r)\, dr\right], \qquad (4.2.3)$$

where

$$A(r) = \frac{1}{2}\sigma^2 \triangle + \sigma^2 \mathbf{b}(r, x) \cdot \nabla + c(r, x)$$

Hence, if

$$u(s,x) = P_{s,b}f(x) = \mathbf{P}_{(s,x)}\left[K_s^b f(X_b)\right],$$

then

$$u(s,x) = \mathbf{P}_{(s,x)}\left[K_s^{s+h}\mathbf{P}_{(s+h, X_{s+h})}\left[K_{s+h}^b f(X_b)\right]\right]$$
$$= \mathbf{P}_{(s,x)}\left[K_s^{s+h} u(s+h, X_{s+h})\right],$$

and so

$$u(s,x) - u(s+h, x) = \mathbf{P}_{(s,x)}\left[K_s^{s+h} u(s+h, X_{s+h})\right] - u(s+h, x).$$

On the right-hand side we use formula (4.2.3) with $f(x) = u(s+h,x)$ and $t = s+h$. Since

$$u(s,x) - u(s+h,x) = \mathbf{P}_{(s,x)}\left[\int_s^{s+h} K_s^r A(r) u(s+h,X_r) dr\right],$$

we have

$$-\lim_{h\downarrow 0}\frac{u(s+h,x) - u(s,x)}{h} = \lim_{h\downarrow 0}\frac{1}{h}\mathbf{P}_{(s,x)}\left[\int_s^{s+h} K_s^r A(r) u(s+h,X_r) dr\right].$$

The right-hand side is equal to

$$A(s)u = \frac{1}{2}\sigma^2\triangle u + \sigma^2\mathbf{b}(s,x)\cdot\nabla u + c(s,x)u.$$

Let us write these conclusions as a theorem.

Theorem 4.2.2 (Kac's formula). *Consider the Kac functional*

$$K_s^t = \exp\left(\int_s^t c(r,X_r)dr\right)$$

and define $u(s,x) = \mathbf{P}_{(s,x)}\left[K_s^b f(X_b)\right]$, *that is,*

$$u(s,x) = \mathbf{P}_{(s,x)}\left[\exp\left(\int_s^b c(r,X_r)dr\right)f(X_b)\right]. \tag{4.2.4}$$

Then $u(s,x)$ *satisfies the equation*

$$\frac{\partial u}{\partial s} + \frac{1}{2}\sigma^2\triangle u + \sigma^2\mathbf{b}(s,x)\cdot\nabla u + c(s,x)u = 0.$$

Formula (4.2.4) is called the "Kac formula".

Comments. 1. Thus, when we want to add a potential function $c(s,x)$ to an equation of the form

$$\frac{\partial u}{\partial s} + \frac{1}{2}\sigma^2\triangle u + \sigma^2\mathbf{b}(s,x)\cdot\nabla u = 0,$$

we use the Kac functional.

2. On Feynman's formula: Consider again equation (1.9.14) appearing in the comment to Theorem 1.9.2. For simplicity, we assume that $\mathbf{b}(t,x) = 0$, $V(t,x) = 0$ and $\psi(s,x,t,y)$ is the fundamental solution of the equation

$$\frac{\partial\psi}{\partial t} = i\frac{1}{2}\sigma^2\triangle\psi.$$

Fix times $s < t_1 \leq t_2 \leq \cdots \leq t_n = b$.

Changing equation (1.9.14) slightly, let

$$F^{(n)}\left[\Gamma\right] = \int \psi\left(s, x, t_1, x_1\right) dx_1 \mathbf{1}_{\Gamma_1}\left(x_1\right) \psi\left(t_1, x_1, t_2, x_2\right) dx_2 \mathbf{1}_{\Gamma_2}(x_2) \times \cdots$$
$$\times \psi\left(t_{n-1}, x_{n-1}, t_n, x_n\right) dx_n \mathbf{1}_{\Gamma_n}\left(x_n\right),$$

which is a well-defined quantity.

This expression involves a complex-valued measure $\psi(s, x, t, y)dy$ instead of the transition probability $Q(s, x, t, dy)$ figuring in the Kolmogorov equation (1.2.10). So one would expect that, the same way a probability measure $\mathbf{P}_{(s,x)}$ was defined by (1.2.10), a complex-valued measure $\mathbf{F}_{(s,x)}$ can be defined on the path space as a limit of $F^{(n)}$ when $n \to \infty$. Feynman called the integral with respect to $\mathbf{F}_{(s,x)}$ a "path integral".

Then for a potential function V he defined $\psi(s, x)$ by

$$\psi(s, x) = \mathbf{F}_{(s,x)}\left[\exp\left(-i\int_s^b V(r, \omega_r)dr\right)f(\omega_b)\right].$$

This is known as Feynman's formula (see Feynman (1948)). Since the stochastic process X_r does not exist, we write ω_r in this formula.

Feynman showed that $\psi(s, x)$ is a solution of the equation

$$\frac{\partial \psi}{\partial s} = i\left(\frac{1}{2}\sigma^2 \triangle - V(s, x)\right)\psi,$$

and expanding the complex-valued exponential function on the right-hand side of Feynman's formula, he invented a successful approximation technique.

Incidentally, Feynman's complex-valued measure $\mathbf{F}_{(s,x)}$ does not exist. Nevertheless, calculations using it are effective. Why is that the case? Most physicists accept Feynman's path integral and Feynman's approximation calculus without question. Even people with doubts thought that the new regorous mathematical theory could be constructed and Feynman's path integral would be justified, like, say, the Dirac delta function was incorporated in the theory of distributions. However, such new mathematical theory has not yet been invented.

Feynman's formula and Kac's formula (4.2.4) are of the same type. The difference is that in Kac's formula $\{X_t, \mathbf{P}_{(s,x)}\}$ is a mathematically well-defined stochastic process, whereas we do not know if $\{\omega_t, \mathbf{F}_{(s,x)}\}$ of Feynman's formula exists as a rigorous mathematical object (or if any justification method exists). So the two formulas should not be confused.

Next, as an application of Kac's formula, we state

Theorem 4.2.3. *Consider the Kac functional*

$$K_s^t = \exp\left(\int_s^t \overline{c}\left(r, X_r\right)dr\right)$$

and for an exit function $\phi_b(x)$ define

$$\phi(s,x) = \mathbf{P}_{(s,x)}\left[K_s^b \phi_b(X_b)\right], \qquad (4.2.5)$$

where

$$\overline{c}(t,x) = c(t,x) + \frac{1}{2}\sigma^2\left(\nabla \cdot \mathbf{b}(t,x) + \mathbf{b}(t,x)^2\right).$$

Then $\phi(s,x)$ is a solution of the equation of motion

$$\frac{\partial \phi}{\partial s} + \frac{1}{2}\sigma^2 \triangle\phi + \sigma^2 \mathbf{b}(s,x)\cdot\nabla\phi + \overline{c}(s,x)\,\phi = 0,$$

that is, $\phi(s,x)$ is an evolution function.

As already noted, the measure $\mathbf{P}_{(s,x)}^K$ defined by

$$\mathbf{P}_{(s,x)}^K[F] = \mathbf{P}_{(s,x)}\left[K_s^b F\right],$$

does not yield directly a stochastic process, because $\mathbf{P}_{(s,x)}^K[1] \neq 1$. This is a drawback when we discuss the Kac functional.

Nevertheless, there is a way to obtain a stochastic process, namely, the renormalization of the measure $\mathbf{P}_{(s,x)}^K$. First, set

$$\xi(s,x) = \mathbf{P}_{(s,x)}^K[1] = \mathbf{P}_{(s,x)}\left[K_s^b\right]. \qquad (4.2.6)$$

Since $0 < \xi(s,x) < \infty$, if we define a measure $\overline{\mathbf{P}}_{(s,x)}$ by

$$\overline{\mathbf{P}}_{(s,x)}[F] = \frac{1}{\xi(s,x)}\mathbf{P}_{(s,x)}^K[F],$$

then

$$\overline{\mathbf{P}}_{(s,x)}[1] = \frac{1}{\xi(s,x)}\mathbf{P}_{(s,x)}^K[1] = 1.$$

That is, the measure $\overline{\mathbf{P}}_{(s,x)}$ satisfies the normalization condition, and hence yields a Markov process $\{X_t, \overline{\mathbf{P}}_{(s,x)}, (s,x) \in [a,b] \times \mathbf{R}^d\}$.

We will call the process $\{X_t, \overline{\mathbf{P}}_{(s,x)}, (s,x) \in [a,b] \times \mathbf{R}^d\}$ defined by means of the measure $\overline{\mathbf{P}}_{(s,x)}$ **renormalized process** (and call the procedure described above **renormalization**).

Moreover, using $\xi(s,x)$ we define the "renormalized" Kac functional by

$$N_s^t = \frac{1}{\xi(s,X_s)}K_s^t\xi(t,X_t). \qquad (4.2.7)$$

Then we have the following theorem.

Theorem 4.2.4 (Renormalization).

(i) *The renormalized multiplicative functional N_s^t defined by recipe (4.2.7) satisfies the normalization condition*

$$\mathbf{P}_{(s,x)}\left[N_s^t\right] = 1.$$

(ii) $\overline{\mathbf{P}}_{(s,x)}[F] = \mathbf{P}_{(s,x)}[N_s^t F]$ *for any $\mathcal{F}_s^t$-measurable function F. That is, the renormalized stochastic process*

$$\{X_t, \overline{\mathbf{P}}_{(s,x)}, (s,x) \in [a,b] \times \mathbf{R}^d\}$$

is obtained by means of the transformation based on the renormalized Kac functional N_s^t.

Proof. First, in view of the formula (4.2.7), the generalized Markov property (4.2.1) and the multiplicativity of K_s^t,

$$\mathbf{P}_{(s,x)}\left[N_s^t\right] = \frac{1}{\xi(s,x)}\mathbf{P}_{(s,x)}\left[K_s^t \xi(t,X_t)\right]$$

$$= \frac{1}{\xi(s,x)}\mathbf{P}_{(s,x)}\left[K_s^t \mathbf{P}_{(t,X_t)}\left[K_t^b\right]\right]$$

$$= \frac{1}{\xi(s,x)}\mathbf{P}_{(s,x)}\left[K_s^t K_t^b\right] = \frac{1}{\xi(s,x)}\mathbf{P}_{(s,x)}\left[K_s^b\right] = 1,$$

that is, the functional N_s^t satisfies the normalization condition.

Next, let $F \geq 0$ be $\mathcal{F}_s^t$-measurable. By the Markov property (4.2.1),

$$\overline{\mathbf{P}}_{(s,x)}[F] = \frac{1}{\xi(s,x)}\mathbf{P}_{(s,x)}^K[F] = \frac{1}{\xi(s,x)}\mathbf{P}_{(s,x)}\left[K_s^b F\right]$$

$$= \frac{1}{\xi(s,x)}\mathbf{P}_{(s,x)}\left[K_s^t K_t^b F\right] = \frac{1}{\xi(s,x)}\mathbf{P}_{(s,x)}\left[K_s^t F \mathbf{P}_{(t,X_t)}\left[K_t^b\right]\right]$$

$$= \frac{1}{\xi(s,x)}\mathbf{P}_{(s,x)}\left[K_s^t \xi(t,X_t) F\right]$$

$$= \mathbf{P}_{(s,x)}\left[\frac{1}{\xi(s,X_s)}K_s^t \xi(t,X_t) F\right]$$

$$= \mathbf{P}_{(s,x)}\left[N_s^t F\right],$$

that is, $\overline{\mathbf{P}}_{(s,x)}[F] = \mathbf{P}_{(s,x)}[N_s^t F]$. Therefore, the renormalized process is obtained by the transformation based on the renormalized Kac functional N_s^t, as claimed.
□

We note here that Theorem 4.2.4 is a special case of Theorem 4.2.8 that will be proved below.

Example 2. The Maruyama-Girsanov-Motoo functional.

Let $\mathbf{a}(s, x)$ be a bounded vector-function.

Definition 4.2.3. The exponential functional

$$M_s^t = \exp\left(\int_s^t \mathbf{a}\left(r, X_r\right) \cdot dB_r - \frac{1}{2}\int_s^t \mathbf{a}^2\left(r, X_r\right) dr\right) \qquad (4.2.8)$$

will be called the Maruyama-Girsanov-Motoo functional (see Maruyama (1954), Girsanov (1960) and Motoo (1960)) (or, often in the literature, just the Maruyama-Girsanov functional).

Consider a stochastic process $\{X_t; \mathcal{F}_s^t, \mathbf{P}_{(s,x)}, (s, x) \in [a, b] \times \mathbf{R}^d\}$ defined by the solution of a stochastic differential equation

$$X_t = X_a + \int_a^t \boldsymbol{\sigma}\left(s, X_s\right) dB_s + \int_a^t \mathbf{b}\left(s, X_s\right) ds, \qquad (4.2.9)$$

where $\boldsymbol{\sigma}(s, x)$ is a symmetric matrix-function.

Theorem 4.2.5. *The Maruyama-Girsanov-Motoo functional M_s^t defined by formula (4.2.8) is a normalized multiplicative functional of the stochastic process $\{X_t, \mathcal{F}_s^t, \mathbf{P}_{(s,x)}\}$.*

Proof. It is enough to show that if

$$W_t = \int_s^t \mathbf{a}\left(r, X_r\right) \cdot dB_r - \frac{1}{2}\int_s^t \mathbf{a}^2\left(r, X_r\right) dr,$$

then the normalization condition $\mathbf{P}_{(s,x)}\left[M_s^t\right] = \mathbf{P}_{(s,x)}\left[e^{W_t}\right] = 1$ is satisfied. We take $f(x) = e^x$ and apply Itô's formula (1.3.10). Then

$$
\begin{aligned}
d\left(e^{W_t}\right) &= e^{W_t} dW_t + \frac{1}{2}e^{W_t}(dW_t)^2 \\
&= e^{W_t}\left(\mathbf{a}\left(t, X_t\right) \cdot dB_t - \frac{1}{2}\mathbf{a}^2\left(t, X_t\right) dt\right) + \frac{1}{2}e^{W_t}(\mathbf{a}\left(t, X_t\right) \cdot dB_t)^2 \\
&= e^{W_t}\mathbf{a}\left(t, X_t\right) \cdot dB_t, \qquad (4.2.10)
\end{aligned}
$$

where we used the fact that, by Lévy's formulas (1.3.8)-(1.3.9),

$$e^{W_t}(\mathbf{a}\left(t, X_t\right) \cdot dB_t)^2 = e^{W_t}(\mathbf{a}\left(t, X_t\right))^2 dt.$$

Integrating both sides of equation (4.2.10) and computing expectations, we obtain

$$\mathbf{P}_{(s,x)}\left[e^{W_t}\right] - 1 = \mathbf{P}_{(s,x)}\left[\int_s^t e^{W_r}\mathbf{a}\left(r, X_r\right) \cdot dB_r\right],$$

where the right-hand side vanishes, being the expectation of a stochastic integral. Therefore, $\mathbf{P}_{(s,x)}\left[e^{W_t}\right] = 1$, as desired. $\qquad\square$

Clearly, Theorem 4.2.1 and Theorem 4.2.5 imply the following result.

Theorem 4.2.6. *Define the measure* $\mathbf{Q}_{(s,x)}$ *by* $\mathbf{Q}_{(s,x)} = \mathbf{P}_{(s,x)}M_s^b$, *where* M_s^t *is the Maruyama-Girsanov-Motoo functional* (4.2.8). *Then*

$$\{X_t; \mathbf{Q}_{(s,x)}, (s,x) \in [a,b] \times \mathbf{R}^d\}$$

is a Markov process.

Theorem 4.2.7 (Maruyama-Girsanov-Motoo Formula).

(i) *Define the process* $\widetilde{B}_t$ *by*

$$\widetilde{B}_{t-a} = B_{t-a} - \int_a^t \mathbf{a}(s, X_s)\, d. \tag{4.2.11}$$

Then $\widetilde{B}_t$ *is the Brownian motion with respect to the probability measure* $\mathbf{Q}_{(s,x)} = \mathbf{P}_{(s,x)}M_s^b$.

(ii) *Under the transformed probability measure* $\mathbf{Q}_{(s,x)} = \mathbf{P}_{(s,x)}M_s^b$, *the stochastic process* $\{X_t; \mathbf{Q}_{(s,x)}\}$ *satisfies the stochastic differential equation*

$$X_t = X_a + \int_a^t \boldsymbol{\sigma}(s, X_s)\, d\widetilde{B}_s + \int_a^t \{\mathbf{b}(s, X_s)\, ds + \boldsymbol{\sigma}(s, X_s)\mathbf{a}(s, X_s)\}ds,$$

where a new drift $\boldsymbol{\sigma}(s, X_s)\mathbf{a}(s, X_s)$ *is added.*

Proof. First, by equation (4.2.10),

$$d\left(M_s^t\right) = M_s^t \mathbf{a}(t, X_t) \cdot dB_t,$$
$$d\left(M_s^t\right)\widetilde{B}_{t-s} = \widetilde{B}_{t-s}M_s^t \mathbf{a}(t, X_t) \cdot dB_t,$$
$$M_s^t d\widetilde{B}_t = M_s^t\left(dB_t - \mathbf{a}(t, X_t)\, dt\right),$$

and

$$d\left(M_s^t\right)d\widetilde{B}_{t-s} = M_s^t \mathbf{a}(t, X_t) \cdot dB_t\left(dB_t - \mathbf{a}(t, X_t)\, dt\right)$$
$$= M_s^t \mathbf{a}(t, X_t)\, dt.$$

Hence,

$$d\left(M_s^t\widetilde{B}_{t-s}\right) = d\left(M_s^t\right)\widetilde{B}_{t-s} + M_s^t d\widetilde{B}_{t-s} + d\left(M_s^t\right)d\widetilde{B}_{t-s}$$
$$= \widetilde{B}_{t-s}M_s^t \mathbf{a}(t, X_t) \cdot dB_t + M_s^t dB_t.$$

Therefore,

$$M_s^t\widetilde{B}_{t-s} = \int_s^t \widetilde{B}_{r-s}M_s^r a(r, X_r) \cdot dB_r + \int_s^t M_s^r dB_r.$$

Since the right-hand side is a sum of stochastic integrals, $M_s^t\widetilde{B}_{t-s}$ is a $\mathbf{P}_{(s,x)}$-martingale. Further, since $\mathbf{Q}_{(s,x)} = \mathbf{P}_{(s,x)}M_s^b$, $\widetilde{B}_{t-s}$ is a $\mathbf{Q}_{(s,x)}$-martingale as well, and as $d\big(\widetilde{B}_{t-s}\big)^2 = dt$, we conclude that $\widetilde{B}_{t-s}$ is a $\mathbf{Q}_{(s,x)}$-Brownian motion (see Theorem 4.7.1.)

Substituting the expression (4.2.11) in

$$X_t = X_a + \int_a^t \boldsymbol{\sigma}\,(s, X_s)\,dB_s + \int_a^t \mathbf{b}\,(s, X_s)\,ds,$$

we obtain the claimed equation

$$X_t = X_a + \int_a^t \boldsymbol{\sigma}\,(s, X_s)\,d\widetilde{B}_s + \int_a^t \{\mathbf{b}\,(s, X_s)\,ds + \boldsymbol{\sigma}\,(s, X_s)\,\mathbf{a}\,(s, X_s)\}ds.$$

The proof is complete. □

Thus, when we want to add a new drift, we use the transformation based on the Maruyama-Girsanov-Motoo formula. We see the added drift by examining the stochastic differential equations of the transformed stochastic process $\{X_t; \mathbf{Q}_{(s,x)} = \mathbf{P}_{(s,x)} M_s^b\}$.

Application 1. Let σ be a constant and let $\{X_t; \mathcal{F}_s^t, \mathbf{P}_{(s,x)}, (s, x) \in [a, b] \times \mathbf{R}^d\}$ be defined by a stochastic differential equation of the form

$$X_t = X_a + \sigma B_{t-a} + \int_a^t \sigma^2 \mathbf{b}\,(s, X_s)ds.$$

Let $\phi(t, x)$ be a solution of the equations of motion (1.4.1) and consider the Maruyama-Girsanov-Motoo functional

$$M_s^t = \exp\left(\int_s^t \sigma \frac{\nabla\phi}{\phi}(r, X_r) \cdot dB_r - \frac{1}{2}\int_s^t \left(\sigma \frac{\nabla\phi}{\phi}\right)^2 (r, X_r)dr\right).$$

Then, by Theorem 4.2.7, with respect to the transformed probability measure $\mathbf{Q}_{(s,x)} = \mathbf{P}_{(s,x)} M_s^b$, the stochastic process X_t satisfies the stochastic differential equation

$$X_t = X_a + \sigma \widetilde{B}_{t-a} + \int_a^t \left(\sigma^2 \mathbf{b}\,(s, X_s) + \sigma^2 \frac{\nabla\phi}{\phi}(s, X_s)\right)ds,$$

where $\widetilde{B}_s$ is Brownian motion with respect to the measure $\mathbf{Q}_{(s,x)} = \mathbf{P}_{(s,x)} M_s^b$. That is, we get the equation of paths that is determined by the solution $\phi(t, x)$ of the equations of motion (1.4.1) (see Theorem 1.5.4).

Application 2. Consider the motion of a charged particle in a magnetic field, treated in Section 2.2. We have shown that the motion of the particle in the (x, y)-plane is governed by the equation

$$\frac{\partial u}{\partial t} + \frac{1}{2}\sigma^2\left(\frac{\partial^2 u}{\partial x^2} + \frac{\partial^2 u}{\partial y^2}\right) + \sigma^2\left(\mathbf{b}_x \frac{\partial u}{\partial x} + \mathbf{b}_y \frac{\partial u}{\partial y}\right)$$

$$+\sigma^2\alpha H\left(-y\frac{\partial u}{\partial x}+x\frac{\partial u}{\partial y}\right)=0, \qquad (4.2.12)$$

which is obtained from the equation

$$\frac{\partial u}{\partial t}+\frac{1}{2}\sigma^2\left(\frac{\partial^2 u}{\partial x^2}+\frac{\partial^2 u}{\partial y^2}\right)+\sigma^2\left(\mathbf{b}_x\frac{\partial u}{\partial x}+\mathbf{b}_y\frac{\partial u}{\partial y}\right)=0 \qquad (4.2.13)$$

by introducing a new drift term (counter-clockwise rotation)

$$\sigma^2\alpha H\left(-y\frac{\partial u}{\partial x}+x\frac{\partial u}{\partial y}\right).$$

If we apply the Maruyama-Girsanov-Motoo formula, the random motion $\{X_t, \mathbf{Q}_x\}$ determined by equation (4.2.12) is obtained from the random motion $\{X_t, \mathbf{P}_x\}$ determined by equation (4.2.13). Since both motions are time-homogeneous stochastic processes, using a two-dimensional drift

$$\mathbf{a}(x,y)=\sigma\alpha H(-y,x),$$

we first define a time-homogeneous Maruyama-Girsanov functional

$$M_0^t=\exp\left(\int_0^t\mathbf{a}\left(X_r\right)\cdot dB_r-\frac{1}{2}\int_0^t\mathbf{a}^2\left(X_r\right)dr\right)$$

where B_t is a two-dimensional Brownian motion.

Take b sufficiently large and denote $\mathbf{Q}_x=\mathbf{P}_x M_0^b$. Let $0\leq t<b$, and let F be a $\mathcal{F}_0^t$-measurable function. Then, by the Markov property (4.2.1),

$$\mathbf{Q}_x\left[F\right]=\mathbf{P}_x\left[M_0^b F\right]=\mathbf{P}_x\left[M_0^t M_t^b F\right]$$
$$=\mathbf{P}_x\left[M_0^t F\mathbf{P}_x\left[M_t^b\right]\right]=\mathbf{P}_x\left[M_0^t F\right].$$

Hence, $\mathbf{Q}_x$ does not depend on b.

By Theorem 4.2.6 and Theorem 4.2.7, we see that the counter-clockwise rotation $\sigma^2\alpha H(-y\partial u/\partial x+x\partial u/\partial y)$ is added to the random motion $\{X_t,\mathbf{Q}_x\}$.

Example 3. The Nagasawa functional.

Assuming that σ is a constant, let $\{X_t;\mathbf{P}_{(s,x)},(s,x)\in[a,b]\times\mathbf{R}^d\}$ be the stochastic process determined by the equation

$$X_t=x+\sigma B_{t-a}+\int_a^t\sigma^2\mathbf{b}\left(s,X_s\right)ds. \qquad (4.2.14)$$

First of all, let us state the following known results:

(i) Let $\phi_b\left(x\right)$ be an exit function and define

$$\phi\left(s,x\right)=\mathbf{P}_{(s,x)}\left[K_s^b\phi_b\left(X_b\right)\right],$$

where K_s^t is the Kac functional

$$K_s^t = \exp \left(\int_s^t \overline{c}(r, X_r) \, dr \right).$$

Then, as shown in Theorem 4.2.3, we obtain an evolution function $\phi(s, x)$ which satisfies the equation of motion

$$\frac{\partial \phi}{\partial s} + \frac{1}{2} \sigma^2 \triangle \phi + \sigma^2 \mathbf{b}(s, x) \cdot \nabla \phi + \overline{c}(s, x) \phi = 0,$$

where

$$\overline{c}(t, x) = c(t, x) + \frac{1}{2} \sigma^2 \left(\nabla \cdot \mathbf{b}(t, x) + \mathbf{b}(t, x)^2 \right).$$

(ii) By Theorem 1.5.1, if with the fundamental solution $p(s, x, t, y)$ of the equations of motions (1.4.1) and an evolution function $\phi(t, x)$ we define

$$q(s, x, t, y) = \frac{1}{\phi(s, x)} p(s, x, t, y) \phi(t, y),$$

then $q(s, x, t, y)$ is a transition probability density, and, by Theorem 1.5.3, it is the fundamental solution of the kinematics equation

$$\frac{\partial u}{\partial s} + \frac{1}{2} \sigma^2 \triangle u + \left(\sigma^2 \mathbf{b}(s, x) + \mathbf{a}(s, x) \right) \cdot \nabla u = 0,$$

where $\mathbf{a}(t, x) = \sigma^2 \nabla \log \phi(t, x)$ is the evolution drift determined by $\phi(t, x)$.

Now, following Nagasawa (1989, a), we introduce a new functional that combines the Kac functional and the evolution function $\phi(t, x)$.

Definition 4.2.4. With the function $\overline{c}(s, x)$ and an evolution function $\phi(s, x)$, define the multiplicative functional N_s^t by

$$N_s^t = \exp \left(\int_s^t \overline{c}(r, X_r) \, dr \right) \frac{\phi(t, X_t)}{\phi(s, X_s)}. \qquad (4.2.15)$$

Let $K_s^t = \exp \left(\int_s^t \overline{c}(r, X_r) \, dr \right)$ be the Kac functional. Using the evolution function $\phi(s, x)$, define the multiplicative functional L_s^t by

$$L_s^t = \frac{\phi(t, X_t)}{\phi(s, X_s)}.$$

Then the Nagasawa functional N_s^t defined by formula (4.2.15) is the product $N_s^t = K_s^t L_s^t$.

Theorem 4.2.8 (Nagasawa (1989, a)).

(i) *The Nagasawa functional $N_s^t = K_s^t L_s^t$ is normalized, that is,*

$$\mathbf{P}_{(s,x)}\left[N_s^t\right] = 1. \tag{4.2.16}$$

(ii) *Define the probability measure $\mathbf{Q}_{(s,x)}$ by $\mathbf{Q}_{(s,x)} = \mathbf{P}_{(s,x)}N_s^b$. Then*

$$\{X_t; \mathbf{Q}_{(s,x)}, (s,x) \in [a,b] \times \mathbf{R}^d\}$$

is a Markov process and X_t satisfies the stochastic differential equation

$$X_t = x + \sigma\widetilde{B}_{t-a} + \int_a^t \left(\sigma^2\mathbf{b}\left(s, X_s\right) + \mathbf{a}\left(s, X_s\right)\right) ds, \tag{4.2.17}$$

where $\widetilde{B}_t$ is the Brownian motion with respect to the probability measure $\mathbf{Q}_{(s,x)}$ and the added drift vector

$$\mathbf{a}\left(t, x\right) = \sigma^2 \frac{\nabla\phi\left(t, x\right)}{\phi\left(t, x\right)}$$

is the evolution drift determined by the evolution function $\phi(s, x)$. That is, the multiplicative functional N_s^t given by formula (4.2.15) induces the evolution drift $\mathbf{a}\left(t, x\right)$ determined by the equation of motion.

Proof. Applying Itô's formula (1.3.12) to the function $f\left(x\right) = \log\phi\left(t, x\right)$ and to the process X_t which satisfies equation (4.2.14), and using Paul Lévy's formulas (1.3.8)-(1.3.9), we have

$$\log\frac{\phi\left(t, X_t\right)}{\phi\left(s, X_s\right)} = \int_s^t \sigma\frac{\nabla\phi\left(r, X_r\right)}{\phi\left(r, X_r\right)} \cdot dB_r - \frac{1}{2}\int_s^t \sigma^2\frac{\nabla\phi \cdot \nabla\phi\left(r, X_r\right)}{\phi^2\left(r, X_r\right)} dr$$
$$+ \int_s^t \left(\frac{\partial\phi}{\partial r} + \frac{1}{2}\triangle\phi + \sigma^2\mathbf{b} \cdot \nabla\phi\right)\left(r, X_r\right)\frac{1}{\phi\left(r, X_r\right)}dr.$$

Next, substituting $\sigma^2\frac{\nabla\phi(t,x)}{\phi(t,x)} = \mathbf{a}\left(t, x\right)$ in the first and the second terms, and substituting

$$\frac{\partial\phi}{\partial r} + \frac{1}{2}\triangle\phi + \sigma^2\mathbf{b} \cdot \nabla\phi = -\bar{c}\phi$$

in the third term, we obtain the equation

$$\log\frac{\phi\left(t, X_t\right)}{\phi\left(s, X_s\right)} = \int_s^t \sigma^{-1}\mathbf{a}\left(r, X_r\right) \cdot dB_r - \frac{1}{2}\int_s^t \left(\sigma^{-1}\mathbf{a}\left(r, X_r\right)\right)^2 dr$$
$$- \int_s^t \bar{c}\left(r, X_r\right) dr.$$

If we now move the third term on the right-hand side and exponentiate bothe both sides of the resulting equality, we obtain

$$\exp \left(\int_s^t \overline{c}\,(r, X_r)\, dr \right) \frac{\phi\,(t, X_t)}{\phi\,(s, X_s)}$$

$$= \exp \left(\int_s^t \sigma^{-1} \mathbf{a}\,(r, X_r) \cdot dB_r - \frac{1}{2} \int_s^t \left(\sigma^{-1} \mathbf{a}\,(r, X_r) \right)^2 dr \right),$$

where $\mathbf{a}(t, x) = \sigma^2 \nabla \log \phi(t, x)$ is the evolution drift.

The left-hand side is the functional $N_s^t = K_s^t L_s^t$, while the right-hand side is the Maruyama-Girsanov-Motoo functional. Therefore, if we set $\mathbf{Q}_{(s,x)} = \mathbf{P}_{(s,x)} N_s^b$, then, by Theorem 4.2.6, the transformed stochastic process $\{X_t; \mathbf{Q}_{(s,x)}, (s, x) \in [a, b] \times \mathbf{R}^d\}$ is a Markov process, and, by Theorem 4.2.7, X_t satisfies the stochastic differential equation (4.2.17) with the drift $\mathbf{a}(t, x) = \sigma^2 \nabla \log \phi(t, x)$. □

Remark to Theorem 4.2.8. The Kac functional K_s^t yields a transformation that adds the potential function term cu to an equation; on the other hand, the Maruyama-Girsanov-Motoo functional M_s^t yields a transformation that adds the drift term $\mathbf{a} \cdot \nabla u$. So for a long time it was believed that the two functionals are not related. However, Theorem 4.2.8 asserts that "the Kac functional K_s^t and the Maruyama-Girsanov-Motoo functional M_s^t are not that different."

Indeed, if we change a little bit and set $N_s^t = K_s^t L_s^t$, then we obtain the same transformation as the one based on the Maruyama-Girsanov-Motoo functional M_s^t. This is the surprising content of Theorem 4.2.8, which is easier to understand if we look at the functional N_s^t in the context of the mechanics of random motion, as shown by the following result.

Theorem 4.2.9 (Nagasawa (1989, a)). *Assign to the Markov process $\{X_t; \mathbf{Q}_{(s,x)}\}$, obtained via the transformation based on the functional $N_s^t = K_s^t L_s^t$ in the preceding theorem, an initial distribution with the density $\mu_a(x) = \phi(a, x)\widehat{\phi}_a(x)$, and denote*

$$\mathbf{Q} = \int \mu_a\,(x)\, dx\, \mathbf{Q}_{(a,x)}.$$

Then $\{X_t; \mathbf{Q}\}$ coincides with the stochastic process constructed by means of the triplet $\{p(a, z, b, y), \widehat{\phi}_a(z), \phi_b(y)\}$ in Section 1.4. In other words, if the stochastic process determined by a solution of the stochastic differential equation (4.2.14) is transformed by means of the multiplicative functional

$$N_s^t = \exp \left(\int_s^t \overline{c}\,(r, X_r)\, dr \right) \frac{\phi(t, X_t)}{\phi(s, X_s)},$$

then it is the random motion determined by the equations of motion (1.4.1).

Theorem 4.2.9, which uses the multiplicative functional $N_s^t = K_s^t L_s^t$, is the direct way to derive the Markov process (4.2.17) determined by the equations of motion (1.4.1) from the stochastic process (4.2.14).

In this method the backward evolution function $\widehat{\phi}(t, x)$ is not used explicitly, but one should note that even in this construction an entrance function $\widehat{\phi}_a(x)$ and an exit function $\phi_b(x)$ are given. As a matter of fact, the function $\phi(a, x)$ which is used to define the initial distribution density is determined by the exit function $\phi_b(x)$.

4.3   Change of Time Scale

In this section we treat again time-homogeneous Markov processes

$$\{X_t; t \geq 0, \mathcal{F}_t, \mathbf{P}_x\}.$$

If we change the time scale of the Markov process X_t, is the result again a Markov process? In general this is not the case, but if the new time variable that replaces t depends only on the past history of the Markov process X_t, then the stochastic process with changed time scale is also a Markov process. Let us show this.

For a given function $z(t) > 0$, we define

$$\tau(t, \omega) = \int_0^t z(X_s(\omega))\, ds, \tag{4.3.1}$$

where for simplicity we assume that $\tau(\infty, \omega) = \infty$. Then

$$\tau(t + r, \omega) = \tau(t, \omega) + \tau(r, \theta_t \omega).$$

We next define the so-called "time-change function" $\tau^{-1}(t, \omega)$ as

$$\tau^{-1}(t, \omega) = \sup\{s; \tau(s, \omega) \leq t\}. \tag{4.3.2}$$

Using $\tau^{-1}(t, \omega)$ as a new time scale we define

$$Y_t(\omega) = X\left(\tau^{-1}(t, \omega), \omega\right),$$

where we write $X_t(\omega)$ as $X(t, \omega)$.

Then *the time-changed stochastic process* $\{Y_t, t \geq 0; \mathcal{F}_t, \mathbf{P}_x\}$ *is a Markov process.*

To prove this, denote the semigroups of the Markov processes $X_t(\omega)$ and $Y_t(\omega)$ by P_t and P_t^Y, respectively. Let

$$G_\alpha f(x) = \int_0^\infty e^{-\alpha t} P_t f(x)\, dt = \mathbf{P}_x\left[\int_0^\infty e^{-\alpha t} f(X_t)\, dt\right],$$

and

$$G_\alpha^Y f(x) = \int_0^\infty e^{-\alpha t} P_t^Y f(x)\, dt = \mathbf{P}_x\left[\int_0^\infty e^{-\alpha t} f(Y_t)\, dt\right],$$

respectively, be the corresponding resolvent operators.

The second and third expressions in each of these definition represent the same object in terms of the operator semi-group P_t and of the probability measure $\mathbf{P}_x$, respectively; the latter representation will prove convenient in calculations later on.

Lemma 4.3.1. *The resolvent operators G_α and G_α^Y are connected by the relations*

$$G_\alpha^Y f = G_\alpha\left(zf\right) - \alpha G_\alpha\left(\left(z-1\right) G_\alpha^Y f\right), \tag{4.3.3}$$

and

$$G_\alpha f = G_\alpha^Y\left(\frac{1}{z}f\right) - \alpha G_\alpha^Y\left(\left(\frac{1}{z}-1\right)G_\alpha f\right). \tag{4.3.4}$$

Proof. First, we have

$$G_\alpha^Y f\left(x\right) = \mathbf{P}_x\left[\int_0^\infty e^{-\alpha t} f\left(Y_t\right) dt\right] = \mathbf{P}_x\left[\int_0^\infty e^{-\alpha t} f\left(X\left(\tau^{-1}\right)\right) dt\right]$$

$$= \mathbf{P}_x\left[\int_0^\infty d\tau(s) e^{-\alpha\tau(s)} f\left(X_s\right)\right].$$

Therefore,

$$G_\alpha\left(zf\right) - G_\alpha^Y f = \mathbf{P}_x\left[\int_0^\infty z\left(X_s\right) ds\, e^{-\alpha s} f\left(X_s\right)\right]$$

$$- \mathbf{P}_x\left[\int_0^\infty d\tau(s) e^{-\alpha\tau(s)} f\left(X_s\right)\right].$$

Rewriting $z\left(X_s\right) ds$ as $d\tau\left(s\right)$ in the first term on the right-hand side, we see that

$$G_\alpha\left(zf\right) - G_\alpha^Y f = \mathbf{P}_x\left[\int_0^\infty d\tau\left(s\right) e^{-\alpha\tau(s)} f\left(X_s\right)\left(e^{-\alpha s+\alpha\tau(s)}-1\right)\right]$$

$$= \mathbf{P}_x\left[\int_0^\infty d\tau\left(s\right) e^{-\alpha\tau(s)} f\left(X_s\right)\int_0^s\left(\alpha z\left(X_t\right)-\alpha\right) dt\, e^{-\alpha t+\alpha\tau(t)}\right],$$

which by changing the order of integration is further equal to the following expression, which we denote by

$$I := \alpha\int_0^\infty dt\,\mathbf{P}_x\left[e^{-\alpha t}\left(z\left(X_t\right)-1\right) e^{\alpha\tau(t)}\int_t^\infty d\tau(s) f\left(X_s\right) e^{-\alpha\tau(s)}\right].$$

Since $\tau\left(t+r,\omega\right) = \tau\left(t,\omega\right) + \tau\left(r,\theta_t\omega\right)$, we have

$$I = e^{\alpha\tau(t)}\int_t^\infty d\tau\left(s\right) f\left(X_s\right) e^{-\alpha\tau(s)}$$

$$= \int_0^\infty d\tau(t+r) f\left(X_{t+r}\right) e^{-\alpha(\tau(t+r)-\tau(t))}$$

$$= \int_0^\infty d\tau(r,\theta_t\omega) f\left(X_r\left(\theta_t\omega\right)\right) e^{-\alpha\tau(r,\theta_t\omega)}.$$

Therefore, putting

$$F(t) = e^{-\alpha t}\left(z\left(X_t\right) - 1\right),$$

we have

$$I = \alpha \int_0^\infty dt\, \mathbf{P}_x \left[F(t) \int_0^\infty d\tau(r, \theta_t \omega) f(X_r(\theta_t \omega)) e^{-\alpha \tau(r, \theta_t \omega)} \right]$$

Here we use the Markov property (4.1.6). Then

$$I = \alpha \int_0^\infty dt\, \mathbf{P}_x \left[F(t)\, \mathbf{P}_{X_t} \left[\int_0^\infty d\tau(r) f(X_r) e^{-\alpha \tau(r)} \right] \right]$$

$$= \alpha \int_0^\infty dt \mathbf{P}_x \left[F(t)\, G_\alpha^Y f(X_t) \right],$$

which in view of the definition of $F(t)$, implies that

$$I = \alpha \int_0^\infty dt\, \mathbf{P}_x \left[e^{-\alpha t}\left(z\left(X_t\right) - 1\right) G_\alpha^Y f(X_t) \right]$$

$$= \alpha G_\alpha \left((z - 1) G_\alpha^Y f \right).$$

Thus relation (4.3.3) is proved. Relation (4.3.4) is established in the same way. □

With the new time scale $\tau^{-1}(t, \omega)$ defined by formula (4.3.2), we set

$$Y_t(\omega) = X\left(\tau^{-1}(t, \omega), \omega\right).$$

Theorem 4.3.1. *The time-changed process $\{Y_t, t \geq 0; \mathcal{F}_t, \mathbf{P}_x\}$ is a Markov process. The generator A^Y of the semigroup of operators $P_t^Y f(x) = \mathbf{P}_x[f(Y_t)]$ is*

$$A^Y = \frac{1}{z(x)} A, \tag{4.3.5}$$

with

$$\alpha - A^Y = \left(G_\alpha^Y\right)^{-1}, \tag{4.3.6}$$

where A is the generator of the Markov process $\{X_t, t \geq 0; \mathcal{F}_t, \mathbf{P}_x\}$.

Proof. We apply the operator $\alpha - A = G_\alpha^{-1}$ to both sides of equation (4.3.3). Then

$$(\alpha - A) G_\alpha^Y f = (G_\alpha)^{-1} G_\alpha(zf) - \alpha(G_\alpha)^{-1} G_\alpha\left((z-1) G_\alpha^Y f\right)$$

$$= zf - \alpha(z-1) G_\alpha^Y f.$$

Therefore,

$$\alpha z G_\alpha^Y f - A G_\alpha^Y f = zf$$

that is,

$$\left(\alpha - \frac{1}{z} A\right) G_\alpha^Y f = f,$$

which establishes both (4.3.5) and (4.3.6). □

Example 1. Let $\{B_t, t \geq 0; \mathbf{P}_x, x \in (-\infty, \infty)\}$ be the one-dimensional Brownian motion and let P_t be the associated semigroup of operators. Then the function

$$u(t, x) = P_t f(x) = \mathbf{P}_x\left[f(B_t)\right]$$

satisfies the equation

$$\frac{\partial u}{\partial t} = \frac{1}{2}\frac{\partial^2 u}{\partial x^2}.$$

That is,

$$A = \frac{1}{2}\frac{d^2}{dx^2}$$

is the generator of the semigroup of operators P_t of the one-dimensional Brownian motion. (See Remark below and Theorem 4.3.2.) Now take a function $z(x) = \sigma^{-2}(x)$ and define

$$\tau(t, \omega) = \int_0^t z(B_s)\,ds = \int_0^t \sigma^{-2}(B_s)\,ds$$

and the new time scale

$$\tau^{-1}(t, \omega) = \sup\{s; \tau(s, \omega) \leq t\},$$

and then set

$$X_t = B\left(\tau^{-1}(t)\right),$$

where we write $B(t) = B_t$.

The Markov process with the new time scale,

$$\{X_t, t \geq 0; \mathbf{P}_x, x \in (-\infty, \infty)\},$$

has the semigroup of operators

$$P_t^X f(x) = \mathbf{P}_x\left[f(X_t)\right].$$

By Theorem 4.3.1, its generator A^X is

$$A^X = \frac{1}{z(x)}A = \sigma^2(x)A = \frac{1}{2}\sigma^2(x)\frac{d^2}{dx^2}.$$

Therefore, $u(t, x) = P_t^X f(x) = P_x\left[f(X_t)\right]$ satisfies

$$\frac{\partial u}{\partial t} = \frac{1}{2}\sigma^2(x)\frac{\partial^2 u}{\partial x^2}.$$

That is, the time-changed Brownian motion $X_t = B\left(\tau^{-1}(t)\right)$ has the randomness intensity coefficient $\sigma^2(x)$. Thus, if the time scale is modified, the randomness intensity changes.

Example 2. Let us consider the radial component of the random motion treated in Section 1.8,

$$r_t = r_0 + \sigma B_t^1 + \int_0^t ds \left(\frac{3}{2}\sigma^2 \frac{1}{r_s} - \sigma \kappa r_s \right). \tag{4.3.7}$$

The radial motion cannot reach the origin because the flow $\frac{3}{2}\sigma^2 \frac{1}{r_s}$ becomes infinite there. Also, the motion considered in Section 2.4,

$$X_t = X_0 + \sigma \int_0^t \sqrt{X_s} dB_s + \int_0^t ds \left(\frac{3}{2}\sigma^2 - \sigma \kappa X_s^2 \right), \tag{4.3.8}$$

cannot reach the origin, because the random component $\sigma \int_0^t \sqrt{X_s} dB_s$ vanishes at the origin.

In Section 2.4, we stated without proof that one motion can be deduced from the other, even though they describe different phenomena. Let us prove this by applying a time change.

The solution r_t of the stochastic differential equation (4.3.7) is a time-homogeneous Markov process $\{r_t, t \geq 0; \mathbf{P}_x, x \in (0, \infty)\}$. Let P_t be the associated semigroup of operators. Then the function

$$u(t, x) = P_t f(x) = \mathbf{P}_x [f(r_t)]$$

satisfies the partial differential equation

$$\frac{\partial u}{\partial t} = \frac{1}{2}\sigma^2 \frac{\partial^2 u}{\partial x^2} + \left(\sigma^2 \frac{3}{2}\frac{1}{x} - \sigma \kappa x \right) \frac{\partial u}{\partial x}.$$

Therefore,

$$A = \frac{1}{2}\sigma^2 \frac{d^2}{dx^2} + \left(\sigma^2 \frac{3}{2}\frac{1}{x} - \sigma \kappa x \right) \frac{d}{dx}$$

is the generator of the semigroup of operators P_t. Taking the function $z(x) = x^{-1}$ and setting

$$\tau(t, \omega) = \int_0^t z(r_s) ds = \int_0^t r_s^{-1} ds$$

and

$$\tau^{-1}(t, \omega) = \sup \{s; \tau(s, \omega) \leq t\},$$

we define, with the new time-scale $\tau^{-1}(t, \omega)$, the Markov process

$$X_t = r\left(\tau^{-1}(t) \right).$$

By Theorem 4.3.1, the generator A^X of the semigroup $P_t^X f(x) = \mathbf{P}_x [f(X_t)]$ of the Markov process $\{X_t, t \geq 0; \mathbf{P}_x, x \in (0, \infty)\}$ with the new time scale is

$$A^X = \frac{1}{z(x)} A = xA = \frac{1}{2}\sigma^2 x \frac{d^2}{dx^2} + \left(\sigma^2 \frac{3}{2} - \sigma \kappa x^2 \right) \frac{d}{dx}.$$

That is, $u(t, x) = P_t^X f(x) = \mathbf{P}_x [f(X_t)]$ satisfies the equation

$$\frac{\partial u}{\partial t} = \frac{1}{2}\sigma^2 x \frac{\partial^2 u}{\partial x^2} + \left(\sigma^2 \frac{3}{2} - \sigma \kappa x^2\right)\frac{\partial u}{\partial x}.$$

Hence, the random motion with the new time scale $X_t = r\left(\tau^{-1}(t)\right)$ satisfies the stochastic differential equation (4.3.8). Thus, the solution r_t of stochastic differential equation (4.3.7) and solution $X_t = r\left(\tau^{-1}(t)\right)$ of equation (4.3.8) differ only by their time scales.

Remark. For time-homogeneous Markov processes Theorem 1.3.1 takes on the following form.

Theorem 4.3.2. *Let B_t be a d-dimensional Brownian motion defined on a probability space $\{\Omega, \mathcal{F}, \mathbf{P}\}$, X_0 be a random variable, σ be a constant, and $\mathbf{b}(x)$ be a d-dimensional drift vector. Then:*

(i) *The stochastic differential equation*

$$X_t = X_0 + \sigma B_t + \int_0^t \mathbf{b}(X_s)\,ds, \quad 0 \le t < \infty,$$

has a unique solution X_t, which is a time-homogeneous Markov process with the transition probability density $p(t, x, y)$. The functions

$$u(t, x) = \int p(t, x, y) f(y)\, dy,$$

and

$$\mu(t, x) = \int \mu_0(y) dy\, p(t, y, x),$$

satisfy the equations

$$\frac{\partial u}{\partial t} = \frac{1}{2}\sigma^2 \triangle u + \mathbf{b}(x)\cdot \nabla u$$

and

$$\frac{\partial \mu}{\partial t} = \frac{1}{2}\sigma^2 \triangle \mu - \nabla \cdot (\mathbf{b}(x)\mu),$$

respectively.

(ii) *Now let σ be not a constant, but a matrix function $\boldsymbol{\sigma}(x) : \mathbf{R}^d \to \mathbf{R}^{d^2}$. Then the stochastic differential equation*

$$X_t = X_0 + \int_0^t \sigma(X_s)\,dB_s + \int_0^t \mathbf{b}(X_s)\,ds \tag{4.3.9}$$

has a unique solution X_t, which is a time-homogeneous Markov process with the transition probability density $p(t, x, y)$. The function

$$u(t, x) = \int p(t, x, y) f(y)\, dy$$

satisfies the evolution equation

$$-\frac{\partial u}{\partial t} + \frac{1}{2} \sum_{i,j=1}^{d} \sigma^2(x)^{ij} \frac{\partial^2 u}{\partial x^i \partial x^j} + \sum_{i=1}^{d} b^i(x) \frac{\partial u}{\partial x^i} = 0. \tag{4.3.10}$$

Moreover, the stochastic process X_t determined by the evolution equation satisfies the stochastic differential equation (4.3.9).

To establish Theorem 4.3.2, it is enough to repeat the proof of Theorem 1.3.1 for the stochastic differential equation (4.3.9). Let $p(t, x, z)$ be the density function of a time-homogeneous transition probability and $q(s, x, t, y)$ be the density function in Theorem 1.3.1. Then $q(s, x, t, y) = p(t - s, x, y)$, by definition, and so

$$\frac{\partial}{\partial s} q(s, x, t, y) = -\frac{\partial}{\partial r} p(r, x, y)|_{r=t-s}.$$

Therefore, in the case of a time-homogeneous Markov process, the time derivative in the evolution equation (4.3.10) has the negative sign.

4.4   Duality and Time Reversal

We discussed the creation of the universe in Section 2.7, and learned that changing the direction of time alters completely the description of the physical phenomenon. This is reflected by the difference between formulas (2.7.9) and (2.7.10). In fact, formula (2.7.9) displays a singularity at $t = 0$, while formula (2.7.10) is singularity-free. According to (2.7.9), the universe started as a big bang, whereas (2.7.10) shows there was no such big bang. What happened to the stochastic process after time was reversed? It was Schrödinger (1931, 1932) who first considered the time reversal of random motion. Kolmogorov (1936, 1937) analyzed it as a duality relation between two transition probabilities.

We explain this next.

Suppose the transition probability has a density $p(t, x, y)$:

$$P(t, x, \Gamma) = \int_{\Gamma} p(t, x, y) \, dy. \tag{4.4.1}$$

Define a family of probability measures $\{\mathbf{P}_x, x \in \mathbf{R}^d\}$ on the sample space Ω by

$$\mathbf{P}_x\left[f\left(X_{t_1}, \ldots, X_{t_n}\right)\right] = \int p(t_1, x, x_1) \, dx_1 p(t_2 - t_1, x_1, x_2) \, dx_2 \times \cdots$$

$$\times p(t_n - t_{n-1}, x_{n-1}, x_n) \, dx_n f(x_1, \ldots, x_n), \tag{4.4.2}$$

for any bounded measurable function $f(x_1, \ldots, x_n)$ on the space $\left(\mathbf{R}^d\right)^n$. Moreover, define a random variable $X_t(\omega)$ on the sample space Ω by $X_t(\omega) = \omega(t)$. Then, by equation (4.4.2),

$$\int p(t, x, y) \, dy f(y) = \mathbf{P}_x[f(X_t)].$$

Therefore, $\{X_t, t \geq 0; \mathbf{P}_x, x \in \mathbf{R}^d\}$ is a time-homogeneous Markov process with the transition probability $p(t, x, y) dy$.

Let $f(x)$ be a bounded continuous function on the space $\mathbf{R}^d$ and define an operator P_t by

$$P_t f(x) = \int p(t, x, y) f(y) \, dy. \tag{4.4.3}$$

Then, by the Chapman-Kolmogorov equation,

$$
\begin{aligned}
P_{t+s} f(x) &= \int p(t+s, x, y) f(y) \, dy \\
&= \int p(t, x, z) \, dz \int p(s, z, y) f(y) \, dy = P_t P_s f(x),
\end{aligned}
\tag{4.4.4}
$$

i.e., $\{P_t; t \geq 0\}$ is a semigroup of operators.

If a distribution density $m(x)$ satisfies the equation

$$m(y) = \int m(x) \, dx \, p(t, x, y), \quad t \geq 0, \tag{4.4.5}$$

then $m(x) \, dx$ is called an **invariant measure**.

Invariant measures does not always exist. However, in what follows we will assume that they do, without affecting the topics addressed in the book.

Using the density function $m(x)$ of the invariant measure, we define

$$\widehat{p}(t, y, x) = m(x) p(t, x, y) \frac{1}{m(y)}. \tag{4.4.6}$$

Note that here the variables x and y are switched in $\widehat{p}(t, y, x)$ and $p(t, x, y)$.

Next, define an operator $\widehat{P}_t$ by

$$\widehat{P}_t f(x) = \int f(y) \, dy \, \widehat{p}(t, y, x), \tag{4.4.7}$$

for any bounded continuous functions $f(x)$ on $\mathbf{R}^d$. Then $\{\widehat{P}_t; t \geq 0\}$ is a semigroup of operators. In fact

$$\widehat{P}_{t+s} f(x) = \int f(y) \, dy \, m(x) p(t+s, x, y) \frac{1}{m(y)}.$$

Applying the Chapman-Kolmogorov equation to $p(t+s, x, y)$, this is equal to

$$\int f(y) \, dy \, m(x) \int p(t, x, z) \, dz \, p(s, z, y) \frac{1}{m(y)},$$

which upon dividing by $m(z)$ and multiplying by $m(z)$ becomes

$$\int f(y) \, dy \, m(x) \int p(t, x, z) \frac{1}{m(z)} dz \, m(z) \, p(s, z, y) \frac{1}{m(y)}.$$

By the defining equation (4.4.6), this last expression is equal to

$$= \int f(y)\, dy \int \widehat{p}(t, z, x)\, dz \widehat{p}(s, y, z)$$

and, by equation (4.4.7), to

$$\int \widehat{P}_s f(z)\, dz \widehat{p}(t, z, x) = \widehat{P}_t \widehat{P}_s f(x).$$

Therefore, the semigroup property $\widehat{P}_{t+s} f(x) = \widehat{P}_t \widehat{P}_s f(x)$ holds.

Now we have a pair of semigroups of operators, P_t and $\widehat{P}_t$. The analytical relationship between the two is characterized by a duality result of Kolmogorov.

Theorem 4.4.1 (Duality of Semigroups of Operators). *The semigroups of operators P_t and $\widehat{P}_t$ are in duality with respect to the invariant measure $m(x)\, dx$, that is, the* **duality relation**

$$\int m(x)\, dx\, g(x)\, P_t f(x) = \int m(y)\, dy\, f(y)\, \widehat{P}_t g(y) \tag{4.4.8}$$

holds for any bounded continuous functions $f(x)$ and $g(x)$.

Proof. The left-hand side of equation (4.4.8) is

$$\int m(x)\, dx\, g(x)\, P_t f(x) = \int m(x)\, dx\, g(x) \int p(t, x, y)\, f(y)\, dy,$$

which upon changing the order of integration, multiplying by $m(y)$ and dividing by $m(y)$, and then using equation (4.4.6), becomes

$$\int m(y)\, dy f(y) \int g(x)\, dx m(x)\, p(t, x, y)\, \frac{1}{m(y)} = \int m(y)\, dy f(y)\, \widehat{P}_t g(y),$$

i.e., the right-hand side of equation (4.4.8). $\qquad\square$

From now on we assume that the randomness intensity σ is a constant, the drift vector $\mathbf{a}(x)$ does not depend on time, and the stochastic process X_t is the solution of a time-homogeneous stochastic differential equation

$$X_t = x + \sigma dB_s + \int_0^t \mathbf{a}(X_s)\, ds, \tag{4.4.9}$$

where $x \in \mathbf{R}^d$ is a starting position and the starting time is $t = 0$.

Then

$$P(t, x, \Gamma) = \mathbf{P}_x[X_t \in \Gamma] \tag{4.4.10}$$

is a time-homogeneous transition probability.

The generators of the semigroup of operators P_t and $\widehat{P}_t$ are

$$A f(x) = \lim_{t \downarrow 0} \frac{P_t f(x) - f(x)}{t}$$

and

$$\widehat{A} g(x) = \lim_{t \downarrow 0} \frac{\widehat{P}_t g(y) - g(y)}{t},$$

respectively. We assume that

$$A = \frac{1}{2}\sigma^2 \triangle + \mathbf{a}(x) \cdot \nabla$$

and

$$\widehat{A} = \frac{1}{2}\sigma^2 \triangle + \widehat{\mathbf{a}}(x) \cdot \nabla$$

where $\widehat{\mathbf{a}}(x)$ denotes the dual drift vector.

Lemma 4.4.1. *If the pair of semigroups of operators P_t and $\widehat{P}_t$ introduce above satisfy the duality relation (4.4.8) with respect to the measure $m(x)\, dx$, then*

$$\int m(x)dx\, g(x)Af(x) = \int m(y)\, dy\, f(y)\widehat{A}g(y). \tag{4.4.11}$$

Proof. Since the semigroups P_t and $\widehat{P}_t$ satisfy the duality relation (4.4.8), we have

$$\int m(x)dx\, g(x)\frac{P_t f(x) - f(x)}{t} = \int m(y)dy\, f(y)\frac{\widehat{P}_t g(y) - g(y}{t}.$$

Letting here $t \downarrow 0$ on both sides, we obtain equation (4.4.11). □

The next lemma is technical, but equation (4.4.12) in it is one of the most important formulas concerning the duality of semigroups of operators.

Lemma 4.4.2 (Nagasawa (1961)). *Let A and $\widehat{A}$ be the generators of the semigroups of operators P_t and $\widehat{P}_t$. Then*

$$\int \{(Af(x))g(x) - f(x)(\widehat{A}g(x))\}m(x)dx$$

$$= \int f(x)g(x)A^0\, m(x)dx \tag{4.4.12}$$

$$+ \int f(x)\{\mathbf{a}(x) + \widehat{\mathbf{a}}(x) - \sigma^2 \nabla \log m(x)\} \cdot \nabla g(x)m(x)dx,$$

where

$$A^0 m(x) = \frac{1}{2}\sigma^2 \triangle m(x) - \nabla \cdot (\mathbf{a}(x)m(x)).$$

Proof. The proof of this statement can be obtained by calculating the expression on the left-hand side of equation (4.4.12) and rearranging it.   □

Formula (4.4.12) connects the drift vector $\mathbf{a}(x)$ and the dual drift vector $\widehat{\mathbf{a}}(x)$ to the density of the invariant measure $m(x)$.

Lemma 4.4.3. *The duality relation* (4.4.11) *between the generators A and $\widehat{A}$ is equivalent to*

$$\mathbf{a}(x) + \widehat{\mathbf{a}}(x) - \sigma^2 \nabla \log m(x) = 0, \tag{4.4.13}$$

$$A^0 m(x) = 0, \quad \widehat{A^0} m(x) = 0, \tag{4.4.14}$$

where

$$\widehat{A^0} m(x) = \frac{1}{2}\sigma^2 \triangle m(x) - \nabla \cdot (\widehat{\mathbf{a}}(x)m(x)).$$

Proof. If the generators A and $\widehat{A}$ satisfy the duality relation (4.4.11), then the left-hand side of equation (4.4.12) vanishes. If we set $g \equiv 1$ in (4.4.11), we get

$$\int A f(x)m(x)dx = 0.$$

Therefore,

$$\int f(x)A^0 m(x)dx = 0,$$

and so $A^0 m(x) = 0$ because $f(x)$ is arbitrary. In the same way, if we set $f \equiv 1$ in (4.4.1), we get

$$\int \widehat{A} g(x)m(x)dx = 0.$$

Therefore

$$\int g(x) \widehat{A^0} m(x)dx = 0,$$

and so $\widehat{A^0} m(x) = 0$, because $g(x)$ is arbitrary. So, (4.4.14) holds.

The first integral on the right-hand side of equation (4.4.12) vanishes. The second integral term on the right-hand side of equation (4.4.12) also vanishes. Therefore,

$$\int f(x) \left\{ \mathbf{a}(x) + \widehat{\mathbf{a}}(x) - \sigma^2 \nabla \log m(x) \right\} \cdot \nabla g(x)m(x)dx = 0.$$

Since the functions $f(x)$ and $g(x)$ are arbitrary, we obtain equation (4.4.13).

Conversely, if (4.4.13) and (4.4.14) hold, then the right-hand side of equation (4.4.12) vanishes. Hence, we get the duality relation (4.4.11) between the generators A and $\widehat{A}$. This completes the proof.   □

We now introduce a new function $R(x)$ and represent $m(x)$ in the form

$$m(x) = e^{2R(x)}. \tag{4.4.15}$$

Then (4.4.13) implies that

$$\mathbf{a}(x) + \widehat{\mathbf{a}}(x) = 2\sigma^2 \nabla R(x). \tag{4.4.16}$$

Moreover, by (4.4.14),

$$A^0 m(x) - \widehat{A^0} m(x) = 0,$$

and since the terms involving the Laplacian compensate, we have

$$\nabla \cdot ((\mathbf{a}(x) - \widehat{\mathbf{a}}(x)) m(x)) = 0.$$

This equation does not determine $\mathbf{a}(x) - \widehat{\mathbf{a}}(x)$ uniquely, but it ensures that there is a function $S(x)$ such that

$$\mathbf{a}(x) - \widehat{\mathbf{a}}(x) = 2\sigma^2 \nabla S(x). \tag{4.4.17}$$

Definition 4.4.1. The function $R(x)$ determined by equation (4.4.15) will be called the **generating function of distribution**. The function $S(x)$ which satisfies equation (4.4.17) will be called the **drift potential**.

The drift potential $S(x)$ appeared first in the context of the duality of Markov processes via equation (4.4.17), so the term "drift potential" for the function $S(x)$ is justified.

The fact that the function $S(x)$ is a fundamental quantity was first clarified and recognized in the context of mechanics of random motion.

The results proved so far are summarized in the following fundamental Theorem 4.4.2, which characterizes the duality of semigroups of operators. It provides a mathematical foundation for the development of the mechanics of random motion.

Theorem 4.4.2. *Let the generators of the semigroups of operators P_t and $\widehat{P}_t$ be*

$$A = \frac{1}{2}\sigma^2 \triangle + \mathbf{a}(x) \cdot \nabla$$

and

$$\widehat{A} = \frac{1}{2}\sigma^2 \triangle + \widehat{\mathbf{a}}(x) \cdot \nabla,$$

respectively, and suppose they satisfy the duality relation with respect to the invariant measure $m(dx) = m(x)\,dx$:

$$\int m(x)dx\, g(x) A f(x) = \int m(y)dy\, f(y) \widehat{A} g(y).$$

Then there exist two functions, the generating function of distribution $R(x)$ and the drift potential $S(x)$, such that:

(i) *The drifts* $\mathbf{a}(x)$ *and* $\widehat{\mathbf{a}}(x)$ *are expressed in terms of* $R(x)$ *and* $S(x)$ *as*

$$\mathbf{a}(x) = \sigma^2 \nabla R(x) + \sigma^2 \nabla S(x) \qquad (4.4.18)$$

and

$$\widehat{\mathbf{a}}(x) = \sigma^2 \nabla R(x) - \sigma^2 \nabla S(x). \qquad (4.4.19)$$

(ii) *The distribution density is given in terms of* $R(x)$ *by the formula*

$$m(x) = e^{2R(x)}. \qquad (4.4.20)$$

Comments to Theorem 4.4.2.

(1) The function $R(x)$ determines the distribution density $m(x)$ in exponential form by formula (4.4.20), so we named $R(x)$ the distribution generation function. This is straightforward. However, in the context of duality, it is not clear what the function $S(x)$ represents.

(2) The Kolmogorov-Itô theory does not contain the decompositions (4.4.18) and (4.4.19) of the drift vectors. Since in that theory the random motion is determined by the total drift vector $\mathbf{a}(x)$, such decompositions play no role. Therefore, the Kolmogorov-Itô theory could not lead to the function $S(x)$ and recognize that it plays an important role in the theory of Markov processes; this became clear only after the elaboration of the mechanics of random motion.

(3) The existence of the pair of drift vectors $\mathbf{a}(x)$ and $\widehat{\mathbf{a}}(x)$ emerged through the study of the duality between pairs of Markov processes. That was in the 1960's. (Cf. Nagasawa (1961), Ikeda, Nagasawa and Sato (1964), Nagasawa (1964), Kunita and T. Watanabe (1966).) This in turn has led to the introduction of the function $S(x)$. We found much later that the solution of the equations of motion (1.4.1) given in Chapter 1 determines the drift potential $S(x)$. Cf. Nagasawa (1993, 2000).

Now, it still remains to clarify what the semigroup P_t and the dual semi-group $\widehat{P}_t$ represent. We explain this next.

Let us take an invariant measure $m(dx) = m(x)\,dx$ as an initial distribution and define

$$\mathbf{P}_m = \int m(x)\,dx\,\mathbf{P}_x.$$

In view of equation (4.4.5), the distribution of the random motion $\{X_t, t \geq 0; \mathbf{P}_m\}$ with the initial distribution $m(dx) = m(x)\,dx$ remains $m(x)\,dx$ at any time, i.e., the random motion $\{X_t, t \geq 0; \mathbf{P}_m\}$ is time-stationary. Therefore, the nature of the random motion is independent of time interval considered.

So let us take a time interval $[0, b]$, $b > 0$, and consider

$$\{X_t, t \in [0, b]; \mathbf{P}_m\}.$$

Let us look at X_t by taking $b > 0$ as the terminal time and reversing the direction of time from time $b > 0$ (that is, play the movie backwards). Then what we "see" is the stochastic process

$$\{\widehat{X}_t, t \in [0, b]\,; \mathbf{P}_m\},$$

where

$$\widehat{X}_t = X_{b-t} \quad (t \uparrow b)\,.$$

The finite-dimensional distribution of the random motion $\widehat{X}_t$, which is what we see when we play the movie backwards, is

$$\mathbf{P}_m[f(\widehat{X}_0, \widehat{X}_{t_1}, \dots, \widehat{X}_{t_{n-1}}, \widehat{X}_b)] = \mathbf{P}_m[f(X_b, X_{b-t_1}, \dots, X_{b-t_{n-1}}, X_0)]\,.$$

When we read the right-hand side, we do it from the right to the left, so that it aligns with the direction of time, that is, $t = 0 \uparrow b$, because what we have on the right-hand side is the original random motion X_t. Then, since the transition probability density of X_t is $p(t, x, y)$ and the distribution density is $m(x)$, the last expression is equal to

$$\int m(x_n)dx_n\, p(b - t_{n-1}, x_n, x_{n-1})dx_{n-1} \times \cdots$$

$$\times\, p(t_2 - t_1, x_2, x_1)dx_1\, p(t_1, x_1, x_0)dx_0\, f(x_0, x_1, \dots, x_n)\,.$$

This expression must be read from the left to the right. If we rewrite it from the right to the left, we obtain

$$\int dx_0\, p(t_1, x_1, x_0)\, dx_1\, p(t_2 - t_1, x_2, x_1)\, dx_2 \times \cdots$$

$$\times\, dx_{n-1}\, p(b - t_{n-1}, x_n, x_{n-1})m(x_n)\, dx_n\, f(x_0, x_1, \dots, x_n)\,.$$

Multiplying here by $m(x_i)$ and also dividing by $m(x_i)$, we obtain

$$\int dx_0\, m(x_0)\frac{1}{m(x_0)}p(t_1, x_1, x_0)m(x_1)dx_1$$

$$\times\, \frac{1}{m(x_1)}p(t_2 - t_1, x_2, x_1)m(x_2)dx_2 \times \cdots$$

$$\times\, \frac{1}{m(x_{n-1})}p(b - t_{n-1}, x_n, x_{n-1})m(x_n)dx_n f(x_0, x_1, \dots, x_n)\,.$$

Since, by (4.4.6), $\widehat{p}(t, y, x) = m(x)p(t, x, y)\dfrac{1}{m(y)}$, we conclude that

$$\mathbf{P}_m\big[f\big(\widehat{X}_0, \widehat{X}_{t_1}, \dots, \widehat{X}_{t_{n-1}}, \widehat{X}_b\big)\big]$$

$$= \int m(x_0)\, dx_0 \widehat{p}(t_1, x_0, x_1)\, dx_1 \widehat{p}(t_2 - t_1, x_1, x_2)\, dx_2 \times \cdots$$

$$\times\, \widehat{p}(b - t_{n-1}, x_{n-1}, x_n)\, dx_n f(x_0, x_1, \dots, x_n)\,.$$

This formula implies that the time-reversed process $\widehat{X}_t = X_{b-t}$, $(t \uparrow b)$ is a Markov process with the time-homogeneous transition probability $\widehat{p}(t, x, y) \, dy$ and the initial distribution $m(x_0) \, dx_0$.

Summarizing, we have

Theorem 4.4.3. *Let $\{X_t, t \in [0, b]; \mathbf{P}_m\}$ be a random motion with an invariant measure $m(dx) = m(x) \, dx$ as the initial distribution. Then $\{X_t, t \in [0, b]; \mathbf{P}_m\}$ is a Markov process with the transition probability density $p(t, x, y)$.*

Take $b > 0$ as the terminal time and look at X_t from $b > 0$ with time reversed (play the movie backwards), that is, consider the random motion $\widehat{X}_t = X_{b-t}$ $(t \uparrow b)$. Then $\{\widehat{X}_t, t \in [0, b]; \mathbf{P}_m\}$ is a Markov process with the initial distribution $m(x) \, dx$ and the time homogeneous transition probability

$$\widehat{p}(t, x, y) = m(y) \, p(t, y, x) \, \frac{1}{m(x)}.$$

In this section we discussed the duality and time reversal of time-homogeneous Markov processes. The theorems proved can be applied also to the duality and time reversal of time-inhomogeneous Markov processes by using the method of space-time Markov processes that will be discussed in Section 4.6.

4.5   Time Reversal, Last Occurrence Time

We consider a time-homogeneous Markov process $\{X_t, t \geq 0; \mathcal{F}_t, \mathbf{P}_x\}$, **but we do not assume beforehand that it admits an invariant measure.**

In the mechanics of random motion, the starting (entrance) time and the terminal (exit) time are predetermined, so to reverse time we should watch the motion from the terminal time to the starting time. In the time-homogeneous Markov process the time continues from zero to infinity, i.e., there is no terminal time. So we may decide to arbitrarily choose a terminal time b $(0 < b < \infty)$ and consider the process

$$\widehat{X}_t = X_{b-t} \quad (t \uparrow b).$$

Calculating as in the previous section, we have

$$\mathbf{P}_m[f(\widehat{X}_0, \widehat{X}_{t_1}, \ldots, \widehat{X}_{t_{n-1}}, \widehat{X}_b)] = \mathbf{P}_m[f(X_b, X_{b-t_1}, \ldots, X_{b-t_{n-1}}, X_0)].$$

We read this in the direction in which the time variable increases from 0 to b, that is, from right to left. Then, in terms of the density function $p(t, x, y)$ of the transition probability of the Markov process X_t and the distribution density $\mu_t(x)$, the last expression is equal to

$$\int \mu_0(x_n) dx_n \, p(b - t_{n-1}, x_n, x_{n-1}) dx_{n-1} \times \cdots$$
$$\times \, p(t_2 - t_1, x_2, x_1) dx_1 \, p(t_1, x_1, x_0) dx_0 \, f(x_0, x_1, \ldots, x_n).$$

Rewriting this expression from right to left, we obtain

$$\int dx_0\, p\,(t_1, x_1, x_0)\, dx_1\, p\,(t_2 - t_1, x_2, x_1)\, dx_2 \times \cdots$$

$$\times\, dx_{n-1}\, p(b - t_{n-1}, x_n, x_{n-1})\mu_0\,(x_n)\, dx_n\, f\,(x_0, x_1, \ldots, x_n)\,.$$

Next, replacing here dx_i by $\mu_{b-t_i}\,(x_i)\, dx_i\, \frac{1}{\mu_{b-t_i}(x_i)}$, we get

$$\int dx_0\, \mu_b\,(x_0)\, \frac{1}{\mu_b\,(x_0)}\, p\,(t_1, x_1, x_0)\, \mu_{b-t_1}\,(x_1)\, dx_1$$

$$\times\, \frac{1}{\mu_{b-t_1}\,(x_1)}\, p\,(t_2 - t_1, x_2, x_1)\, \mu_{b-t_2}\,(x_2)\, dx_2 \times \cdots$$

$$\times\, \frac{1}{\mu_{b-t_{n-1}}\,(x_{n-1})}\, p\,(b - t_{n-1}, x_n, x_{n-1})\mu_0\,(x_n)\, dx_n\, f(x_0, x_1, \ldots, x_n)\,.$$

Defining a time-inhomogeneous transition probability density by

$$\widehat{p}\,(s, x, t, y) = \frac{1}{\mu_{b-s}\,(x)}\, p\,(t - s, y, x)\, \mu_{b-t}\,(y)\,, \qquad (4.5.1)$$

and rewriting, the last expression becomes

$$\int \mu_b\,(x_0)\, dx_0\, \widehat{p}\,(0, x_0, t_1, x_1)\, dx_1\, \widehat{p}\,(t_1, x_1, t_2, x_2)\, dx_2 \times \cdots$$

$$\times\, \widehat{p}\,(t_{n-1}, x_{n-1}, b, x_n)\, dx_n\, f\,(x_0, x_1, \ldots, x_n)\,.$$

Thus, for the time-reversed process $\widehat{X}_t = X_{b-t},\ (t \uparrow b)$ we have

$$\mathbf{P}[f(\widehat{X}_0, \widehat{X}_{t_1}, \ldots, \widehat{X}_{t_{n-1}}, \widehat{X}_b)]$$

$$= \int \mu_b\,(x_0)\, dx_0\, \widehat{p}\,(0, x_0, t_1, x_1)\, dx_1\, \widehat{p}\,(t_1, x_1, t_2, x_2)\, dx_2 \times \cdots$$

$$\times\, \widehat{p}\,(t_{n-1}, x_{n-1}, b, x_n)\, dx_n\, f\,(x_0, x_1, \ldots, x_n)\,.$$

Hence, the time-reversed process $\widehat{X}_t = X_{b-t}\ (t \uparrow b)$ is a time-inhomogeneous Markov process with the initial distribution $\mu_b\,(x_0)\, dx_0$ and time-inhomogeneous transition probability $\widehat{p}\,(s, x, t, y)\, dy$.

Note that on the right-hand side of the defining equation (4.5.1) the positions of x and y in $p\,(t - s, y, x)$ are interchanged .

The time-reversed process $\widehat{X}_t = X_{b-t},\, (t \uparrow b)$ becomes again a time-homogeneous Markov process, if and only if in the right-hand side of equation (4.5.1) one has

$$\mu_{b-s}\,(x) = \mu_{b-t}\,(x) = \mu\,(x)\,,$$

that is, if and only if the distribution density $\mu\,(y)$ does not depend on time. This case reduces to the case of the preceding section:

Theorem 4.5.1. *Suppose the distribution* $\mu(y)\,dy$ *of a Markov process*

$$\{X_t, t \geq 0; \mathcal{F}_t, P_\mu\}$$

is an invariant measure. Then:

(i) *The time-reversed process* $\widehat{X}_t = X_{b-t}, (t \uparrow b)$ *is a time-homogeneous Markov process with the time-homogeneous transition probability density*

$$\widehat{p}(t, x, y) = \frac{1}{\mu(x)}\, p(t, y, x)\, \mu(y), \qquad (4.5.2)$$

(ii) *The semigroups of operators*

$$P_t f(x) = \int p(t, x, y)\, f(y)\, dy$$

and

$$\widehat{P}_t f(x) = \int \widehat{p}(t, y, x)\, f(y)\, dy$$

satisfy the duality relation

$$\int g(x)\, \mu(x)\, dx\, P_t f(x) = \int f(y)\, \mu(y)\, dy\, \widehat{P}_t\, g(y)$$

with respect to the invariant measure $\mu(y)\,dy$.

Markov processes do not always admit an invariant measure. So, the question arises: "what should we do if an invariant measure does not exist?"

In that case, we use measures with properties similar to invariant measures.

Definition 4.5.1. If

$$\int P_t f(x)\, m(dx) \leq \int f(x)\, m(dx), \quad t \geq 0,$$

for any bounded positive function $f(x)$, the measure $m(dx)$ is called **excessive**.

In the following we assume that there exists the Green function

$$g(x, y) = \int_0^\infty ds\, p(s, x, y) < \infty. \qquad (4.5.3)$$

Lemma 4.5.1. *Let the Green function* $g(x, y)$ *be defined by formula* (4.5.3). *Let* $\mu(dx)$ *be an arbitrary initial distribution and set*

$$m(dy) = \left(\int \mu(dx)\, g(x, y) \right) dy. \qquad (4.5.4)$$

Then $m(dx)$ *is an excessive measure.*

Proof. By the definition (4.5.4),

$$\int m\,(dy)\,f\,(y) = \int \mu\,(dx) \int_0^\infty ds\, P_s f\,(x).$$

for any bounded positive function $f(y)$. Hence, by the semigroup property,

$$\int P_t f\,(y)\,m\,(dy) = \int \mu\,(dx) \int_0^\infty ds\, P_s P_t f\,(x)$$

$$= \int \mu\,(dx) \int_0^\infty ds\, P_{t+s} f\,(x),$$

which after a change of variables becomes equal to

$$\int \mu\,(dx) \int_t^\infty ds\, P_s f\,(x) \le \int \mu\,(dx) \int_0^\infty ds\, P_s f\,(x)$$

$$= \int f\,(x)\,m\,(dx).$$

Hence, the measure $m(dx)$ defined by formula (4.5.4) is excessive. □

Now using the Green function (4.5.3) we define

$$m\,(y) = \int \mu\,(dx)\,g\,(x,y).$$

Moreover, we define

$$\widehat{p}\,(t,x,y) = \frac{1}{m\,(x)} p\,(t,y,x)\,m\,(y)$$

and set

$$\widehat{P}_t h(y) = \int \widehat{p}\,(t,y,z)\,h\,(z)\,dz. \qquad (4.5.5)$$

Lemma 4.5.2. *The semigroups P_t and $\widehat{P}_t$ satisfy the duality relation*

$$\int m\,(dx)\,h\,(x)\,P_t f\,(x) = \int m\,(dy)\,f\,(y)\,\widehat{P}_t h(y), \qquad (4.5.6)$$

with respect to the excessive measure $m(dx)$.

Proof. Indeed,

$$\int m\,(dy)\,f\,(y)\,\widehat{P}_t h(y) = \int m\,(y)\,dy f\,(y) \int \widehat{p}\,(t,x,y)\,h\,(y)\,dy$$

$$= \int m\,(y)\,dy f\,(y)\,\frac{1}{m\,(y)} \int p\,(t,z,y)\,m\,(z)\,h\,(z)\,dz$$

$$= \int m\,(z)\,dz h\,(z) \int p\,(t,z,y)\,f\,(y)\,dy = \int h\,(x)\,m\,(dx)\,P_t f\,(x),$$

as claimed. □

Consider now two time-homogeneous Markov processes such that the corresponding semigroups of operators P_t and $\widehat{P}_t$ are in duality with respect to the excessive measure $m(dx)$. Are P_t and $\widehat{P}_t$ time-reversed versions of one another?

If we select the point of time reversal, for example, $t = b$, then, as already mentioned, the time-reversed Markov process becomes time-inhomogeneous, which is not good for our purposes.

Fortunately, there is a way to solve this issue. We select the moment of time reversal appropriately for each individual sample path ω. Then the duality relation (4.5.6) can be interpreted as time reversal. Let us explain this.

Definition 4.5.2. If a random variable $L(\omega) \geq 0$ satisfies

$$L(\theta_t \omega) = (L(\omega) - t)^+, \tag{4.5.7}$$

then $L(\omega)$ is called the **"last occurrence time"** or "final time".

As the name indicates, the last occurrence time $L(\omega)$ is the time after which nothing else occurs (no train runs after the last train). Looking at the defining formula (4.5.7), we see that:

(i) If $t < L(\omega)$, then

$$L(\omega) = t + L(\theta_t \omega),$$

so there is still a time lag $L(\theta_t \omega)$ before the last occurrence time $L(\omega)$.

(ii) If $t \geq L(\omega)$, then

$$L(\theta_t \omega) = 0,$$

so nothing happens after time t.

Example. Let

$$L_B(\omega) = \sup \{t : X_t(\omega) \in B\}$$

be the last moment of time the Markov process X_t leaves a subset B. Then in view of (4.5.7), $L_B(\omega)$ is the last occurrence time. (If the set $\{t : X_t(\omega) \in B\}$ is empty, then $\sup \{t : X_t(\omega) \in B\} = 0$.)

For convenience, we give another definition of the last occurrence time $L(\omega)$ which is equivalent to the defining equation (4.5.7).

Lemma 4.5.3. *The defining equation (4.5.7) of the last occurrence time $L(\omega)$ is equivalent to the relation*

$$\{\omega : s < L(\omega) - t\} = \{\omega : s < L(\theta_t \omega)\}, \quad s, t \geq 0. \tag{4.5.8}$$

Proof. Clearly, if (4.5.7) holds, then so does (4.5.8). Conversely, if (4.5.8) holds, then taking $s = 0$ and considering the complement subset, we have

$$\{\omega : L(\omega) \leq t\} = \{\omega : L(\theta_t \omega) = 0\};$$

hence, if $L(\omega) \leq t$, then $L(\theta_t \omega) = 0$.

Now let $t < L(\omega)$. If $L(\omega) < t + L(\theta_t \omega)$, take r such that

$$t < L(\omega) < r < t + L(\theta_t \omega),$$

i.e., $L(\omega) - t < r - t < L(\theta_t \omega)$. Then, by (4.5.8),

$$0 < L(\omega) - t < r - t < L(\omega) - t.$$

which is impossible. In the same way one shows that $L(\omega) > t + L(\theta_t \omega)$ is also impossible. Therefore, if $t < L(\omega)$, then $t < L(\omega) = t + L(\theta_t \omega)$.   □

From now on we consider a time-homogeneous Markov process

$$\{X_t, t \geq 0; \mathcal{F}_t, \mathbf{P}_\mu\}$$

with an initial distribution $\mu(dx)$.

We define the time reversal of the Markov process $\{X_t, t \geq 0; \mathcal{F}_t, \mathbf{P}_\mu\}$ from the last occurrence time $L(\omega)$ for $0 < t < L(\omega)$ by

$$Y_t(\omega) = X_{L(\omega)-t}(\omega), \tag{4.5.9}$$

where

$$Y_0(\omega) = \lim_{t \downarrow 0} X_{L(\omega)-t}(\omega),$$

and assume that $L(\omega) < \infty$. When $L(\omega) \leq t$, $Y_t(\omega)$ is at an extra point δ. Note that the time $L(\omega)$ at which time is reversed depends on the path ω.

Now we define the time-reversed stochastic process by

$$\{Y_t, t \geq 0; \widehat{\mathcal{F}}_t, \mathbf{P}_\mu\},$$

where $\widehat{\mathcal{F}}_t$ is the σ-algebra generated by $\{Y_r; 0 \leq r \leq t\}$.

Theorem 4.5.2 (Nagasawa (1964)). *Let $\{X_t, t \geq 0; \mathcal{F}_t, \mathbf{P}_\mu\}$ be a time-homogeneous Markov process. Then the time-reversed process $\{Y_t, t \geq 0; \widehat{\mathcal{F}}_t, \mathbf{P}_\mu\}$ from the last occurrence time $L(\omega)$ is itself a time-homogeneous Markov process. Its semigroup of operators $\widehat{P}_t$ is given by formula (4.5.5). Moreover, $\widehat{P}_t$ and the semigroup of operators P_t of the Markov process X_t are in duality with respect to the excessive measure $m(dx)$ defined by formula (4.5.4), that is, equality (4.5.6) holds.*

Proof. We show that the time-reversed process $\{Y_t, t \geq 0; \widehat{\mathcal{F}}_t, \mathbf{P}_\mu\}$ from the last occurrence time $L(\omega)$ is a Markov process with the semigroup of operators $\widehat{P}_t$ given by formula (4.5.5).

To this end, taking continuous functions $f, f_j, \ j = 1, 2, \ldots, n$ with bounded supports and setting

$$F(\omega) = \prod_{n=1}^{n} f_j(Y_{t_j}), \quad 0 \leq t_1 \leq \cdots \leq t_n = r \leq s,$$

it is enough to show that

$$\mathbf{P}_\mu\left[Ff\left(Y_{t+s}\right)\right] = \mathbf{P}_\mu\left[F\widehat{P}_t f\left(Y_s\right)\right], \qquad (4.5.10)$$

where $f\left(\delta\right) = 0$, $f_j\left(\delta\right) = 0$, $j = 1, 2, \ldots, n$.

Since both sides of equation (4.5.10) are uniquely determined by their Laplace transforms, it suffices to prove that for any $\alpha, \beta > 0$, the following equality of double Laplace transforms of the two sides of (4.5.10) holds:

$$\int_r^\infty e^{-(\alpha+\beta)s}ds \int_0^\infty e^{-\beta t}dt\, \mathbf{P}_\mu\left[Ff\left(Y_{t+s}\right)\right]$$
$$= \int_r^\infty e^{-(\alpha+\beta)s}ds \int_0^\infty e^{-\beta t}dt\, \mathbf{P}_\mu\left[F\widehat{P}_t f\left(Y_s\right)\right], \qquad (4.5.11)$$

We first prepare a lemma.

Lemma 4.5.4. *Let $F\left(\omega\right)$ be as defined above and let $\alpha, \gamma > 0$. Then*

$$\int_r^\infty e^{-\alpha t}dt\, \mathbf{P}_\mu\left[e^{-\gamma L}Ff\left(Y_t\right)\right]$$
$$= \int m_\gamma\left(dx\right)f\left(x\right)\mathbf{P}_x\left[e^{-(\alpha+\gamma)L}F\right], \qquad (4.5.12)$$

where

$$m_\gamma\left(dy\right) = \int_0^\infty \mu\left(dx\right)g_\gamma\left(x,y\right)dy,$$
$$\text{with } g_\gamma\left(x,y\right) = \int_0^\infty dt\, e^{-\gamma t}p\left(t,x,y\right). \qquad (4.5.13)$$

Proof. Let $r < L(\omega) - u$. Since, by definition, $L(\omega) - u = L(\theta_u\omega)$, we have

$$Y_{t_j}(\omega) = X_{L(\omega)-t_j}(\omega) = X_{L(\omega)-u-t_j+u}(\omega)$$
$$= X_{L(\theta_u\omega)-t_j}(\theta_u\omega) = Y_{t_j}(\theta_u\omega).$$

Therefore,

$$F(\omega) = \prod_{n=1}^n f_j(Y_{t_j}(\omega)) = \prod_{n=1}^n f_j(Y_{t_j}(\theta_u\omega)) = F(\theta_u\omega). \qquad (4.5.14)$$

On the left-hand side of equation (4.5.12), $f(Y_t) = f(\delta) = 0$ if $L \leq t$. Hence, if for the sake of the computations we denote

$$J := \int_r^\infty e^{-\alpha t}dt\, \mathbf{P}_\mu\left[e^{-\gamma L}Ff(Y_t)\right],$$

then

$$J = \mathbf{P}_\mu\left[\int_r^L dt\, e^{-\alpha t-\gamma L}Ff(Y_t)\right] = \mathbf{P}_\mu\left[\int_r^L dt\, e^{-\alpha t-\gamma L}Ff\left(X_{L-t}\right)\right].$$

Changing variables by $u = L - t$, we see that

$$J = \mathbf{P}_\mu \left[\int_0^{L-r} du\, e^{\alpha u - (\alpha + \gamma) L} F f(X_u) \right]$$

$$= \mathbf{P}_\mu \left[\int_0^\infty du\, e^{\alpha u - (\alpha + \gamma) L} F f(X_u); u < L - r \right].$$

Continuing, since, $F(\omega) = F(\theta_u \omega)$ by (4.5.14), and since $\{u < L - r\} = \{0 < L(\theta_u \omega) - r\}$ by (4.5.8) and $L(\omega) = u + L(\theta_u \omega)$, we see that

$$J = \int_0^\infty e^{-\gamma u} du\, \mathbf{P}_\mu \left[f(X_u) e^{-(\alpha + \gamma) L(\theta_u \omega)} F(\theta_u \omega); 0 < L(\theta_u \omega) - r \right].$$

Moreover, the Markov property of $\{X_t, t \geq 0; \mathcal{F}_t, \mathbf{P}_\mu\}$ implies that

$$J = \int_0^\infty e^{-\gamma u} du\, \mathbf{P}_\mu \left[f(X_u)\, \mathbf{P}_{X_u} \left[e^{-(\alpha + \gamma) L} F; 0 < L - r \right] \right]$$

$$= \int \mu(dx) \int_0^\infty du\, e^{-\gamma u} p(u, x, y) f(y)\, \mathbf{P}_y \left[e^{-(\alpha + \gamma) L} F; 0 < L - r \right],$$

where if $L - r \leq 0$, then $F = 0$. Hence, we may omit the condition $\{0 < L - r\}$. We conclude that

$$J = \int \mu(dx)\, g_\gamma(x, y) f(y)\, \mathbf{P}_y \left[e^{-(\alpha + \gamma) L} F \right]$$

$$= \int m_\gamma(dy) f(y)\, \mathbf{P}_y \left[e^{-(\alpha + \gamma) L} F \right],$$

which is the right-hand side of equation (4.5.11), as we needed to show.   □

Let us return to the proof of the theorem.

For the sake of convenience, let us denote by L the left-hand side of equation (4.5.11), i.e.,

$$L = \int_r^\infty e^{-(\alpha + \beta) s} ds \int_0^\infty e^{-\beta t} dt\, \mathbf{P}_\mu \left[F f(Y_{t+s}) \right]$$

$$= \int_r^\infty e^{-\alpha s} ds \int_s^\infty e^{-\beta t} dt\, \mathbf{P}_\mu \left[F f(Y_t) \right].$$

In view of (4.5.12) of Lemma 4.5.4 with $\gamma = 0$,

$$L = \int_r^\infty e^{-\alpha s} ds \int m(dx) f(x)\, \mathbf{P}_x \left[e^{-\beta L} F \right].$$

Changing the order of integration and using again (4.5.12) with $\mu(dy) = \delta_x(dy)$, $\gamma = \beta$, we see that

$$L = \int m(dx) f(x) \int m_\beta(dy)\, \mathbf{P}_y \left[e^{-(\alpha + \beta) L} F \right].$$

Here

$$m_\beta\,(dy) = \int_0^\infty \delta_x\,(dx')\,g_\beta\,(x',y)\,dy = g_\beta\,(x,y)\,dy$$

(set $\mu(dx) = \delta_x(dx)$ in (4.5.13)). Hence, substituting this expression, we further have

$$L = \int m\,(dx)\,f\,(x) \int g_\beta\,(x,y)\,dy\,\mathbf{P}_y\left[e^{-(\alpha+\beta)L}F\right].$$

Denoting $h\,(y) = \mathbf{P}_y\left[e^{-(\alpha+\beta)L}F\right]$, we rewrite L as

$$L = \int m\,(dx)\,f\,(x) \int g_\beta\,(x,y)\,dy\,h\,(y)$$

$$= \int m\,(dx)\,f\,(x)\,G_\beta h\,(x)$$

$$= \int m\,(dy)\,h\,(y)\,\widehat{G}_\beta f(y).$$

Here we used the fact that the duality of semigroups of operators,

$$\int m\,(dx)\,f\,(x)\,P_t h\,(x) = \int m\,(dy)\,h\,(y)\,\widehat{P}_t f(y),$$

implies the duality of resolvent operators,

$$\int m\,(dx)\,f\,(x)\,G_\beta h\,(x) = \int m\,(dy)\,h\,(y)\,\widehat{G}_\beta f(y).$$

Next, returning to the expression of L in terms of h, we have

$$L = \int m\,(dy)\,\widehat{G}_\beta f(y)\mathbf{P}_y\left[e^{-(\alpha+\beta)L}F\right].$$

Finally, using the equation (4.5.12) of Lemma 4.5.4 with $\alpha + \beta$ instead of α, and applying it for $\widehat{G}_\beta f$ instead of f and with $\gamma = 0$, we see that

$$L = \int_r^\infty e^{-(\alpha+\beta)s}ds\,\mathbf{P}_\mu\left[F\widehat{G}_\beta f\,(Y_s)\right]$$

$$= \int_r^\infty e^{-(\alpha+\beta)s}ds \int_0^\infty e^{-\beta t}dt\,\mathbf{P}_\mu\left[F\widehat{P}_t f\,(Y_s)\right],$$

which is the right-hand side of equation (4.5.11). This completes the proof of the theorem. □

Remark. The relationship between duality with respect to an invariant measure and time reversal, expressed by Theorem 4.5.1, was established by Kolmogorov (1936, 1937). The relationship between the duality of semigroups of operators P_t and $\widehat{P}_t$ and the time reversal from the last occurrence (final) time, expressed by Theorem 4.5.2, was established later in Nagasawa (1964). The notion of the "last occurrence time" was introduced in Nagasawa (1964).

4.6 Time Reversal, Equations of Motion

Time reversal plays an indispensable role when we discuss the equation of motion (1.4.1). Let the fundamental solution $p(s, x, t, y)$ of the equations of motion (1.4.1), an entrance function $\widehat{\phi}_a(z)$ and an exit function $\phi_b(y)$, i.e., a "triplet" $\{p(s, x, t, y), \widehat{\phi}_a(x), \phi_b(y)\}$, be given. If the "triplet normalization condition"

$$\int dz\, \widehat{\phi}_a(z)\, p(a, z, b, y)\, \phi_b(y)\, dy = 1$$

is satisfied, then there exists a stochastic process $\{X_t(\omega)\,; t \in [a, b]\,, \mathbf{Q}\}$, with its finite-dimensional distribution given by

$$\mathbf{Q}\left[f\left(X_a, X_{t_1}, \ldots, X_{t_{n-1}}, X_b\right)\right]$$
$$= \int dx_0\, \widehat{\phi}_a(x_0)\, p(a, x_0, t_1, x_1)\, dx_1\, p(t_1, x_1, t_2, x_2)\, dx_2 \times \cdots \qquad (4.6.1)$$
$$\times\, p(t_{n-1}, x_{n-1}, b, x_n)\, \phi_b(x_n)\, dx_n\, f(x_0, x_1, \ldots, x_n)\,.$$

We wrote the right-hand side of equation (4.6.1) from the left to the right, and certainly this is how we read it. That is,

"the stochastic process $X_t(\omega)$ starts with the entrance function $\widehat{\phi}_a(z)$ and ends with the exit function $\phi_b(y)$."

Here we consider that time flows from the past to the future.

However, if we look closely at formula (4.6.1) for the finite-dimensional distribution, we find that there is another way of reading it. Since the right-hand side of equation (4.6.1) is symmetric with respect to the direction of time, we may also read it from the right to the left.

In other words, assuming that time flows from the future to the past, we can read it as

"the stochastic process $X_t(\omega)$ starts with the exit function $\phi_b(y)$ snd ends with the entrance function $\widehat{\phi}_a(z)$".

There is no difference between the two interpretations. So when we want to specify the direction of the time flow, we use two symbols

$$\uparrow t \in [a, b] \quad \text{and} \quad \downarrow t \in [a, b], \quad \text{respectively,}$$

i.e., if we write a stochastic process as $\{X_t(\omega)\,; \uparrow t \in [a, b]\,, \mathbf{Q}\}$ (resp., $\{X_t(\omega)\,; \downarrow t \in [a, b]\,, \mathbf{Q}\}$), time flows from the past to the future (resp., from the future to the past).

Looking at things in the reversed time direction, in many cases they evolve in a completely different way, as when reversing the direction of a movie. Only in a special case the motion looks the same when the flow of time is reversed.

If we consider the time-reversed process $\{X_t(\omega); \downarrow t \in [a, b], \mathbf{Q}\}$, we can rewrite equation (4.6.1) in terms of the time-reversed evolution function

$$\widehat{\phi}_t(x) = \int dz\, \widehat{\phi}_a(z)\, p(a, z, t, x),$$

as follows:

$$\mathbf{Q}\left[f\left(X_a, X_{t_1}, \ldots, X_{t_{n-1}}, X_b\right)\right]$$
$$= \int dx_0\, \widehat{\phi}_a(x_0)\, p(a, x_0, t_1, x_1)\, dx_1 \times \cdots$$
$$\times\, p(t_{n-1}, x_{n-1}, b, x_n)\, \phi_b(x_n)\, dx_n\, f(x_0, x_1, \ldots, x_n).$$

Replacing here dx_i by $\frac{1}{\widehat{\phi}_{t_i}(x_1)}dx_i\, \widehat{\phi}_{t_i}(x_i)$, we obtain

$$\int dx_0\, \widehat{\phi}_a(x_0) p(a, x_0, t_1, x_1) \frac{1}{\widehat{\phi}_{t_1}(x_1)} dx_1 \widehat{\phi}_{t_1}(x_1) p(t_1, x_1, t_2, x_2) \frac{1}{\widehat{\phi}_{t_2}(x_2)} \times \cdots$$
$$\times\, dx_{n-1}\, \widehat{\phi}_{t_{n-1}}(x_{n-1})\, p(t_{n-1}, x_{n-1}, b, x_n)\, \frac{1}{\widehat{\phi}_b(x_n)} \times \cdots$$
$$\times\, \widehat{\phi}_b(x_n)\, \phi_b(x_n)\, dx_n f(x_0, x_1, \ldots, x_n).$$

If we introduce the function

$$\widehat{q}(s, x, t, y) = \widehat{\phi}_s(x)\, p(s, x, t, y)\, \frac{1}{\widehat{\phi}_t(y)}$$

(the time-reversed transition probability density), the last expression above takes on the form

$$\int dx_0 \widehat{q}(a, x_0, t_1, x_1)\, dx_1\, \widehat{q}(t_1, x_1, t_2, x_2)\, dx_2 \times \cdots$$
$$\times\, dx_{n-1}\, \widehat{q}(t_{n-1}, x_{n-1}, b, x_n)\, \widehat{\phi}_b(x_n)\, \phi_b(x_n)\, dx_n\, f(x_0, x_1, \ldots, x_n)$$

which in turn, upon setting $\mu_b(x) = \widehat{\phi}_b(x)\, \phi_b(x)$, becomes

$$\int dx_0\, \widehat{q}(a, x_0, t_1, x_1)\, dx_1 \widehat{q}(t_1, x_1, t_2, x_2)\, dx_2 \times \cdots$$
$$\times\, dx_{n-1}\, \widehat{q}(t_{n-1}, x_{n-1}, b, x_n)\, \mu_b(x_n)\, dx_n f(x_0, x_1, \ldots, x_n).$$

Thus,

$$\mathbf{Q}\left[f\left(X_a, X_{t_1}, \ldots, X_{t_{n-1}}, X_b\right)\right]$$
$$= \int dx_0\, \widehat{q}(a, x_0, t_1, x_1)\, dx_1\, \widehat{q}(t_1, x_1, t_2, x_2)\, dx_2 \times \cdots \qquad (4.6.2)$$
$$\times\, dx_{n-1}\, \widehat{q}(t_{n-1}, x_{n-1}, b, x_n)\mu_b(x_n)dx_n f(x_0, x_1, \ldots, x_n).$$

Here we read the right-hand side from the right to the left, that is to say, we start at time b with the distribution $\mu_b(x_n)dx_n$ and read in the direction of decreasing time up to time a.

Theorem 4.6.1 (Time Reversal of Random Motion Determined by the Equations of Motion).

(i) *The stochastic process $\{X_t(\omega); t \in [a,b], \mathbf{Q}\}$ which is constructed by means of the fundamental solution $p(s,x,t,y)$ of the equations of motion (1.4.1) and prescribed entrance and exit functions $\widehat{\phi}_a(z)$ and $\phi_b(y)$ is a time-reversed stochastic process $\{X_t(\omega); \downarrow t \in [a,b], \mathbf{Q}\}$ which has the density function of terminal distribution*

$$\mu_b(x) = \widehat{\phi}_b(x)\phi_b(x). \tag{4.6.3}$$

This process departs in the opposite direction with the terminal distribution (4.6.3), where $\widehat{\phi}_b(x)$ is the function defined by

$$\widehat{\phi}_b(x) = \int dz\, \widehat{\phi}_a(z)p(a,z,b,x).$$

(ii) *The time-reversed stochastic process $\{X_t(\omega); \downarrow t \in [a,b], \mathbf{Q}\}$ is a Markov process with the time-reversed transition probability density*

$$\widehat{q}(s,x,t,y) = \widehat{\phi}_s(x)\,p(s,x,t,y)\,\frac{1}{\widehat{\phi}_t(y)}. \tag{4.6.4}$$

(iii) *The finite-dimensional distribution of the time-reversed Markov process $\{X_t(\omega); \downarrow t \in [a,b], \mathbf{Q}\}$ is given by formula (4.6.2).*

Proof. The function $\widehat{q}(s,x,t,y)$ given by formula (4.6.4) satisfies the Chapman-Kolmogorov equation. Moreover, since

$$\widehat{\phi}_t(y) = \int dx\, \widehat{\phi}_s(x)\,p(s,x,t,y),$$

we have

$$\int dx\, \widehat{q}(s,x,t,y) = \int dx\, \widehat{\phi}_s(x)p(s,x,t,y)\frac{1}{\widehat{\phi}_t(y)} = \widehat{\phi}_t(y)\frac{1}{\widehat{\phi}_t(y)} = 1,$$

that is, it satisfies the time-reversed normalization condition. So $\widehat{q}(s,x,t,y)$ is the time-reversed transition probability density. Assertions (i), (ii) and (iii) of the theorem follow from (4.6.2). $\qquad\square$

If we look at the time flow in the opposite order, the picture evolves in a completely different manner, which is manifested by the difference of the evolution drift vectors. Specifically, if we look at the stochastic process $\{X_t(\omega); t \in [a,b], \mathbf{Q}\}$ by reversing the time flow, namely, towards the departure time a starting from the terminal time b, we arrive at the following theorem.

Theorem 4.6.2 (Nagasawa (1989, a)) (Time-Reversed Kinematics).

(i) Let $p(s, x, t, y)$ be the fundamental solution of the equations of motion, let $\widehat{\phi}(t, x)$ be the reversed evolution function $\widehat{\phi}(t, x)$, and define $\widehat{q}(s, x, t, y)$ by formula (4.6.4). Then $\widehat{q}(s, x, t, y)$ is the time-reversed transition probability density.

(ii) The function $\widehat{q}(s, x, t, y)$ is the fundamental solution of the reversed kinematic equations

$$\begin{cases} -\dfrac{\partial \widehat{u}}{\partial t} + \dfrac{1}{2}\sigma^2 \triangle \widehat{u} + \left(-\sigma^2 \mathbf{b}(t, y) + \widehat{\mathbf{a}}(t, y)\right) \cdot \nabla \widehat{u} = 0, \\ \dfrac{\partial \mu}{\partial s} + \dfrac{1}{2}\sigma^2 \triangle \mu - \nabla \cdot \left(\left(-\sigma^2 \mathbf{b}(s, x) + \widehat{\mathbf{a}}(s, x)\right) \mu\right) = 0. \end{cases} \tag{4.6.5}$$

(iii) The vector $\widehat{\mathbf{a}}(t, x)$ added in equations (4.6.5) is given by

$$\widehat{\mathbf{a}}(t, x) = \sigma^2 \frac{\nabla \widehat{\phi}(t, x)}{\widehat{\phi}(t, x)} = \sigma^2 \nabla \left(R(t, x) - S(t, x)\right),$$

where $\widehat{\phi}(t, x)$ is the time-reversed evolution function.

(iv) The distribution density of the stochastic process X_t is

$$\mu_t(x) = \phi(t, x)\,\widehat{\phi}(t, x), \quad t \in [a, b].$$

The proof of Theorem 4.6.2 can be carried out in much the same way as that of Theorem 1.5.3 of kinematics by using Lemma 1.5.1 and Lemma 1.5.2 (but using a time-reversed evolution function $\widehat{\phi}(t, x)$ instead of $\phi(t, x)$). This is left to the reader as an exercise.

Theorem 4.6.2 of time-reversed kinematics does not provide the time-reversed path equation (stochastic differential equation). In order to discuss stochastic differential equations with time reversal, we need to deal with the time-reversed stochastic integral. Since we want to avoid it here, Theorem 4.6.2 lacks that part.

The stochastic process $\{X_t(\omega)\,;\ \uparrow t \in [a, b], \mathbf{Q}\}$, in which time flows from the past to the future, is a Markov process and its transition probability density is

$$q(s, x, t, y) = \frac{1}{\phi_s(x)} p(s, x, t, y)\,\phi_t(y). \tag{4.6.6}$$

The stochastic process $\{X_t(\omega)\,;\ \downarrow t \in [a, b], \mathbf{Q}\}$, in which time flows from the future to the past, is a time-reversed Markov process and has the time-reversed transition probability density

$$\widehat{q}(s, x, t, y) = \widehat{\phi}_s(x)\,p(s, x, t, y)\,\frac{1}{\widehat{\phi}_t(y)}.$$

In order to check the duality relation of the Markov processes $\{X_t, \uparrow t \in [a, b]\}$ and $\{X_t, \downarrow t \in [a, b]\}$, it is better to consider the space-time Markov process

$$\{(t, X_t), \uparrow t \in [a, b]\},$$

and the time-reversed space-time Markov process

$$\{(t, X_t), \downarrow t \in [a, b]\}.$$

This is because the space-time Markov process is a time-homogeneous Markov process running in the (space-time) state space $[a, b] \times \mathbf{R}^d$.

For time-homogeneous Markov processes, we have at our disposal a powerful tool, the "theory of semigroups of operators". Therefore, when developing the general theory of the Markov processes, we use time-homogeneous Markov processes.

The semigroup P_r associated to the space-time Markov process

$$\{(t, X_t), \uparrow t \in [a, b]\}$$

is defined by

$$P_r f (s, x) = \begin{cases} \displaystyle\int\!\!\int q(s, x, s + r, y) \, dy \, f(s + r, y) & \text{for } a \leq s < s + r \leq b, \\ 0 & \text{otherwise,} \end{cases} \tag{4.6.7}$$

for any bounded continuous function $f(s, x)$ on the state space $[a, b] \times \mathbf{R}^d$; we set $P_0 = I$.

The variables (s, x) in $P_r f(s, x)$ are the starting time and position of a particle. On the other hand, $r \geq 0$ represents the time elapsed since the particle started at $t \in [a, b]$. If the variable $s \in [a, b]$ in the pair (s, x) is measured using standard time, then the density function $q(s, x, s + r, y)$ shows that the particle moves from the space-time point (s, x) to the space-time point $(s + r, y)$. That is, the particle is moving in space-time as the standard time increases.

The time-reversed space-time process $\{(t, X_t), \downarrow t \in [a, b]\}$ is a time-homogeneous Markov process on the state space $[a, b] \times \mathbf{R}^d$. Its semigroup $\widehat{P}_r$ is defined by

$$\widehat{P}_r g(t, y) = \begin{cases} \displaystyle\int\!\!\int g(t - r, x) \, dx \widehat{q}(t - r, x, t, y) & \text{for } a \leq t - r < t \leq b, \\ 0 & \text{otherwise,} \end{cases} \tag{4.6.8}$$

for any bounded continuous function $g(t, x)$ on the space-time state space $[a, b] \times \mathbf{R}^d$, where $\widehat{P}_0 = I$.

The variables (t, y) in $\widehat{P}_r g\,(t, y)$ are the time and location at the particle's starting point. On the other hand, $r \geq 0$ represents the time elapsed from the moment the particle started at $t \in [a, b]$ of the space-time variable (t, y).

If the variable $t \in [a, b]$ in the space-time pair (t, y) is measured using standard time, then the density function $\widehat{q}\,(t - r, x, t, y)$ shows that the particle moves from the space-time point (t, y) to the space-time point $(t - r, x)$. That is, the particle is moving in space-time as the standard time decreases. So, the elapse of standard time is reversed.

Theorem 4.6.3. (Time Reversal of Space-Time Markov Processes and Space-Time Duality).

(i) *The space-time processes* $\{(t, X_t), \uparrow t \in [a, b]\}$ *and* $\{(t, X_t), \downarrow t \in [a, b]\}$ *are mutually time-reversed Markov processes.*

(ii) *The semigroup* P_r *defined by equation* (4.6.7) *and the semigroup* $\widehat{P}_r$ *defined by equation* (4.6.8) *are in space-time duality, that is,*

$$\int g(t, x) P_r f(t, x) \mu(dtdx) = \int f(t, x) \widehat{P}_r g(t, x) \mu(dtdx), \qquad (4.6.9)$$

for any bounded continuous functions $f(s, x)$ *and* $g(s, x)$ *on the space-time state space* $[a, b] \times \mathbf{R}^d$.

We note here that the duality is with respect to the measure $\mu\,(dtdx) = \widehat{\phi}_t(x)\phi_t(x)dtdx$ on the space-time state space $[a, b] \times \mathbf{R}^d$.

Proof. Assertion (i) was already established. Let us prove assertion (ii). We have

$$\int g\,(t, x)\, P_r f\,(t, x)\, \mu\,(dtdx)$$

$$= \int g\,(t, x)\, dx\, \widehat{\phi}_t\,(x)\, \phi_t\,(x)\, q\,(t, x, t + r, y)\, dy f\,(t + r, y)\, dt.$$

Using here equation (4.6.6), changing variables, dividing by $\widehat{\phi}_s\,(y)$, and then multiplying by $\widehat{\phi}_s\,(y)$, we transform the last expression to

$$\int g\,(s - r, x)\, dx \widehat{\phi}_{s-r}\,(x)\, p\,(s - r, x, s, y)\, \frac{1}{\widehat{\phi}_s\,(y)} \widehat{\phi}_s\,(y)\, \phi_s\,(y)\, dy f\,(s, y)\, ds$$

and then, with the help of equation (4.6.4), to

$$\int g\,(s - r, x)\, dx \widehat{q}\,(s - r, x, s, y)\, \widehat{\phi}_s\,(y)\, \phi_s\,(y)\, dy\, f\,(s, y)\, ds$$

$$= \int f\,(s, y)\, \widehat{P}_r g\,(s, y)\, \mu\,(dsdy)\,.$$

This is equation (4.6.9), where $\mu(dtdx) = \widehat{\phi}(t, x)\phi(t, x)dtdx$. $\qquad\square$

4.7 Conditional Expectation

In this section we define the concept of the conditional expectation and investigate its properties. As an application, we will discuss briefly the stochastic processes called martingales.

Let $\{\Omega, \mathcal{F}, \mathbf{P}\}$ be a probability space, where $\mathcal{F}_t \subset \mathcal{F}$, $t \in [a, b]$ is a family of σ-algebras. Consider the function space $L^2(\mathcal{F}, \mathbf{P})$, which is a Hilbert space. Since $L^2(\mathcal{F}_t, \mathbf{P})$ is a subspace of $L^2(\mathcal{F}, \mathbf{P})$, for every element $X \in L^2(\mathcal{F}, \mathbf{P})$, there exists its projection on the subspace $L^2(\mathcal{F}_t, \mathbf{P})$. We denote it by

$$\mathbf{P}[X|\mathcal{F}_t]$$

and call it the *conditional expectation* of the random variable $X \in L^2(\mathcal{F}, \mathbf{P})$.

Since the conditional expectation $\mathbf{P}[X|\mathcal{F}_t]$ belongs to the subspace $L^2(\mathcal{F}_t, \mathbf{P})$ by definition, it is an $\mathcal{F}_t$-measurable random variable.

Let us show that

$$\mathbf{P}[YX] = \mathbf{P}[Y\mathbf{P}[X|\mathcal{F}_t]], \qquad (4.7.1)$$

for all $Y \in L^2(\mathcal{F}_t, \mathbf{P})$.

First we decompose the random variable $X \in L^2(\mathcal{F}_t, \mathbf{P})$ as

$$X = \mathbf{P}[X|\mathcal{F}_t] + X_0,$$

where X_0 is orthogonal to the subspace $L^2(\mathcal{F}_t, \mathbf{P})$. Then

$$\mathbf{P}[YX] = \mathbf{P}[Y(\mathbf{P}[X|\mathcal{F}_t] + X_0)] = \mathbf{P}[Y\mathbf{P}[X|\mathcal{F}_t]] + \mathbf{P}[YX_0],$$

for all $Y \in L^2(\mathcal{F}_t, \mathbf{P})$ Since $\mathbf{P}[YX_0] = 0$,

$$\mathbf{P}[YX] = \mathbf{P}[Y\mathbf{P}[X|\mathcal{F}_t]]$$

for all $Y \in L^2(\mathcal{F}_t, \mathbf{P})$.

We defined the conditional expectation on $L^2(\mathcal{F}, \mathbf{P})$, but it can be extended to $L^1(\mathcal{F}, \mathbf{P})$.

For any event $\Lambda \in \mathcal{F}$, the conditional probability $\mathbf{P}[\Lambda|\mathcal{F}_t]$ of Λ given $\mathcal{F}_t$ is defined by

$$\mathbf{P}[\Lambda|\mathcal{F}_t] = \mathbf{P}[\mathbf{1}_\Lambda|\mathcal{F}_t].$$

The conditional probability $\mathbf{P}[\Lambda|\mathcal{F}_t]$ is an $\mathcal{F}_t$-measurable random variable.

Let us express the Markov property in terms of conditional expectation.

In Section 1.2 we defined the Markov property by equation (1.2.14), that is,

$$\mathbf{P}_{(r,x)}[g(X_{t_1}, \ldots, X_s) f(X_t)] = \mathbf{P}_{(r,x)}[g(X_{t_1}, \ldots, X_s) \mathbf{P}_{(s,X_s)}[f(X_t)]],$$

where $g(X_{t_1}, \ldots, X_s)$ is an arbitrary $\mathcal{F}_s$-measurable random variable. Due to the arbitrariness of g, for any $\mathcal{F}_s$-measurable random variable Y we have

$$\mathbf{P}_{(r,x)}[Yf(X_t)] = \mathbf{P}_{(r,x)}[Y\mathbf{P}_{(s,X_s)}[f(X_t)]],$$

where in view of equation (4.7.1) the left-hand side is

$$\mathbf{P}_{(r,x)}[Yf(X_t)] = \mathbf{P}_{(r,x)}[Y\mathbf{P}_{(r,x)}[f(X_t)|\mathcal{F}_s]].$$

Hence we can rewrite the equation (1.2.14) referred to above in terms of conditional expectation as

$$\mathbf{P}_{(r,x)}[Y\mathbf{P}_{(r,x)}[f(X_t)|\mathcal{F}_s]] = \mathbf{P}_{(r,x)}\left[Y\mathbf{P}_{(s,X_s)}\left[f(X_t)\right]\right]. \tag{4.7.2}$$

Since this holds for any $\mathcal{F}_s$-measurable random variable Y, we have

$$\mathbf{P}_{(r,x)}[f(X_t)|\mathcal{F}_s] = \mathbf{P}_{(s,X_s)}[f(X_t)], \quad \mathbf{P}_{(r,x)}\text{-a.e.} \tag{4.7.3}$$

This is the expression of the Markov property in terms of conditional expectations.

The left-hand side of equation (4.7.3) is the conditional expectation of $f(X_t)$ given that we know the motion until time s. The right-hand side is the conditional expectation when a particle starts X_s at time s. The assertion is that they are equal.

Thus, if we know the behavior of a particle up to time s, its behavior after the time s is determined by the particle's position X_s at the time s only, i.e, it does not depend on the behavior of the motion up to time s. This is the Markov property. In computations concerning the Markov property we often use equations (4.7.2) or (4.7.3), because they are convenient.

Remark. "$\mathbf{P}_{(r,x)}$-a.e." in equation (4.7.3) means that it does not hold for all $\omega \in \Omega$, but the probability of the set of all ω for which it fails is equal to zero, so this set can be ignored.

On martingales

For Markov processes, we relate X_s and X_t by means of the transition probability; here we will consider stochastic processes for which the relationship between X_s and X_t is characterized in terms of conditional expectations.

Let $\{\Omega, \mathcal{F}, \mathbf{P}\}$ be a probability space, and let the subsets $\mathcal{F}_t$, $t \in [a,b]$, of $\mathcal{F}$ form an increasing family of σ-algebras; we assume that the family $\mathcal{F}_t$, $t \in [a,b]$ is complete, i.e., it contains all sets of $\mathbf{P}$-measure zero.

If a stochastic process X_t is $\mathcal{F}_t$-measurable and $\mathbf{P}$-integrable, i.e., $\mathbf{P}\left[|X_t|\right] < \infty$, and if

$$\mathbf{P}[X_t|\mathcal{F}_s] = X_s, \quad \mathbf{P}\text{-a.e.}, \quad s \le t,$$

then X_t is called a **martingale**.

Example 1. One-dimensional Brownian motion $\{B_t, t \ge 0; \mathbf{P}\}$ is a martingale. Moreover, the process $B_t^2 - t$ is also a martingale.

Proof. Thanks to the Markov property (4.7.3),

$$\mathbf{P}[f(B_t)|\mathcal{F}_s] = \mathbf{P}_{(s,B_s)}[f(B_t)], \quad \mathbf{P}-\text{a.e.}.$$

If term of the transition probability density $q(s,x,t,y)$ given by formula (1.1.5), the right-hand side becomes

$$\mathbf{P}[f(B_t)|\mathcal{F}_s] = \int q(s,B_s,t,y)f(y)dy, \quad \mathbf{P}-\text{a.e.}. \tag{4.7.4}$$

Since $\int q(s,x,t,y)ydy = x$, if we set $f(y) = y$ in (4.7.4), the right-hand side becomes B_s. Hence,

$$\mathbf{P}[B_t|\mathcal{F}_s] = B_s, \quad \mathbf{P}-\text{a.e.},$$

so the Brownian motion $\{B_t, t \geq 0; \mathbf{P}\}$ is a martingale. Moreover,

$$\int q(s,x,t,y)y^2 dy = \int q(s,x,t,y)(x-y)^2 dy$$
$$+ 2x \int q(s,x,t,y)y\,dy - x^2 \int q(s,x,t,y)dy$$
$$= t - s + x^2,$$

where in the first integral we used formula (1.2.2). Taking $f(y) = y^2$ in (4.7.4), we have

$$\mathbf{P}[B_t{}^2|\mathcal{F}_s] = \int q(s,B_s,t,y)y^2 dy = t - s + B_s{}^2, \quad \mathbf{P}-\text{a.e.}.$$

Hence

$$\mathbf{P}\left[B_t^2 - t|\mathcal{F}_s\right] = B_s^2 - s, \quad \mathbf{P}-\text{a.e.},$$

that is, $B_t^2 - t$ is a martingale, as claimed. $\square$

The converse also holds, i.e., we have the following theorem of P. Lévy.

Theorem 4.7.1. *A necessary and sufficient condition for a continuous stochastic process X_t to be a one-dimensional Brownian motion is that X_t and $X_t^2 - t$ are continuous martingales.*

Proof. We need to show that the condition of the theorem is sufficient. Suppose X_t and $X_t^2 - t$ are continuous martingales. Then $(dX_t)^2 = dt$, and Itô's formula (1.3.12) for Brownian motion holds also for the martingale X_t.

Hence, applying Itô's formula (1.3.12) to the function $f(x) = e^{i\lambda x}$ and the martingale X_t, we have

$$e^{i\lambda X_t} - e^{i\lambda X_s} = \int_s^t i\lambda e^{i\lambda X_r} dX_r + \frac{1}{2}\int_s^t -\lambda^2 e^{i\lambda X_r} dr.$$

Multiplying both sides by $e^{-i\lambda X_s}$ and computing the conditional expectations, we have

$$\mathbf{P}\left[e^{i\lambda(X_t - X_s)}|\mathcal{F}_s\right] = 1 - \frac{\lambda^2}{2}\int_s^t \mathbf{P}\left[e^{i\lambda(X_r - X_s)}|\mathcal{F}_s\right]dr.$$

Then observing that the equation

$$f(t) = 1 - \frac{\lambda^2}{2}\int_s^t f(r)dr$$

has the unique solution

$$f(t) = \exp\left(-\lambda^2(t - s)/2\right),$$

we have

$$\mathbf{P}\left[e^{i\lambda(X_t - X_s)}|\mathcal{F}_s\right] = \exp\left(-\lambda^2(t - s)/2\right),$$

whence

$$\mathbf{P}\left[X_t \in dy|\mathcal{F}_s\right] = \frac{1}{\sqrt{2\pi(t - s)}}\exp\left(-(y - X_s)^2/2(t - s)\right)dy.$$

Therefore, X_t is Brownian motion. □

Example 2. The stochastic integral

$$Y_t = \int_a^t f(s, X_s)\,dB_s,$$

where $f(t, x)$ is a bounded continuous function and X_t is continuous and $\mathcal{F}_t$-measurable, is a martingale.

Proof. Let $a = s_0 < \cdots < s_k < \cdots < s_n = t$. Under the above assumptions on $f(t, x)$, the stochastic integral is the limit of the sequence

$$Y_t^n = \sum_{k=0}^n f(s_k, X_{s_k})\left(B_{s_{k+1}} - B_{s_k}\right),$$

in the function space $L^2(F, \mathbf{P})$. Consequently,

$$\mathbf{P}\left[Y_t^n|\mathcal{F}_s\right] = \sum_{k=0}^n \mathbf{P}\left[f(s_k, X_{s_k})\left(B_{s_{k+1}} - B_{s_k}\right)|\mathcal{F}_s\right]. \tag{4.7.5}$$

Here, if $s_k \geq s$, then $\mathcal{F}_{s_k} \supset \mathcal{F}_s$, and hence

$$\mathbf{P}\left[f(s_k, X_{s_k})\left(B_{s_{k+1}} - B_{s_k}\right)|\mathcal{F}_s\right]$$
$$= \mathbf{P}\left[\mathbf{P}\left[f(s_k, X_{s_k})\left(B_{s_{k+1}} - B_{s_k}\right)|\mathcal{F}_{s_k}\right]|\mathcal{F}_s\right]$$
$$= \mathbf{P}\left[f(s_k, X_{s_k})\mathbf{P}\left[B_{s_{k+1}} - B_{s_k}|\mathcal{F}_{s_k}\right]|\mathcal{F}_s\right],$$

which in view of the fact that the Brownian motion is a martingale is equal to

$$\mathbf{P}\left[f\left(s_k, X_{s_k}\right)\left(B_{s_k} - B_{s_k}\right)|\mathcal{F}_s\right] = 0.$$

Therefore, what remains on the right-hand side of equality (4.7.5) are the terms with $s_k \leq s$, and so

$$\mathbf{P}\left[Y_t^n|\mathcal{F}_s\right] = \sum_{k=0}^{n} f\left(s_k, X_{s_k}\right)\left(B_{s_{k+1}\wedge s} - B_{s_k\wedge s}\right) = Y_s^n.$$

Letting $n \to \infty$, we conclude that

$$\mathbf{P}\left[Y_t|\mathcal{F}_s\right] = Y_s,$$

i.e., the stochastic integral Y_t is a martingale. □

For further properties of martingales, see, e.g., Section 16.2 of Nagasawa (2000).

4.8 Paths of Brownian Motion

In this section, we will show that the paths of Brownian motion can be constructed explicitly. This was discovered by N. Wiener and P. Lévy.

Einstein (1905) thought that Brownian motion is caused by countless collisions of small particles with the particle that is being observed. Let us construct a path of the Brownian motion, which is "a non-differentiable function at all values of time". Looking at how to describe such path, we first note that a path of the Brownian motion $B\left(t, \omega\right)$ is obviously continuous as a function of time t. We adopt McKean's approach (1969), which is slightly different from that of N. Wiener and P. Lévy.

Let the time interval be $[0, 1]$ and consider the motion of a particle that was at the origin at time $t = 0$ and collided with many particles, one after another, on the left and on the right.

If we allow such collisions and successively halve the interval, then we get a function of the form

$$B\left(t\right) = g_0 f_0\left(t\right) + \sum_{n=1}^{\infty} \sum_{\text{odd } k < 2^n} g_{k,2^{-n}} f_{k,2^{-n}}\left(t\right), \quad t \in [0, 1],$$

where

$$\left\{f_{k,2^{-n}}\left(t\right), \ k \text{ odd}, \ k < 2^n, \ n \geq 1\right\}$$

are tent functions given by

$$f_{k,2^{-n}}\left(t\right) = \int_0^t h_{k,2^{-n}}\left(t\right) dt,$$

with

$$h_{k,2^{-n}}(t) = \begin{cases} 2^{\frac{n-1}{2}}, & \text{if } (k-1)\,2^{-n} \le t < k2^{-n}, \\ -2^{\frac{n-1}{2}}, & \text{if } k2^{-n} \le t < (k+1)\,2^{-n}, \\ = 0, & \text{otherwise.} \end{cases}$$

The functions $h_{k,2^{-n}}(t)$ used here to define $B(t)$ are called Haar functions. The system of Haar functions

$$\left\{ h_0(t) \equiv 0,\ h_{k,2^{-n}}(t),\ k \text{ odd},\ k < 2^n,\ n \ge 1 \right\}$$

belongs to the function space $L^2([0,1])$ and enjoys the following properties:

Completeness: If $q \in L^2([0,1])$ satisfies

$$\int_0^1 q(t)\, h_0(t)\, dt = 0$$

and

$$\int_0^1 q(t)\, h_{k,2^{-n}}(t)\, dt = 0, \quad k \text{ odd},\ k < 2^n,\ n \ge 1,$$

then $q(t) \equiv 0$.

Normality:

$$\int_0^1 h_0^2(t)\, dt = 1, \quad \int_0^1 h_{k,2^{-n}}^2(t)\, dt = 1.$$

Orthogonality:

$$\int_0^1 h_0(t)\, h_{k,2^{-n}}(t)\, dt = 0$$

and

$$\int_0^1 h_{k,2^{-n}}(t)\, h_{j,2^{-m}}(t)\, dt = 0, \quad k2^{-n} \ne j2^{-m}.$$

Since the Haar functions form a complete orthonormal system in $L^2([0,1])$, the Parseval equality holds:

For any $p, q \in L^2([0,1])$,

$$\int_0^1 p(t)\, q(t)\, dt = \int_0^1 p(t)\, h_0(t)\, dt \int_0^1 q(t)\, h_0(t)\, dt$$

$$+ \sum_{n=1}^{\infty} \sum_{\text{odd } k < 2^n} \int_0^1 p(t)\, h_{k,2^{-n}}(t)\, dt \int_0^1 q(t)\, h_{k,2^{-n}}(t)\, dt.$$

Now as the coefficients

$$\left\{ g_0,\ g_{k,2^{-n}};\ k \text{ odd},\ k < 2^n,\ n \ge 1 \right\},$$

in the definition of the function $B(t)$ we choose independent normally distributed random variables, i.e., with probability distributions

$$\mathbf{P}[g_0 \in dx] = \mathbf{P}[g_{k,2^{-n}} \in dx] = \frac{1}{\sqrt{2\pi}} \exp\left(-\frac{x^2}{2}\right) dx.$$

Then $B(t)$ becomes a function of the variables $t \in [0,1]$ and ω:

$$B(t,\omega) = g_0(\omega) f_0(t) + \sum_{n=1}^{\infty} \sum_{\text{odd } k < 2^n} g_{k,2^{-n}}(\omega) f_{k,2^{-n}}(t). \qquad (4.8.1)$$

However, if we do not need to emphasize the variable ω, we can omit ω in $B(t,\omega)$ and henceforth simply denote the random variable as $B(t)$ in many cases.

Since $B(t)$ is a sum of normally distributed independent random variables, it is also normally distributed.

Now let us take two times $t_1, t_2 \in [0,1]$ and compute the expectation

$$\mathbf{P}[B(t_1) B(t_2)].$$

Since $\{g_0, \ g_{k,2^{-n}} \ k \text{ odd}, \ k < 2^n, \ n \geq 1\}$ are independent random variables,

$$\mathbf{P}[g_{k,2^{-n}} g_{j,2^{-m}}] = 0, \quad k2^{-n} \neq j2^{-m}.$$

Hence,

$$\mathbf{P}[B(t_1) B(t_2)] = \mathbf{P}[g_0 f_0(t_1) g_0 f_0(t_2)]$$
$$+ \sum_{n=1}^{\infty} \sum_{\text{odd} k < 2^n} \mathbf{P}\left[g_{k,2^{-n}} f_{k,2^{-n}}(t_1) g_{k,2^{-n}} f_{k,2^{-n}}(t_2)\right].$$

Moreover, since

$$\mathbf{P}\left[|g_0|^2\right] = 1 \quad \text{and} \quad \mathbf{P}\left[|g_{k,2^{-n}}|^2\right] = 1,$$

we have

$$\mathbf{P}[B(t_1) B(t_2)] = f_0(t_1) f_0(t_2) + \sum_{n=1}^{\infty} \sum_{\text{odd } k < 2^n} f_{k,2^{-n}}(t_1) f_{k,2^{-n}}(t_2)$$

$$= \int_0^{t_1} h_0(t) \, dt \int_0^{t_2} h_0(s) \, ds$$

$$+ \sum_{n=1}^{\infty} \sum_{\text{odd } k < 2^n} \int_0^{t_1} h_{k,2^{-n}}(t) \, dt \int_0^{t_2} h_{k,2^{-n}}(s) \, ds,$$

which by the Parseval equality is equal to

$$\int_0^1 \mathbf{1}_{[0,t_1]}(s) \mathbf{1}_{[0,t_2]}(s) \, ds = t_1 \wedge t_2.$$

Thus, we obtained the formula

$$\mathbf{P}\left[B\left(t_1\right)B\left(t_2\right)\right]=t_1\wedge t_2, \tag{4.8.2}$$

and, as a special case,

$$\mathbf{P}\left[\left|B\left(t\right)\right|^2\right]=t. \tag{4.8.3}$$

That is, the variance of $B\left(t\right)$ is equal to t.

Now, preparing independent copies of $\{B\left(t,\omega\right),\ t\in[0,1]\}$, we can extend the range of the time variable as

$$B\left(t,\omega_1\right),\quad t\in[0,1]\,,$$

$$B\left(1,\omega_1\right)+B\left(t-1,\omega_2\right),\quad t\in[1,2]\,,$$

$$B\left(1,\omega_1\right)+B\left(1,\omega_2\right)+B\left(t-2,\omega_3\right),\quad t\in[2,3]\,,$$

$$\dots\dots\dots\dots\dots\dots\dots\dots\dots\dots\dots\dots$$

and then define, for all values of time $t\in[0,\infty)$, ,

$$B\left(t,\varpi\right),\quad\varpi=\left(\omega_1,\omega_2,\omega_3,\dots\right),\quad\in[0,\infty)\,. \tag{4.8.4}$$

The random variable $B\left(t\right)$ constructed in this manner is normally distributed, since it is a sum of normally distributed random variables. By formula (4.8.3), its variance is t. Therefore:

(i) The distribution of the stochastic process $B\left(t\right)$ is

$$\mathbf{P}\left[B\left(t\right)\in dx\right]=\frac{1}{\sqrt{2\pi t}}\exp\left(-\frac{x^2}{2t}\right)dx. \tag{4.8.5}$$

(ii) $B\left(0,\varpi\right)=0$, and $B\left(t,\varpi\right)$ is a continuous function of t.

(iii) Moreover, for $r<s\le s'<t$, formula (4.8.2) implies that

$$\mathbf{P}\left[\left(B\left(t\right)-B\left(s'\right)\right)\left(B\left(s\right)-B\left(r\right)\right)\right]=0,\quad\left(t,s'\right)\cap\left(s,r\right)=\emptyset.$$

Hence, the variables $B\left(t\right)-B\left(s'\right)$ and $B\left(s\right)-B\left(r\right)$ are independent.

Therefore, the function $B\left(t,\varpi\right)$ defined in (4.8.4) is a path of the Brownian motion, i.e., $\{B\left(t,\varpi\right),\mathbf{P}\}$ is the one-dimensional Brownian motion started from the origin at time $t=0$.

Further, preparing d one-dimensional Brownian motions

$$\{B_i\left(t\right);\Omega_i,\mathcal{F}_i,\mathbf{P}_i\}\,,\ i=1,2,\dots,d,$$

we first introduce the product space

$$\Omega = \prod_{1=1}^{d} \Omega_i, \quad \mathcal{F} = \prod_{i=1}^{d} \mathcal{F}_i, \quad \mathbf{P} = \prod_{i=1}^{d} \mathbf{P}_i.$$

Then if for $\boldsymbol{\omega} = (\omega_1, \omega_2, \ldots, \omega_d) \in \Omega$ we set

$$B(t, \boldsymbol{\omega}) = (B_1(t, \omega_1), B_2(t, \omega_2), \ldots, B_d(t, \omega_d)),$$

we obtain a d-dimensional Brownian motion $\{B(t, \omega); \Omega, \mathcal{F}, \mathbf{P}\}$.

A further noteworthy fact is that the path of the Brownian movement is continuous, but it cannot be differentiated everywhere with respect to the time variable t.

Theorem 4.8.1. *For almost every fixed* $\boldsymbol{\omega}$ *the Brownian motion* $B(t, \boldsymbol{\omega})$ *is not differentiable at time* t.

Proof. Suppose, for simplicity, that the space is one-dimensional and the time interval is $[0, 1]$. First, note that if we set

$$A_{\ell n} = \bigcup_{0 < i \leq n+2} \bigcap_{i < k \leq i+3} \left\{ \omega : \left| B\left(\frac{k}{n}, \omega\right) - B\left(\frac{k-1}{n}, \omega\right) \right| < \frac{7\ell}{n} \right\},$$

then

$$\mathbf{P}[A_{\ell n}] \leq (n+2)\, \mathbf{P}\left[\left| B\left(\frac{1}{n}\right) \right| < \frac{7\ell}{n} \right]^3$$
$$= (n+2) \left(\int_{|x| < \frac{7\ell}{\sqrt{n}}} \frac{1}{\sqrt{2\pi}} e^{-\frac{x^2}{2}}\, dx \right)^3 < \frac{\text{const.}}{\sqrt{n}} \downarrow 0, \quad n \uparrow \infty. \tag{4.8.6}$$

However, if $B(t, \boldsymbol{\omega})$ is differentiable, then there exist $0 \leq s \leq 1$ and $m \geq 1$ such as

$$|B(t, \boldsymbol{\omega}) - B(s, \boldsymbol{\omega})| < \ell(t - s), \quad 0 < |t - s| < \frac{5}{n}, \quad n \geq m.$$

That is,

$$\{\boldsymbol{\omega} : B(t, \boldsymbol{\omega}) \text{ is differentiable at } t \in [0, 1]\} \subset \bigcup_{\ell \geq 1} \bigcup_{m \geq 1} \bigcap_{n \geq m} A_{\ell n}.$$

Therefore, by inequality (4.8.6)

$$\mathbf{P}[\{\boldsymbol{\omega} : B(t, \boldsymbol{\omega}) \text{ is differentiable at some time point } t \in [0, 1]\}] = 0.$$

Hence,

$$\mathbf{P}[\{\boldsymbol{\omega} : B(t, \boldsymbol{\omega}) \text{ is not differentiable everywhere in } [0, 1]\}] = 1,$$

which completes the proof. $\qquad \square$

Chapter 5

Applications of Relative Entropy

In this chapter we first introduce the concept of the relative entropy and prove the entropy projection theorem for probability measures. Then, as an application of the projection theorem, we establish the existence of a unique solution of the variational principle (principle of least action). (Incidentally, the existence of the solution was assumed when we discussed the principle of least action in Section 1.14.) Further, applying the projection theorem we show that the entrance function and the exit function can be constructed for given initial and terminal distributions. When we constructed the random motion in Section 1.4, we assumed that entrance and exit functions were given. In the preceding chapters we considered that the complex evolution equation describes the motion of a single particle. In Section 5.6, we will show that it actually describes also the motion of a group (cloud) of particles. We will discuss the "Schrödinger cloud" and Kac's phenomenon of "propagation of chaos", and note that the Schrödinger equation can be regarded as a kind of Boltzmann equation for a set of electrons.

5.1 Relative Entropy

Relative entropy is a quantity that measures how different two given probability distributions are.

Let $\mathbf{Q}$ and $\mathbf{P}$ be probability measures. If $\mathbf{P}(\Gamma) = 0$ implies $\mathbf{Q}(\Gamma) = 0$, we say that the probability measure $\mathbf{Q}$ is absolutely continuous with respect to the probability measure $\mathbf{P}$, and write $\mathbf{Q} \ll \mathbf{P}$. In this case there exists a density function q such that

$$\mathbf{Q}(\Gamma) = \int_{\Gamma} q \, d\mathbf{P}.$$

M. Nagasawa, *Markov Processes and Quantum Theory*, Monographs in Mathematics 109, https://doi.org/10.1007/978-3-030-62688-4_5

This density function is denoted by $q = \frac{d\mathbf{Q}}{d\mathbf{P}}$ and is called the Radon-Nikodym derivative of the measure $\mathbf{Q}$ with respect to the measure $\mathbf{P}$.

Definition 5.1.1. The **relative entropy** $\mathrm{H}\left(\mathbf{Q}|\mathbf{P}\right)$ of a probability measure $\mathbf{Q}$ with respect to a probability measure $\mathbf{P}$ is defined as

$$\mathrm{H}\left(\mathbf{Q}|\mathbf{P}\right) = \begin{cases} \int q \log q\, d\mathbf{P}, & \text{if } \mathbf{Q} \ll \mathbf{P}, \\ \infty, & \text{otherwise.} \end{cases} \tag{5.1.1}$$

Equivalently,

$$\mathrm{H}\left(\mathbf{Q}|\mathbf{P}\right) = \int \log \frac{d\mathbf{Q}}{d\mathbf{P}}\, d\mathbf{Q}. \tag{5.1.2}$$

Let us state a number of important properties of the relative entropy.

Property 1.

$$\mathrm{H}\left(\mathbf{Q}|\mathbf{P}\right) \geq 0, \textit{with equality if and only if } \mathbf{P} = \mathbf{Q}. \tag{5.1.3}$$

Proof. Set

$$g\left(x\right) = x \log x - x + 1$$

for $x \geq 0$. Then the function $g\left(x\right) \geq 0$ and is strictly convex with the minimum $g(1) = 0$. Therefore

$$0 \leq \int g\left(q\right) d\mathbf{P} = \int q \log q\, d\mathbf{P} = \mathrm{H}\left(\mathbf{Q}|\mathbf{P}\right),$$

and the minimum 0 is attained when $q = \frac{d\mathbf{Q}}{d\mathbf{P}} = 1$ $\mathbf{P}$-a.e.. $\qquad\qquad\square$

We denote the set of all probability measures on the space Ω by $\mathbf{M}_1\left(\Omega\right)$. The distance between two elements $\mathbf{Q}, \mathbf{P} \in \mathbf{M}_1\left(\Omega\right)$ is defined by

$$\|\mathbf{Q} - \mathbf{P}\| = \sup_{|f| \leq 1} |\mathbf{Q}[f] - \mathbf{P}[f]|, \tag{5.1.4}$$

where the right-hand side is the total variation of the difference $\mathbf{Q} - \mathbf{P}$.

Property 2. *For* $\mathbf{Q}, \mathbf{P} \in \mathbf{M}_1\left(\Omega\right)$,

$$\|\mathbf{Q} - \mathbf{P}\|^2 \leq 2\mathrm{H}\left(\mathbf{Q}|\mathbf{P}\right). \tag{5.1.5}$$

Proof. First note that if $\mathbf{Q} \ll \mathbf{P}$, then

$$\|\mathbf{Q} - \mathbf{P}\| = \int |q - 1|\, d\mathbf{P}, \quad q = \frac{d\mathbf{Q}}{d\mathbf{P}}. \tag{5.1.6}$$

Indeed, by (5.1.4)

$$\|\mathbf{Q} - \mathbf{P}\| \leq \sup_{|f| \leq 1} \int |q - 1|\, |f|\, d\mathbf{P} \leq \int |q - 1|\, d\mathbf{P}. \tag{5.1.7}$$

On the other hand,

$$|\mathbf{Q}\,[f] - \mathbf{P}\,[f]| = \left| \int (q-1)\,f\,d\mathbf{P} \right|.$$

In particular, if

$$f = \mathbf{1}_{\{q-1 \geq 0\}} - \mathbf{1}_{\{q-1 < 0\}},$$

then

$$|\mathbf{Q}\,[f] - \mathbf{P}\,[f]| = \int |q-1|\,d\mathbf{P}.$$

Hence, by (5.1.4),

$$\|\mathbf{Q} - \mathbf{P}\| \geq \int |q-1|\,d\mathbf{P},$$

which in conjunction with (5.1.7) yields formula (5.1.6).

Next, set for $x \geq 0$,

$$h\,(x) = 4 + 2x \quad \text{and} \quad g\,(x) = x \log x - x + 1.$$

Then $h\,(x), g\,(x) \geq 0$. Moreover, since

$$F\,(x) = h\,(x)\,g\,(x) - 3(x-1)^2$$

is a convex function with the minimum value $F(1) = 0$, we have

$$h\,(x)\,g\,(x) \geq 3(x-1)^2.$$

Consequently,

$$3\|\mathbf{Q} - \mathbf{P}\|^2 = 3\left(\int |q-1|\,d\mathbf{P} \right)^2 = \left(\int \sqrt{3|q-1|^2}\,d\mathbf{P} \right)^2$$

$$\leq \left[\int \sqrt{h\,(q)}\sqrt{g\,(q)}\,d\left(\int \sqrt{3|q-1|^2}\,d\mathbf{P} \right)^2 \right]^2$$

$$\leq \int h\,(q)\,d\mathbf{P} \int g\,(q)\,d\mathbf{P},$$

where we used the Cauchy-Schwarz inequality. Since the first integral on the right-hand side is

$$\int h\,(q)\,d\mathbf{P} = \int \left(4 + 2\frac{d\mathbf{Q}}{d\mathbf{P}} \right)\,d\mathbf{P} = 6,$$

we conclude that

$$3\|\mathbf{Q} - \mathbf{P}\|^2 \leq 6\int g\,(q)\,d\mathbf{P} = 6\int q \log q\,d\mathbf{P} = 6\mathrm{H}\,(\mathbf{Q}|\mathbf{P}),$$

i.e., (5.1.5) holds. $\qquad\qquad\qquad\square$

Definition 5.1.2. A sequence of probability measures $\mathbf{Q}_n \in \mathbf{M}_1(\Omega)$ is said to **converge to $\mathbf{P} \in \mathbf{M}_1(\Omega)$ in entropy** if

$$\mathrm{H}(\mathbf{Q}_n|\mathbf{P}) \to 0 \quad as\ n \to \infty.$$

Property 3. *If a sequence $\{\mathbf{Q}_n\} \subset \mathbf{M}_1(\Omega)$ converges to $\mathbf{P} \in \mathbf{M}_1(\Omega)$ in entropy, then it converges in the total variation distance, i.e.,*

$$\|\mathbf{Q}_n - \mathbf{P}\| \to \infty \quad as\ n \to \infty.$$

Proof. This is an immediate consequence of inequality (5.1.5).   □

Property 4. *The following parallelogram identity holds for any three probability measures $\mathbf{P}$, $\mathbf{Q}$ and $\mathbf{R}$:*

$$\mathrm{H}(\mathbf{P}|\mathbf{R}) + \mathrm{H}(\mathbf{Q}|\mathbf{R})$$
$$= 2\mathrm{H}\left(\frac{\mathbf{P}+\mathbf{Q}}{2}\middle|\mathbf{R}\right) + \mathrm{H}\left(\mathbf{P}\middle|\frac{\mathbf{P}+\mathbf{Q}}{2}\right) + \mathrm{H}\left(\mathbf{Q}\middle|\frac{\mathbf{P}+\mathbf{Q}}{2}\right) \qquad (5.1.8)$$

Proof. Consider the Radon-Nikodym derivatives with respect to $\mathbf{R}$,

$$p = \frac{d\mathbf{P}}{d\mathbf{R}} \quad \text{and} \quad q = \frac{d\mathbf{Q}}{d\mathbf{R}}.$$

Then

$$2\mathrm{H}\left(\frac{\mathbf{P}+\mathbf{Q}}{2}\middle|\mathbf{R}\right) = \int (p+q)\log\left(\frac{p+q}{2}\right)d\mathbf{R},$$

$$\mathrm{H}\left(\mathbf{P}\middle|\frac{\mathbf{P}+\mathbf{Q}}{2}\right) = \int p\log\left(\frac{2p}{p+q}\right)d\mathbf{R},$$

$$\mathrm{H}\left(\mathbf{Q}\middle|\frac{\mathbf{P}+\mathbf{Q}}{2}\right) = \int q\log\left(\frac{2q}{p+q}\right)d\mathbf{R}.$$

Adding these three equalities, we obtain the claimed parallelogram identity (5.1.8).
                                                                                    □

The following theorem due to Csiszár (Csiszár (1975)) holds in the space of probability measures $\mathbf{M}_1(\Omega)$.

Theorem 5.1.1 (Entropy Projection Theorem).

(i) *Let $\mathbf{A} \subset \mathbf{M}_1(\Omega)$ be a closed convex subset. Assume that for a point $\overline{\mathbf{P}} \in \mathbf{M}_1(\Omega)$ there exists at least one point $\mathbf{P} \in \mathbf{A}$ such that $\mathrm{H}(\mathbf{P}|\overline{\mathbf{P}}) < \infty$. Then there exists a projection $\mathbf{Q} \in \mathbf{A}$ of $\overline{\mathbf{P}}$ on the subset $\mathbf{A}$ such that*

$$\inf_{\mathbf{P} \in \mathbf{A}} \mathrm{H}(\mathbf{P}|\overline{\mathbf{P}}) = \mathrm{H}(\mathbf{Q}|\overline{\mathbf{P}}). \qquad (5.1.9)$$

The projection $\mathbf{Q} \in \mathbf{A}$ is uniquely determined.

(ii) *A necessary and sufficient condition for* $\mathbf{Q} \in \mathbf{A}$ *to be the projection of* $\overline{\mathbf{P}} \in$ $\mathbf{M_1}(\Omega)$ *on the set* $\mathbf{A} \subset \mathbf{M_1}(\Omega)$ *is that* $\mathrm{H}(\mathbf{Q}|\overline{\mathbf{P}}) < \infty$ *and*

$$\mathrm{H}(\mathbf{P}|\overline{\mathbf{P}}) \geq \mathrm{H}(\mathbf{P}|\mathbf{Q}) + \mathrm{H}(\mathbf{Q}|\overline{\mathbf{P}}) \tag{5.1.10}$$

for all $\mathbf{P} \in \mathbf{A}$.

Proof. Let us prove assertion (i). Denote

$$\rho = \inf_{\mathbf{P} \in \mathbf{A}} \mathrm{H}(\mathbf{P}|\overline{\mathbf{P}}),$$

and take a sequence $\{\mathbf{P}_n\} \subset \mathbf{A}$ such that $\mathrm{H}(\mathbf{P}_n|\overline{\mathbf{P}}) < \infty$ and

$$\lim_{n \to \infty} \mathrm{H}(\mathbf{P}_n|\overline{\mathbf{P}}) = \rho.$$

Since the set $\mathbf{A}$ is convex,

$$\frac{\mathbf{P}_m + \mathbf{P}_n}{2} \in \mathbf{A}.$$

Observing that

$$\mathrm{H}\left(\frac{\mathbf{P}_m + \mathbf{P}_n}{2}\middle|\overline{\mathbf{P}}\right) \geq \rho,$$

we have

$$\mathrm{H}(\mathbf{P}_m|\overline{\mathbf{P}}) + \mathrm{H}(P_n|\overline{\mathbf{P}}) - 2\rho \geq \mathrm{H}(\mathbf{P}_m|\overline{\mathbf{P}}) + \mathrm{H}(\mathbf{P}_n|\overline{\mathbf{P}}) - 2\mathrm{H}\left(\frac{\mathbf{P}_m + \mathbf{P}_n}{2}\middle|\overline{\mathbf{P}}\right),$$

which by the parallelogram identity (5.1.8) is equal to

$$\mathrm{H}\left(\mathbf{P}_m\middle|\frac{\mathbf{P}_m + \mathbf{P}_n}{2}\right) + \mathrm{H}\left(\mathbf{P}_n\middle|\frac{\mathbf{P}_m + \mathbf{P}_n}{2}\right) \geq 0,$$

where the left-hand side vanishes when $m, n \to \infty$.

Therefore, when $m, n \to \infty$,

$$\mathrm{H}\left(\mathbf{P}_m\middle|\frac{\mathbf{P}_m + \mathbf{P}_n}{2}\right) \to 0, \quad \mathrm{H}\left(\mathbf{P}_n\middle|\frac{\mathbf{P}_m + \mathbf{P}_n}{2}\right) \to 0. \tag{5.1.11}$$

Moreover, by inequality (5.1.5),

$$\|\mathbf{P}_m - \mathbf{P}_n\| \leq \left\|\mathbf{P}_m - \frac{\mathbf{P}_m + \mathbf{P}_n}{2}\right\| + \left\|\mathbf{P}_n - \frac{\mathbf{P}_m + \mathbf{P}_n}{2}\right\|$$

$$\leq \sqrt{2\mathrm{H}\left(\mathbf{P}_m\middle|\frac{\mathbf{P}_m + \mathbf{P}_n}{2}\right)} + \sqrt{2\mathrm{H}\left(\mathbf{P}_n\middle|\frac{\mathbf{P}_m + \mathbf{P}_n}{2}\right)}.$$

Since, by (5.1.11), the right-hand side vanishes in the limit $m, n \to \infty$, we have $\|\mathbf{P}_m - \mathbf{P}_n\| \to 0$. Since the set $\mathbf{A}$ is closed, there exists $\mathbf{Q} \in \mathbf{A}$ such that $\|\mathbf{P}_n - \mathbf{P}\| \to 0$.

Since $\mathbf{Q} \in \mathbf{A}$, clearly

$$\inf_{\mathbf{P} \in \mathbf{A}} \mathrm{H}\left(\mathbf{P}|\overline{\mathbf{P}}\right) \leq \mathrm{H}\left(\mathbf{Q}|\overline{\mathbf{P}}\right). \tag{5.1.12}$$

On the other hand, let

$$q = \frac{d\mathbf{Q}}{d\overline{\mathbf{P}}} \quad \text{and} \quad p_n = \frac{d\mathbf{P}_n}{d\overline{\mathbf{P}}},$$

and set

$$\eta_n = \inf_{m \geq n} p_m \log p_m.$$

Then the sequence $\{\eta_n\}$ is monotonically increasing and converges to $q \log q$. Hence,

$$\mathrm{H}\left(\mathbf{Q}|\overline{\mathbf{P}}\right) = \int q \log q \, d\overline{\mathbf{P}} = \lim_{n \to \infty} \int \eta_n \, d\overline{\mathbf{P}}.$$

Moreover, since $\eta_n \leq p_m \log p_m$ by definition,

$$\int \eta_n \, d\overline{\mathbf{P}} \leq \inf_{m \geq n} \int p_m \log p_m \, d\overline{\mathbf{P}}.$$

Therefore,

$$\lim_{n \to \infty} \int \eta_n \, d\overline{\mathbf{P}} \leq \lim_{n \to \infty} \inf_{m \geq n} \int p_m \log p_m \, d\overline{\mathbf{P}}$$
$$= \lim_{n \to \infty} \inf_{m \geq n} \mathrm{H}\left(\mathbf{P}_m|\overline{\mathbf{P}}\right) = \rho = \inf_{\mathbf{P} \in \mathbf{A}} \mathrm{H}\left(\mathbf{P}|\overline{\mathbf{P}}\right).$$

(The above inequality is known as Fatou's lemma in measure theory.)

Thus, together with inequality (5.1.12), we have

$$\inf_{\mathbf{P} \in \mathbf{A}} \mathrm{H}\left(\mathbf{P}|\overline{\mathbf{P}}\right) = \mathrm{H}\left(\mathbf{Q}|\overline{\mathbf{P}}\right).$$

To show that the projection $\mathbf{Q}$ is uniquely determined, suppose, by contradiction, that there are two projections, $\mathbf{Q}_1$ and $\mathbf{Q}_2$. Since the set $\mathbf{A}$ is convex, $\frac{\mathbf{Q}_1 + \mathbf{Q}_2}{2} \in \mathbf{A}$. In view of the parallelogram identity (5.1.8),

$$\mathrm{H}\left(\mathbf{Q}_1|\overline{\mathbf{P}}\right) + \mathrm{H}\left(\mathbf{Q}_2|\overline{\mathbf{P}}\right) - 2\rho \geq \mathrm{H}\left(\mathbf{Q}_1|\overline{\mathbf{P}}\right) + \mathrm{H}\left(\mathbf{Q}_2|\overline{\mathbf{P}}\right) - 2\mathrm{H}\left(\frac{\mathbf{Q}_1 + \mathbf{Q}_2}{2}\bigg|\overline{\mathbf{P}}\right)$$
$$= \mathrm{H}\left(\mathbf{Q}_1\bigg|\frac{\mathbf{Q}_1 + \mathbf{Q}_2}{2}\right) + \mathrm{H}\left(\mathbf{Q}_2\bigg|\frac{\mathbf{Q}_1 + \mathbf{Q}_2}{2}\right) \geq 0,$$

since the left-hand side vanishes

$$H\left(\mathbf{Q}_1\middle|\frac{\mathbf{Q}_1+\mathbf{Q}_2}{2}\right)=0.$$

Therefore

$$\mathbf{Q}_1=\frac{\mathbf{Q}_1+\mathbf{Q}_2}{2},$$

whence $\mathbf{Q}_1=\mathbf{Q}_2$. This completes the proof of (i).

Now let us prove assertion (ii). Let

$$\bar{p}=\frac{d\mathbf{P}}{d\overline{\mathbf{P}}} \quad\text{and}\quad \bar{q}=\frac{d\mathbf{Q}}{d\overline{\mathbf{P}}}.$$

Then

$$H\left(\mathbf{P}|\overline{\mathbf{P}}\right)-H\left(\mathbf{P}|\mathbf{Q}\right)=\int\left(\bar{p}\log\bar{p}-\bar{p}\log\frac{\bar{p}}{\bar{q}}\right)d\overline{\mathbf{P}}$$

$$=\int\bar{p}\log\bar{q}\,d\overline{\mathbf{P}}=\int\log\bar{q}\,d\mathbf{P}. \qquad(5.1.13)$$

Therefore, to establish the inequality (5.1.10), it is enough to prove the following statement. $\square$

Lemma 5.1.1. *A probability measure* $\mathbf{Q}\in\mathbf{A}$ *is the projection of the measure* $\overline{\mathbf{P}}\in M_1(\Omega)$ *on a subset* $\mathbf{A}\subset\mathbf{M_1}(\Omega)$ *if and only if* $H\left(\mathbf{Q}|\overline{\mathbf{P}}\right)<\infty$ *and*

$$\int\log\bar{q}\,d\mathbf{P}\geq H\left(\mathbf{P}|\overline{\mathbf{P}}\right) \quad\text{for all } \mathbf{P}\in\mathbf{A}. \qquad(5.1.14)$$

Proof. Define $y(\alpha)$, $0\leq\alpha\leq1$, by

$$y(\alpha)=p_\alpha\log p_\alpha,\quad p_\alpha=\alpha\bar{p}+(1-\alpha)\bar{q}.$$

The function $y(\alpha)$ is convex, as the composition of the convex function $g(x)=x\log x$ and the linear function p_α of α. Therefore, if $\alpha\downarrow$, then $y'(\alpha)\downarrow$; moreover,

$$y'(0)=(\bar{p}-\bar{q})(\log\bar{q}+1).$$

Let $\mathbf{P}_\alpha=p_\alpha\overline{\mathbf{P}}$. Then

$$\lim_{\alpha\downarrow0}\frac{1}{\alpha}\left\{H\left(\mathbf{P}_\alpha|\overline{\mathbf{P}}\right)-H\left(\mathbf{P}_0|\overline{\mathbf{P}}\right)\right\}=\lim_{\alpha\downarrow0}\frac{1}{\alpha}\int(y(\alpha)-y(0))\,d\overline{\mathbf{P}}$$

$$=\int y'(0)\,d\overline{\mathbf{P}}=\int(\bar{p}-\bar{q})(\log\bar{q}+1)\,d\overline{\mathbf{P}}$$

$$=\int\log\bar{q}\,d\mathbf{P}-H\left(\mathbf{P}|\overline{\mathbf{P}}\right). \qquad(5.1.15)$$

If inequality (5.1.14) does not hold, then the right-hand side of equation (5.1.15) is negative, and hence there is an $\alpha > 0$ such as $H\left(\mathbf{P}_\alpha|\overline{\mathbf{P}}\right) < H\left(\mathbf{P}_0|\overline{\mathbf{P}}\right)$. Since $\mathbf{P}_\alpha \in \mathbf{A}$ and $\mathbf{P}_0 = \mathbf{Q}$, $\mathbf{Q}$ is not a projection. Therefore, if $\mathbf{Q}$ is a projection, then (5.1.14) holds.

Conversely, suppose that (5.1.14) holds. Then in view of equality (5.1.13), inequality (5.1.10) holds. Consequently,

$$\inf_{\mathbf{P} \in \mathbf{A}} H\left(\mathbf{P}|\overline{\mathbf{P}}\right) \geq H\left(\mathbf{Q}|\overline{\mathbf{P}}\right),$$

which in view of (5.1.12) implies that equality holds, so $\mathbf{Q}$ is a projection.   □

Comments to Theorem 5.1.1. Since the square root of the relative entropy does not satisfy the triangle inequality, it is not a distance function. Nevertheless, the proof of Theorem 5.1.1 uses the relative entropy like the square of a distance. The square root of the relative entropy has properties similar to the square of a distance: (5.1.3), inequality (5.1.5), and the parallelogram identity (5.1.7) hold. Just like the projection theorem for metric spaces in functional analysis, the Entropy Projection Theorem plays an important role in the theory of stochastic processes and statistics.

5.2   Variational Principle

Let us apply the Entropy Projection Theorem to the variational principle.

As before, $\mathbf{M}_1\left(\Omega\right)$ denotes the set of probability measures on the space Ω. Let $\mathbf{A}_{a,b}$ denote the subset of $\mathbf{M}_1\left(\Omega\right)$ consisting of all measures with prescribed initial distribution density $\mu_a\left(x\right)$ and terminal distribution density $\mu_b\left(x\right)$.

Choosing appropriately a probability measure $\overline{\mathbf{P}}$, we project it on $\mathbf{A}_{a,b}$, and denote the projection by $\mathbf{Q}$. It can be shown that $\mathbf{Q} \in \mathbf{A}_{a,b}$ determines a random motion with the vector potential $\mathbf{b}(t,x)$ and the scalar potential $c(t,x)$. In other words, one has a kind of variational principle for the equations of motion (1.4.1) as an application of the Entropy Projection Theorem.

First of all, we must specify the probability measure $\overline{\mathbf{P}}$. To this end, we consider the Markov process

$$\{X_t; \mathbf{P}_{(s,x)}, (s,x) \in [a,b] \times \mathbf{R}^d\},$$

defined by the equation

$$\frac{\partial u}{\partial t} + \frac{1}{2}\sigma^2 \triangle u + \sigma^2 \mathbf{b}\left(t,x\right) \cdot \nabla u = 0. \tag{5.2.1}$$

Further, we consider the Kac functional

$$K_s^t = \exp\left(\int_s^t \overline{c}\left(r, X_r\right) dr\right)$$

with

$$\overline{c}(t,x) = c(t,x) + \frac{1}{2}\sigma^2\left(\nabla \cdot \mathbf{b}(t,x) + \mathbf{b}(t,x)^2\right).$$

In Section 4.2, we proved that the function

$$\phi(t,x) = \mathbf{P}_{(s,x)}\left[K_t^b f(b, X_b)\right],$$

satisfies the equation

$$\frac{\partial \phi}{\partial t} + \frac{1}{2}\sigma^2 \triangle\phi + \sigma^2 \mathbf{b}(t,x) \cdot \nabla\phi + \overline{c}(t,x)\phi = 0. \tag{5.2.2}$$

Let $p(s,x,t,y)$ denote the fundamental solution of equation (5.2.2).

We would like to construct a stochastic process determined by the Kac functional. As we already know, the measure defined by $\mathbf{P}_{(s,x)}^K[F] = \mathbf{P}_{(s,x)}\left[K_s^b F\right]$ does not define a stochastic process directly, because $\mathbf{P}_{(s,x)}^K[1] \neq 1$. To circumvent this difficulty, we apply the renormalization discussed in Section 4.2. Namely, using the function

$$\xi(s,x) = \mathbf{P}_{(s,x)}^K[1] = \mathbf{P}_{(s,x)}\left[K_s^b\right]$$

we define the renormalized Kac functional K_s^t by

$$N_s^t = \frac{1}{\xi(s,X_s)}K_s^t\xi(t,X_t) \tag{5.2.3}$$

and set

$$\overline{\mathbf{P}}_{(s,x)}[F] = \mathbf{P}_{(s,x)}\left[N_s^t F\right].$$

We thus get the renormalized stochastic process

$$\{X_t; \overline{\mathbf{P}}_{(s,x)}, (s,x) \in [a,b] \times \mathbf{R}^d\}.$$

Given an initial distribution $\mu_a(x)\,dx$ for the renormalized process, we set

$$\overline{\mathbf{P}} = \int \mu_a(x)\,dx\,\overline{\mathbf{P}}_{(a,x)} \tag{5.2.4}$$

and define a stochastic process $\{X_t, \overline{\mathbf{P}} \in \mathbf{M}_1(\Omega)\}$.

Theorem 5.2.1 (Variational Principle of Random Motion). *Let* $\mu_a(x)$ *and* $\mu_b(x)$ *be an initial distribution density and a terminal distribution density* $\mu_b(x)$, *respectively, and let*

$$\mathbf{A}_{a,b} = \{\mathbf{R} \in \mathbf{M}_1(\Omega) : \mathbf{R}[X_r \in dx] = \mu_r(x)\,dx, \ r = a,b\}. \tag{5.2.5}$$

Assume that there is at least one probability measure $\mathbf{R} \in \mathbf{A}_{a,b}$ *such that* $\mathrm{H}(\mathbf{R}|\overline{\mathbf{P}}) < \infty$. *Then the following holds true:*

(i) *Define* $\overline{\mathbf{P}} \in \mathbf{M}_1$ *by formula (5.2.4). Then there exist the projection* $\mathbf{Q} \in \mathbf{A}_{a,b}$
of $\overline{\mathbf{P}}$ *on* $\mathbf{A}_{a,b}$ *such that*

$$\inf_{\mathbf{P} \in \mathbf{A}_{a,b}} \mathrm{H}\left(\mathbf{P}|\overline{\mathbf{P}}\right) = \mathrm{H}\left(\mathbf{Q}|\overline{\mathbf{P}}\right). \tag{5.2.6}$$

(ii) *Moreover, there exist an entrance function* $\widehat{\phi}_a(x)$ *and an exit function* $\phi_b(x)$
such that

$$\widehat{\phi}_a(x)\left(\int p(a, x, b, y)\, \phi_b(y)\, dy\right) = \mu_a(x)$$

and

$$\left(\int \widehat{\phi}_a(x)\, dx_1 p(a, x, b, y)\right) \phi_b(y) = \mu_b(y),$$

where $p(s, x, t, y)$ *is the fundamental solution of the equation (5.2.2).*

(iii) *The finite-dimensional distribution of* $\mathbf{Q}$ *is given by*

$$\begin{aligned}
\mathbf{Q}&\left[f\left(X_a, X_{t_1}, \ldots, X_{t_{n-1}}, X_b\right)\right] \\
&= \int dx_0 \widehat{\phi}_a(x_0)\, p(a, x_0, t_1, x_1)\, dx_1 p(t_1, x_1, t_2, x_2)\, dx_2 \times \cdots \tag{5.2.7} \\
&\quad \times\ p(t_{n-1}, x_{n-1}, b, x_n)\, \phi_b(x_n)\, dx_n f(x_0, x_1, \ldots, x_n).
\end{aligned}$$

(iv) $\{X_t, \mathbf{Q}\}$ *is a Markov process with the transition probability density*

$$q(s, x, t, y) = \frac{1}{\phi_s(x)} p(s, x, t, y)\, \phi_t(y).$$

Proof. Assertion (i) was already proved in Theorem 5.1.1. Assertion (ii) is called a "difficult" problem in Schrödinger's paper (1931). We will prove it later as Theorem 5.4.3.

Let us prove assertion (iii). To distinguish the probability measure defined by the right-hand side of equation (5.2.7) from the projection $\mathbf{Q} \in \mathbf{A}_{a,b}$ of $\overline{\mathbf{P}} \in \mathbf{M}_1$ on $\mathbf{A}_{a,b}$, we denote it by $\mathbf{Q}_0$.

Let

$$\mathbf{A}_{\mu_a \mu_b} = \{\mathbf{P} \in \mathbf{M}_1(S_a \times S_b) : \mathbf{P}_{S_a} = \mu_a,\ \mathbf{P}_{S_b} = \mu_b\} \subset \mathbf{M}_1(S_a \times S_b).$$

If we look at the definitions of the probability measures $\overline{\mathbf{P}} \in \mathbf{M}_1$ and $\mathbf{Q}_0$, the only difference between them is the marginal distribution on $S_a \times S_b$, with the relationship between marginal distributions given by assertion (ii).

So let $\overline{p}$ denote the marginal distribution of $\overline{\mathbf{P}} \in \mathbf{M}_1$ on $S_a \times S_b$ and let q denote the projection of the marginal distribution of $\mathbf{Q}_0$ on $S_a \times S_b$ on the subset $\mathbf{A}_{\mu_a \mu_b}$. Then

$$\mathrm{H}\left(\mathbf{Q}_0|\overline{\mathbf{P}}\right) = \mathrm{H}\left(\mathbf{Q}|\overline{\mathbf{P}}\right). \tag{5.2.8}$$

On the other hand, by the definition of the relative entropy, with $f(x) = x \log x$, $x \geq 0$,

$$H(\mathbf{Q}|\overline{\mathbf{P}}) = \overline{\mathbf{P}}\left[f\left(\frac{d\mathbf{Q}}{d\overline{\mathbf{P}}}\right)\right],$$

which by the definition of conditional expectation is equal to

$$\overline{\mathbf{P}}\left[\overline{\mathbf{P}}\left[f\left(\frac{d\mathbf{Q}}{d\overline{\mathbf{P}}}\right)\Big|\sigma(X_a, X_b)\right]\right]$$

where $\sigma(X_a, X_b)$ is the σ-algebra generated by X_a and X_b. Since $f(x)$ is a convex function, Jensen's inequality (see Remark after the proof) implies that the last expression is

$$\geq \overline{\mathbf{P}}\left[f\left(\overline{\mathbf{P}}\left[\frac{d\mathbf{Q}}{d\overline{\mathbf{P}}}\Big|\sigma(X_a, X_b)\right]\right)\right].$$

This in turn is equal to

$$\overline{\mathbf{P}}\left[f\left(\frac{d\mathbf{Q}}{d\overline{\mathbf{P}}}\right)\right] = \overline{p}\left[f\left(\frac{d\mathbf{Q}}{d\overline{\mathbf{P}}}\right)\right] = H(\mathbf{Q}|\overline{\mathbf{P}}),$$

the reason being that

$$\overline{\mathbf{P}}\left[\frac{d\mathbf{Q}}{d\overline{\mathbf{P}}}\Big|\sigma(X_a, X_b)\right] = \frac{d\mathbf{Q}}{d\overline{\mathbf{P}}}$$

by the definition of $\mathbf{Q}$ and $\overline{\mathbf{P}}$, because the marginal distributions of $\mathbf{Q}$ and $\mathbf{Q}_0$ on $S_a \times S_b$ coincide. Thus,

$$H(\mathbf{Q}|\overline{\mathbf{P}}) \leq H(\mathbf{Q}|\overline{\mathbf{P}}). \tag{5.2.9}$$

Combining (5.2.9) and (5.2.8), we have

$$H(\mathbf{Q}_0|\overline{\mathbf{P}}) \leq H(\mathbf{Q}|\overline{\mathbf{P}}).$$

However, since $\mathbf{Q}_0 \in \mathbf{A}_{a,b}$ and $\mathbf{Q} \in \mathbf{A}_{a,b}$ is the projection of $\overline{\mathbf{P}}$, one necessarily has $H(\mathbf{Q}_0|\overline{\mathbf{P}}) = H(\mathbf{Q}|\overline{\mathbf{P}})$. We conclude that $\mathbf{Q}_0 = \mathbf{Q}$. Assertion (iii) is proved.

Assertion (iv) was established in the kinematics Theorem 1.5.2. □

Comment to Theorem 5.2.1. In Section 1.4, the stochastic process with the finite-dimensional distributions given by formula (5.2.7) was constructed with the triplet $\{p(s, x, t, y), \widehat{\phi}_a(x), \phi_b(y)\}$. Theorem 5.2.1 provides another way, which uses the triplet $\{p(s, x, t, y), \mu_a(x), \mu_b(y)\}$. (See Nagasawa (1990).)

Remark on Jensen's inequality. Let $f(x)$ be a downward convex function, and let X be an integrable random variable, i.e., $\mathbf{P}[\|X\|] < \infty$. Then

$$f(\mathbf{P}[X]) \leq \mathbf{P}[f(X)].$$

This is just Jensen's inequality. Indeed, for any x,

$$f(x_0) + \alpha(x_0) \cdot (x - x_0) \leq f(x).$$

Replacing here x and x_0 by a random variable X and the expectation $\mathbf{P}[X]$, respectively, we have

$$f(\mathbf{P}[X]) + \alpha(\mathbf{P}[X]) \cdot (X - \mathbf{P}[X]) \leq f(X).$$

If we take the expectation of both sides, then since the second term on the left-hand side vanishes, we obtain $f(\mathbf{P}[X]) \leq \mathbf{P}[f(X)]$, as claimed. Moreover, a Jensen inequality holds for conditional expectations:

$$f(\mathbf{P}[X|\mathcal{F}]) \leq \mathbf{P}[f(X)|\mathcal{F}].$$

The subset $\mathbf{A}_{a,b} \subset \mathbf{M}_1(\Omega)$ defined in (5.2.5) consists of the stochastic processes with the initial distribution density $\mu_a(x)$ and the terminal distribution density $\mu_b(x)$. Therefore, Theorem 5.2.1 is a variational principle formulated in terms of relative entropy.

On the other hand, in Section 1.14 we discussed the principle of least action for random motion, which states that

$$\int_a^b L(X_t)\, dt = \inf_Y \int_a^b L(Y_t)\, dt. \tag{5.2.10}$$

Actually, the principle of least action (5.2.10) and the variational principle (5.2.6) say the same thing. However, to see this, it is necessary to clarify the relationship between relative entropy and the action integral. This will be done below.

Consider the Markov process $\{X_t; \mathbf{P}_{(s,x)}, (s,x) \in [a,b] \times \mathbf{R}^d\}$ defined by the equation

$$\frac{\partial u}{\partial t} + \frac{1}{2}\sigma^2 \triangle u + \sigma^2 \mathbf{b}(t,x) \cdot \nabla u = 0. \tag{5.2.11}$$

Define the corresponding Maruyama-Girsanov functional by

$$M_s^t = \exp\left(\int_s^t \sigma^{-1}\mathbf{a}(r) \cdot dB_{r-s} - \frac{1}{2}\int_s^t \left(\sigma^{-1}\mathbf{a}(r)\right)^2 dr\right), \tag{5.2.12}$$

where the vector $\mathbf{a}(t,\omega)$ is $\mathcal{F}_a^t$-measurable. Next, consider the transformed probability measure $\mathbf{R}_{(s,x)}[F] = \mathbf{P}_{(s,x)}[M_s^b F]$. Then

$$\widetilde{B}_{t-a} = B_{t-a} - \sigma^{-1}\int_a^t \mathbf{a}(r)\, dr$$

is an $\mathbf{R}_{(s,x)}$-local martingale (Revuz and Yor (1991), Chapter VIII).

We prescribe an initial distribution $\mu_a(x)\, dx$ for $\mathbf{P}$, for $\mathbf{R}_{(s,x)}[F] = \mathbf{P}_{(s,x)}[M_s^b F]$, and for $\overline{\mathbf{P}}_{(s,x)}[F] = \mathbf{P}_{(s,x)}[N_s^b F]$, where N_s^b is the renormalized Kac functional, by

$$\mathbf{P} = \int \mu_a(x)\, dx\, \mathbf{P}_{(a,x)},$$

$$\mathbf{R} = \int \mu_a(x)\, dx\, \mathbf{R}_{(a,x)},$$

and

$$\overline{\mathbf{P}} = \int \mu_a(x)\, dx\, \overline{\mathbf{P}}_{(a,x)}.$$

Lemma 5.2.1. *Let* $\mathbf{R}$ *be the measure transformed by means of the Maruyama-Girsanov functional and let* $\overline{\mathbf{P}}$ *be the renormalized measure. Then*

$$\frac{d\mathbf{R}}{d\overline{\mathbf{P}}} = \xi(a, X_a) \tag{5.2.13}$$

$$\times \exp\left(\int_a^b \sigma^{-1}\mathbf{a}(r) \cdot d\widetilde{B}_{r-a} + \frac{1}{2}\int_a^b \left(\sigma^{-1}\mathbf{a}(r)\right)^2 dr - \int_a^b c(r, X_r)\, dr\right).$$

Proof. By the definitions given above,

$$\frac{d\mathbf{R}}{d\overline{\mathbf{P}}} = \frac{\frac{d\mathbf{R}}{d\mathbf{P}}}{\frac{d\overline{\mathbf{P}}}{d\mathbf{P}}} = \frac{M_a^b}{N_a^b} = \xi(a, X_a) \times$$

$$\times \exp\left(\int_a^b \sigma^{-1}\mathbf{a}(r) \cdot dB_{r-a} - \frac{1}{2}\int_a^b \left(\sigma^{-1}\mathbf{a}(r)\right)^2 dr - \int_a^b c(r, X_r)\, dr\right).$$

Substituting here

$$\widetilde{B}_{t-a} = B_{t-a} - \sigma^{-1}\int_a^t \mathbf{a}(r)\, dr,$$

we obtain (5.2.13). $\qquad\square$

Theorem 5.2.2. *Let*

$$L_t(\mathbf{R}) = \mathbf{R}\left[\frac{1}{2}\left(\sigma^{-1}\mathbf{a}(t)\right)^2 - c(t, X_t)\right]$$

be the Lagrangian of the stochastic process $\{X_t, \mathbf{R}\}$ *transformed by means of the Maruyama-Girsanov functional* M_s^b *(5.2.12). Then the relative entropy is equal to*

$$H(\mathbf{R}|\overline{\mathbf{P}}) = \int_a^b L_t(\mathbf{R})\, dt + \kappa, \tag{5.2.14}$$

where

$$\kappa = \mathbf{R}\left[\log \xi(a, X_a)\right] = \int \mu_a(x)\, dx \log \xi(a, x).$$

Proof. By the definition of relative entropy and formula (5.2.13),

$$H(\mathbf{R}|\overline{\mathbf{P}}) = \int \left(\log \frac{d\mathbf{R}}{d\overline{\mathbf{P}}}\right) d\mathbf{R}$$

$$= \int_a^b dt\, \mathbf{R}\left[\frac{1}{2}\left(\sigma^{-1}\mathbf{a}(t)\right)^2 - c(t, X_t)\right] + \mathbf{R}\left[\log \xi(a, X_a)\right]$$

$$= \int_a^b L_t(\mathbf{R})\, dt + \kappa,$$

where we used the fact that

$$\mathbf{R}\left[\sigma^{-1}\int_a^b \mathbf{a}\,(r)\cdot d\widetilde{B}_{r-a}\right]=0.$$

The proof is complete.   □

When we discussed the principle of least action (see Section 1.14), we first changed the additional drift vector $\mathbf{a}(t,\omega)$ and constructed the stochastic processes $\{Y_t, t\in[a,b]\,;\mathbf{P}\}$ defined via the stochastic differential equation

$$Y_t = Y_a + \sigma B_{t-a} + \int_a^t \left(\sigma^2\mathbf{b}\,(s,Y_s)+\mathbf{a}\,(s,\omega)\right)ds,$$

where $\mathbf{a}(t,\omega)$ is $\mathcal{F}_a^t$-measurable and the distribution density of Y_a is $\mu_a(x)$. Then the random motion X_t defined by the principle of least action

$$\int_a^b L\,(X_t)\,dt = \inf_Y \int_a^b L\,(Y_t)\,dt$$

satisfies the equation

$$X_t = X_a + \sigma B_{t-a} + \int_a^t \left(\sigma^2\mathbf{b}\,(s,X_s)+\mathbf{a}\,(s,X_s)\right)ds,$$

where $\mathbf{a}\,(s,x)=\sigma^2\nabla\log\phi\,(s,x)$ is the evolution drift vector determined by $\phi\,(s,x)$. Thus, Section 1.14 characterizes the stochastic process that minimizes the action integral.

However, therein we did not discuss whether such a stochastic process actually exists; rather, existence was assumed to avoid facing a very difficult problem. Here we will employ the Entropy Projection Theorem 5.1.1 to prove the existence of the requisite stochastic process.

Theorem 5.2.3.

(i) *Let $\mu_a\,(x)$ and $\mu_b\,(x)$ be distribution densities on $\mathbf{R}^d$ and let*

$$\mathbf{A}_{a,b}=\{\mathbf{R}\in\mathbf{M}_1\,(\Omega):\mathbf{R}\,[X_r\in dx]=\mu_r\,(x)\,dx, r=a,b\}\subset\mathbf{M}_1\,(\Omega)\,.$$

Then there exists a stochastic process $\{X_t,\}$ such that the principle of least action

$$\int_a^b L\,(X_t)\,dt = \inf_{\mathbf{R}\in\mathbf{A}_{a,b}} \int_a^b L\,(\mathbf{R})\,dt. \qquad (5.2.15)$$

holds. In terms of the evolution function $\phi\,(t,x)$, i.e., the solution of the equations of motion, the stochastic process X_t satisfies the equation

$$X_t = X_a + \sigma B_{t-a} + \int_a^t \left(\sigma^2\mathbf{b}\,(s,X_s)+\sigma^2\frac{\nabla\phi\,(s,X_s)}{\phi\,(s,X_s)}\right)ds, \qquad (5.2.16)$$

where the distribution density of X_a is $\mu_a(x) = \phi(a,x)\widehat{\phi}(a,x)$.

The Lagrangian of the stochastic process $\{X_t, \mathbf{Q}\}$ on the left-hand side of equation (5.2.15) is

$$L(X_t) = \mathbf{Q}\left[\frac{1}{2}\left(\sigma^{-1}\mathbf{a}(t,X_t)\right)^2 - c(t,X_t)\right],$$

where $\mathbf{a}(t,x) = \sigma^2 \nabla \log \phi(t,x)$.

(ii) Moreover, if an initial distribution density $\mu_a(x)$ and a terminal distribution density $\mu_b(x)$ are given, then there exist an entrance function $\widehat{\phi}_a(x)$ and an exit function $\phi_b(x)$ such that

$$\phi(t,x) = \int p(t,x,b,y)\,\phi_b(y)\,dy, \quad \widehat{\phi}(t,x) = \int \widehat{\phi}_a(z)\,dz\,p(a,z,t,x),$$

and the distribution density of X_a is $\mu_a(x) = \phi(a,x)\widehat{\phi}(a,x)$.

Proof. Consider the renormalized measure

$$\overline{\mathbf{P}} = \int \mu_a(x)\,dx\overline{\mathbf{P}}_{(a,x)}.$$

By Theorem 5.1.1, there exists a measure $\mathbf{Q} \in \mathbf{A}_{a,b}$ such that

$$\inf_{\mathbf{R} \in \mathbf{A}_{a,b}} \mathrm{H}\left(\mathbf{R}|\overline{\mathbf{P}}\right) = \mathrm{H}\left(\mathbf{Q}|\overline{\mathbf{P}}\right). \tag{5.2.17}$$

Since in formula (5.2.14) κ is invariant on the subset $\mathbf{A}_{a,b}$, the measure $\mathbf{Q} \in \mathbf{A}_{a,b}$ that satisfies (5.2.17) minimizes the right-hand side of (5.2.14). Therefore, equality (5.2.15) holds, and hence, by Theorem 1.14.1, so does equation (5.2.16).

Assertion (ii) will be proved in Section 5.4 as Theorem 5.4.3. $\qquad\square$

Comment to Theorem 5.2.3. Since the probability measure $\mathbf{Q} \in \mathbf{A}_{a,b}$, the variational principle guarantees the existence of the process $\{X_t(\omega), t \in [a,b]; \Omega, \mathcal{F}, \mathbf{Q}\}$ that starts with the entrance function $\widehat{\phi}_a(x)$ and terminates with the exit function $\phi_b(x)$. This process coincides with the stochastic process constructed using the triplet in Section 1.4.

5.3 Exponential Family of Distributions

In this section, we apply the Entropy Projection Theorem to prove an exponential density theorem (see Douglas (1964), Lindenstrauss (1965), Csiszár (1975)). In the next section, this theorem will be used to establish the existence of entrance and exit functions for given initial and terminal distributions.

Let $\mathbf{P}_1, \mathbf{P}_2, \mathbf{Q} \in \mathbf{A} \subset \mathbf{M}_1(\Omega)$. If

$$\mathbf{Q} = \alpha \mathbf{P}_1 + (1 - \alpha) \mathbf{P}_2, \quad 0 < \alpha < 1, \quad \mathbf{Q} \neq \mathbf{P}_1, \quad \mathbf{Q} \neq \mathbf{P}_2 \tag{5.3.1}$$

then we say that "the point $\mathbf{Q}$ lies on the segment joining the two points $\mathbf{P}_1, \mathbf{P}_2$".

Lemma 5.3.1. *If the projection $\mathbf{Q} \in \mathbf{A}$ of a point $\overline{\mathbf{P}} \in \mathbf{M}_1(\Omega)$ lies on the segment joining the two points $\mathbf{P}_1, \mathbf{P}_2 \in \mathbf{A}$, then the Radon-Nikodym derivative $\overline{q} = d\mathbf{Q}/d\overline{\mathbf{P}}$ satisfies*

$$\int \log \overline{q} \, d\mathbf{P}_i = \mathrm{H}\left(\mathbf{Q} | \overline{\mathbf{P}}\right), \quad i = 1, 2, \tag{5.3.2}$$

and equality holds in (5.1.10) for $\mathbf{P} = \mathbf{P}_1$ or $\mathbf{P} = \mathbf{P}_2$.

Proof. By Lemma 5.1.1, (5.3.2) holds as an inequality $\geq$. Hence, it suffices to show that the assumption

$$\int \log \overline{q} \, d\mathbf{P}_i > \mathrm{H}\left(\mathbf{Q} | \overline{\mathbf{P}}\right), \quad i = 1, 2, \tag{5.3.3}$$

leads to a contradiction. Indeed, if (5.3.3) holds, then (5.3.1) implies that

$$\int \log \overline{q} \, d\mathbf{Q} > \mathrm{H}\left(\mathbf{Q} | \overline{\mathbf{P}}\right).$$

However, the integral on the left is $\mathrm{H}\left(\mathbf{P} | \overline{\mathbf{P}}\right)$, by definition. This is the desired contradiction, proving that (5.3.2) holds.

Moreover, if we use (5.3.2) on the right-hand side of equation (5.1.13), we see that (5.1.10) holds as an equality for $\mathbf{P} = \mathbf{P}_1$ or $\mathbf{P} = \mathbf{P}_2$. $\square$

Theorem 5.3.1 (Exponential Family of Distributions). *Let f_j be bounded measurable functions and a_j, $j = 1, 2, \ldots, k$ be constants. Let*

$$\mathbf{A} = \left\{ \mathbf{P} \in \mathbf{M}_1(\Omega) : \int f_j \, d\mathbf{P} = a_j, \ j = 1, 2, \ldots, k \right\} \subset \mathbf{M}_1(\Omega). \tag{5.3.4}$$

Assume that there exists at least one probability measure $\mathbf{P} \in \mathbf{A}$ such that $\mathrm{H}\left(\mathbf{P} | \overline{\mathbf{P}}\right) < \infty$. Moreover, let $\mathbf{Q} \in \mathbf{A}$ be a probability measure that is absolutely continuous with respect to $\overline{\mathbf{P}}$, with density $\overline{q} = d\mathbf{Q}/d\overline{\mathbf{P}}$. Then $\mathbf{Q}$ is the projection of $\overline{\mathbf{P}}$ on $\mathbf{A}$ if and only if the density $\overline{q}(\omega)$ is an exponential function, i.e.,

$$\overline{q}(\omega) = c \exp \left(\sum_{j=1}^{k} \alpha_j f_j(\omega) \right), \tag{5.3.5}$$

except for the case when $\overline{q}(\omega) \equiv 0$.

Proof. The subset $\mathbf{A} \subset \mathbf{M}_1(\Omega)$ defined in (5.3.4) is closed and convex. Let $\mathbf{P} \in \mathbf{A}$ be the projection of $\overline{\mathbf{P}}$ on $\mathbf{A}$. Let

$$\mathbf{A_Q} = \left\{ \mathbf{P} \in \mathbf{A} : \frac{d\mathbf{P}}{d\mathbf{Q}} \leq 2 \right\} \subset \mathbf{A}. \tag{5.3.6}$$

Since obviously $\mathbf{Q} \in \mathbf{A_Q}$,

$$\mathrm{H}\left(\mathbf{Q}|\overline{\mathbf{P}}\right) = \inf_{\mathbf{P} \in \mathbf{A_Q}} \mathrm{H}\left(\mathbf{P}|\overline{\mathbf{P}}\right) = \inf_{\mathbf{P} \in \mathbf{A}} \mathrm{H}\left(\mathbf{P}|\overline{\mathbf{P}}\right).$$

Take an arbitrary $\mathbf{P} \in \mathbf{A_Q}$ $(\mathbf{P} \neq \mathbf{Q})$ and set $\mathbf{P}' = 2\mathbf{Q} - \mathbf{P} \in \mathbf{A_Q}$. Then $\mathbf{Q} = (\mathbf{P} + \mathbf{P}')/2$, i.e., $\mathbf{Q} \in \mathbf{A_Q}$ lies on the segment joining the points $\mathbf{P}$ and $\mathbf{P}'$ in $\mathbf{A_Q}$. Therefore, by Lemma 5.3.1,

$$\int \log \overline{q} \, d\mathbf{P} = \mathrm{H}\left(\mathbf{Q}|\overline{\mathbf{P}}\right), \quad \forall \mathbf{P} \in \mathbf{A_Q}, \tag{5.3.7}$$

where $\overline{q} = d\mathbf{Q}/d\overline{\mathbf{P}}$.

Moreover, since $\mathrm{H}\left(\mathbf{Q}|\overline{\mathbf{P}}\right) = \int \log \overline{q} \, d\mathbf{Q}$, (5.3.7) implies that

$$\int (p - 1) \log \overline{q} \, d\mathbf{Q} = 0, \quad \forall \mathbf{P} \in \mathbf{A_Q}, \tag{5.3.8}$$

with $p = d\mathbf{P}/d\mathbf{Q}$.

Now define a subspace $\mathbf{H}$ of $L^\infty(\mathbf{Q})$ by

$$\mathbf{H} = \left\{ h : \int h \, d\mathbf{Q} = 0, \int f_j h \, d\mathbf{Q} = 0, \ 1 \leq j \leq k \right\}.$$

Take an arbitrary $h \in \mathbf{H}$ and denote $\overline{h} = h/\|h\|$. Then, by (5.3.6),

$$\mathbf{P} = \left(1 + \overline{h}\right) \mathbf{Q} \in \mathbf{A_Q}.$$

Therefore,

$$\frac{d\mathbf{P}}{d\mathbf{Q}} = p = 1 + \overline{h},$$

and then, since $p - 1 = \overline{h}$, (5.3.8) implies that,

$$\int \overline{h} \log \overline{q} \, d\mathbf{Q} = 0, \quad \forall h \in \mathbf{H}.$$

Since $L^1(\mathbf{Q})$ is the dual space of $L^\infty(\mathbf{Q})$, this implies that $\log \overline{q}$ belongs to the closed subspace of $L^\infty(\mathbf{Q})$ spanned by the functions 1 and f_j, $1 \leq j \leq k$, i.e., there are constants $\alpha_0, \alpha_1, \ldots, \alpha_k$ such that

$$\log \overline{q} = \alpha_0 + \sum_{j=1}^k \alpha_j f_j(\omega),$$

so formula (5.3.5) holds.

Conversely, if $\overline{q}\,(\omega)$ is given by formula (5.3.5), then

$$\int \log \overline{q}\, d\mathbf{P} = \log c + \sum_{j=1}^{k} \alpha_j \int f_j \, d\mathbf{P} = \log c + \sum_{j=1}^{k} \alpha_j a_j,$$

where the right-hand side is a constant that does not depend on $\mathbf{P} \in \mathbf{A_Q}$. Hence, this last equality holds when on the left-hand side $\mathbf{P}$ is replaced by $\mathbf{Q} \in \mathbf{A_Q}$. Since, $\int \log \overline{q}\, d\mathbf{Q} = \mathrm{H}\left(\mathbf{Q}|\overline{\mathbf{P}}\right)$ by definition, we have

$$\mathrm{H}\left(\mathbf{Q}|\overline{\mathbf{P}}\right) = \int \log \overline{q}\, d\mathbf{P}, \quad \forall \mathbf{P} \in \mathbf{A_Q}.$$

Therefore, by Lemma 5.3.1, $\mathbf{Q} \in \mathbf{A_Q}$ is the projection of $\overline{\mathbf{P}} \in \mathbf{M}_1\,(\Omega)$ on the set $\mathbf{A_P}$ (and on $\mathbf{A}$).   □

5.4   Existence of Entrance and Exit Functions

When we constructed the random motion determined by the equations of motion in Section 1.4, we used the triplet $\{p(s,z,t,y), \widehat{\phi}_a(z), \phi_b(y)\}$, where $p(s,z,t,y)$ is the fundamental solution of the equations of motion (1.4.1).

We then showed that, under the assumption that the given triplet satisfies the normalization condition

$$\int dz\, \widehat{\phi}_a(z) p(a,z,b,y) \phi_b(y) dy = 1,$$

there exists a stochastic process $\{X_t\,(\omega)\,, t \in [a,b]\,; \Omega, \mathcal{F}, \mathbf{Q}\}\,,\ -\infty < a < b < \infty$ starting with the entrance function $\widehat{\phi}_a\,(z)$ and terminating with the exit function $\phi_b\,(y)$, and with the distribution given by

$$\mathbf{Q}\,[X_t \in dx] = \widehat{\phi}_t(x)\phi_t(x)dx.$$

At that time, we assumed, as a fundamental and natural feature of the mechanics of random motion, that the entrance and exit functions functions $\widehat{\phi}_a(z)$ and $\phi_b\,(y)$ are known.

Indeed, when we discussed the double-slit experiment by Walborn et al. in Section 1.13, it was natural to prescribe an entrance function $\widehat{\phi}_a(z)$ and an exit function $\phi_b(y)$. In fact, it is difficult to think about this experiment in other ways.

However, there are many other cases where one prepares and observes the initial distribution density $\mu_a(x)$ and the terminal distribution density $\mu_b(x)$. In these cases there one runs into the serious problem of whether for given $\mu_a\,(x)$ and $\mu_b\,(x)$ there exist an entrance function $\widehat{\phi}_a\,(z)$ and an exit function $\phi_b\,(y)$ such that

$$\widehat{\phi}_a\,(x) \left(\int p\,(a,x,b,y)\,\phi_b\,(y)\,dy \right) = \mu_a\,(x),$$

$$\left(\int \widehat{\phi}_a \left(x \right) dx p \left(a, x, b, y \right) \right) \phi_b \left(y \right) = \mu_b \left(y \right).$$

This is known as *Schrödinger's problem*, which deals with stochastic processes satisfying the equations of motion (1.1.3). In his paper of 1931, Schrödinger asserted that this was an extremely difficult problem to solve because these are nonlinear simultaneous equations. He assumed the existence of solutions and left the discussion open at that time. This problem was subsequently discussed by Bernstein (1932), Fortet (1940), Beurling (1960), Föllmer (1988), Nagasawa (1990), Aebi (1995).

In this section, we will show that Schrödinger's problem can be solved by applying the exponential density theorem proved in the preceding section. Namely, we will show that, given the densities of the initial and terminal distributions $\mu_a \left(x \right)$ and $\mu_b \left(x \right)$, we can obtain an entrance function $\widehat{\phi}_a \left(z \right)$ and an exit function $\phi_b \left(y \right)$.

First, let denote the underlying d-dimensional space as $S = \mathbf{R}^d$ and denote the set of all probability measure on S by $\mathbf{M}_1 \left(S \right)$. Let S_1 an S_2 be two copies of S and denote the set of all probability measures on $S_1 \times S_2$ by $\mathbf{M}_1 \left(S_1 \times S_2 \right)$. For a probability measure $\mathbf{P} \in \mathbf{M}_1 \left(S_1 \times S_2 \right)$, we let $\mathbf{P}_{S_i}$ denote the marginal distribution on S_k, $k = 1, 2$. That is,

$$\mathbf{P}_{S_1} \left(dx \right) = \mathbf{P} \left(dx \times S_2 \right), \quad \mathbf{P}_{S_2} \left(dx \right) = \mathbf{P} \left(S_1 \times dx \right).$$

Theorem 5.4.1. *Given probability measures* $\mathbf{P}_i \in \mathbf{M} \left(S_k \right)$, $k = 1, 2$, *let*

$$\mathbf{A}_{\mathbf{P}_1 \mathbf{P}_2} = \{ \mathbf{P} \in \mathbf{M}_1 \left(S_1 \times S_2 \right) : \mathbf{P}_{S_k} = \mathbf{P}_k, \ k = 1, 2 \} \qquad (5.4.1)$$

be the set of all $\mathbf{P} \in \mathbf{M}_1 \left(S_1 \times S_2 \right)$ *with the given marginal distributions. Take a probability measure* $\overline{\mathbf{P}} \in \mathbf{M}_1 \left(S_1 \times S_2 \right)$, *and assume that there exists at least one* $\mathbf{P} \in \mathbf{A}_{\mathbf{P}_1 \mathbf{P}_2}$ *such that* $\mathrm{H} \left(\mathbf{P} | \overline{\mathbf{P}} \right) < \infty$. *Let* $\mathbf{Q} \in \mathbf{A}_{\mathbf{P}_1 \mathbf{P}_2}$ *be the projection of* $\overline{\mathbf{P}}$ *on* $\mathbf{A}_{\mathbf{P}_1 \mathbf{P}_2}$. *Then there exist functions* $g \left(x_1 \right)$ *and* $h \left(x_2 \right)$ *such that*

$$\frac{d\mathbf{Q}}{d\overline{\mathbf{P}}} = \overline{q} \left(x_1, x_2 \right) = g \left(x_1 \right) h \left(x_2 \right),$$

and $\log g \in L^1 \left(\mathbf{P}_1 \right)$ *and* $\log h \in L^1 \left(\mathbf{P}_2 \right)$.

Proof. Assume that the set of functions $\{ f_{kj} : j = 1, 2, \ldots \}$ in $L^\infty \left(\mathbf{P}_k \right)$ defines $\mathbf{P}_k$, $k = 1, 2$. Let

$$f_{kj} \left(x_1, x_2 \right) = f_{kj} \left(x_k \right) : \ k = 1, 2; \ j = 1, 2, \ldots,$$

$$a_{kj} = \int f_{kj} \left(x_k \right) d\mathbf{P}_k, \ k = 1, 2; \ j = 1, 2, \ldots.$$

Then the set $\mathbf{A}_{\mathbf{P}_1 \mathbf{P}_2}$ in (5.4.1) can be described as

$$\left\{ \mathbf{P} \in \mathbf{M}_1 \left(S_1 \times S_2 \right) : a_{kj} = \int f_{kj} \left(x_1, x_2 \right) d\mathbf{P}, \ k = 1, 2; \ j = 1, 2, \ldots \right\}.$$

Let $\mathbf{F}$ be the closed subspace spanned by the set

$$\{1, f_{kj}(x_1, x_2) : k = 1, 2;\ j = 1, 2, \ldots\}$$

and define a subspace $\mathbf{H}$ of $L^\infty(\mathbf{Q})$ by

$$\mathbf{H} = \left\{h : \int h\, d\mathbf{Q} = 0,\ \int f_{kj} h\, d\mathbf{Q} = 0,\ k = 1, 2;\ j = 1, 2, \ldots\right\}.$$

Arguing as in the proof of Theorem 5.3.1, we see that if a probability measure $\mathbf{Q} \in \mathbf{A}_{\mathbf{P}_1 \mathbf{P}_2}$ is the projection of $\overline{\mathbf{P}}$ on $\mathbf{A}_{\mathbf{P}_1 \mathbf{P}_2}$, then the Radon-Nikodym derivative $\overline{q} = d\mathbf{Q}/d\overline{\mathbf{P}}$ satisfies

$$\int h\,(\log \overline{q})\, d\mathbf{Q} = 0,\ \forall h \in \mathbf{H}.$$

Therefore, $\log \overline{q} \in \mathbf{F}$. Since the subspace $\mathbf{F}$ can be alternatively written as

$$\{f_1(x_1) + f_2(x_2) : f_1 \in L^1(\mathbf{P}_1),\ f_2 \in L^1(\mathbf{P}_2)\},$$

there exist functions $f_1 \in L^1(\mathbf{P}_1)$ and $f_2 \in L^1(\mathbf{P}_2)$ such that

$$\log \overline{q}(x_1, x_2) = f_1(x_1) + f_2(x_2).$$

Hence, defining $g(x)$ and $h(x)$ by

$$\log g(x_1) = f_1(x_1) \in L^1(\mathbf{P}_1) \quad \text{and} \quad \log h(x_2) = f_2(x_2) \in L^1(\mathbf{P}_2),$$

we have

$$\frac{d\mathbf{Q}}{d\overline{\mathbf{P}}} = \overline{q}(x_1, x_2) = g(x_1) h(x_2),$$

as claimed.   □

As an application of Theorem 5.4.1, we have

Theorem 5.4.2. *Let $p(x_1, x_2)$ be a non-negative measurable function on $S_1 \times S_2$, and define the probability measure $\overline{\mathbf{P}}$ by*

$$\overline{\mathbf{P}}(dx_1 dx_2) = p(x_1, x_2)\, dx_1 dx_2.$$

Let $\mu_k(x_k)\, dx_k$ be given distributions on S_k, $k = 1, 2$. Then there exist functions $g(x_1)$ on S_1 and $h(x_2)$ on S_2 such that

$$g(x_1)\left(\int p(x_1, x_2) h(x_2)\, dx_2\right) = \mu_1(x_1), \tag{5.4.2}$$

and

$$\left(\int g(x_1)\, dx_1 p(x_1, x_2)\right) h(x_2) = \mu_2(x_2). \tag{5.4.3}$$

Proof. By Theorem 5.4.1, if $\mathbf{Q}$ is the projection of $\overline{\mathbf{P}}$, then there exist functions $g(x_1)$ and $h(x_2)$ such as

$$\frac{d\mathbf{Q}}{d\overline{\mathbf{P}}} = g(x_1) h(x_2),$$

i.e., $\mathbf{Q} = g(x_1) h(x_2) \overline{\mathbf{P}}$. Moreover, since the marginal distribution of $\mathbf{Q}$ on S_1,

$$\mathbf{Q}_{S_1} = \left(g(x_1) h(x_2) \overline{\mathbf{P}}\right)_{S_1} = g(x_1) dx_1 \left(\int p(x_1, x_2) h(x_2) dx_2\right)$$

is prescribed to be $\mu_1(x_1) dx_1$, we have

$$g(x_1) dx_1 \left(\int p(x_1, x_2) h(x_2) dx_2\right) = \mu_1(x_1) dx_1.$$

Therefore, (5.4.2) holds. In the same way, since the marginal distribution $\mathbf{P}_{S_2}$ of $\mathbf{Q}$ on S_2 is $\mu_1(x_2) dx_2$, (5.4.3) holds. $\qquad\square$

Theorem 5.4.3 (Schrödinger's Problem). *Let $p(s, x, t, y)$ be the fundamental solution of the equations of motion (1.4.1). If a density $\mu_a(x)$ of initial distribution and a density $\mu_b(x)$ of terminal distribution are given, then there exist an entrance function $\widehat{\phi}_a(z)$ and an exit function $\phi_b(y)$, such that*

$$\widehat{\phi}_a(x) \left(\int p(a, x, b, y) \phi_b(y) dy\right) = \mu_a(x)$$

and

$$\left(\int \widehat{\phi}_a(x) dx p(a, x, b, y)\right) \phi_b(y) = \mu_b(y).$$

Proof. Choosing the functions $k_1(x) > 0$ and $k_2(y) > 0$ properly and consider the probability measure

$$\overline{\mathbf{P}}(dxdy) = k_1(x) p(a, x, b, y) k_2(y) dxdy$$

be a probability measure. Then (5.4.2) and (5.4.3) in Theorem 5.4.2 hold with this measure $\overline{\mathbf{P}}(dxdy)$ in place of $\mathbf{P}(x_1, x_2)$. To complete the proof, it remains to set $\widehat{\phi}_a(x) = g(x) k_1(x)$ and $\phi_b(y) = h(y) k_2(y)$. $\qquad\square$

Comments to Theorem 5.4.3. In 1930's it was thought that only an initial distribution and a terminal distribution can be given beforehand. Therefore, in order to find the entrance function $\widehat{\phi}_a(z)$ and the exit function $\phi_b(y)$, it was theoretically necessary to solve the simultaneous equations of Theorem 5.4.3.

Moreover, as we explained in Section 1.13, in Walborn et al.'s double-slit experiment the entrance and exit functions were given. In 1930's it was almost impossible to imagine that an experiment such as Walborn et al.'s (2002) is possible. The contents of this section are based on Nagasawa (1990). Chapter 7 of Aebi (1996) generalizes the discussion in Beurling (1960).

5.5 Cloud of Paths

When Born gave the statistical interpretation of Schrödinger function $\psi_t(x, y, z)$, he regarded $|\psi_t(x, y, z)|^2$ as the distribution density of one particle generated by scattering, because he considered scattering problem. Indeed it was one possible "interpretation", but there was no theoretical backing. We note that we proved in Theorem 1.10.1 that the complex evolution equation (Schrödinger equation) described the motion of a particle through the equations of motion which is equivalent to the Schrödinger equation and $|\psi_t(x, y, z)|^2$ coincided with the distribution density $\mu(t, x) = \widehat{\phi}(t, x)\phi(t, x)$ of the random motion X_t. On the other hand, there was also an another idea that $|\psi_t(x, y, z)|^2$ was the distribution density of the population of particles. In other words, there were two ways of thinking.

In the present and the next sections we consider a population of particles and examine what happens when one lets the number of particles forming the population go to infinity. We then show that the Schrödinger equation (complex evolution equation) can be used to describes the evolution a population of particles. In other words, we will show that both viewpoints formulated above are relevant.

First, we consider that the motion of the system consisting of n-particles under consideration is described by a Markov process

$$\left\{ (X_t(\omega_1), \ldots, X_t(\omega_n)),\ t \in [a, b] \,;\ (\omega_1, \ldots, \omega_n) \in \Omega^n, \mathbf{P}^{(n)} \right\}$$

and investigate what happens in the limit $n \to \infty$.

To single out one path ω_k in the path space Ω, we use the point measure δ_{ω_k}, i.e, the probability measure on the Ω given by

$$\delta_{\omega_k}\left[f\left(X_t \right) \right] = f\left(X_t\left(\omega_k \right) \right), \quad \omega_k \in \Omega.$$

Then to single out n paths $(\omega_1, \ldots, \omega_n) \in \Omega^n$ we use the probability measure on Ω given by

$$L_n(\overline{\omega}) = \sum_{k=1}^{n} \frac{1}{n} \delta_{\omega_k}, \quad \overline{\omega} = (\omega_1, \ldots, \omega_n) \in \Omega^n.$$

We call $L_n(\overline{\omega})$ **cloud of n particle paths** $\overline{\omega} = (\omega_1, \ldots, \omega_n) \in \Omega^n$.

This is the term we use in the mechanics of random motion; in statistics, $L_n(\overline{\omega}) = \sum_{k=1}^{n} \frac{1}{n} \delta_{\omega_k}$ is known as the empirical distribution.

To understand what $L_n(\overline{\omega})$ is, let $f(x) = \mathbf{1}_B(x)$. Then since

$$L_n(\overline{\omega})[\mathbf{1}_B(X_t)] = \sum_{k=1}^{n} \frac{1}{n} \delta_{\omega_k}[\mathbf{1}_B(X_t)] = \sum_{k=1}^{n} \frac{1}{n} \mathbf{1}_B\left(X_t\left(\omega_k \right) \right),$$

the quantity $L_n(\overline{\omega})[\mathbf{1}_B(X_t)]$ specifies, for a given $\overline{\omega} = (\omega_1, \ldots, \omega_n) \in \Omega^n$, how many of the n particles $X_t(\omega_1), X_t(\omega_2), \ldots, X_t(\omega_n)$ are in the set B that we observe at time t. That is, it indicates how thick is the cloud in B.

If we denote the expectation of $L_n(\overline{\omega})$ with respect to a probability measure $\mathbf{P}^{(n)}$ on Ω^n by

$$\mathbf{K}^{(n)} = \mathbf{P}^{(n)}[L_n],$$

then $\mathbf{K}^{(n)}$ is a probability measure on the path space Ω, and $\{X_t, \mathbf{K}^{(n)}\}$ represents the averaged random motion of a particle in the cloud.

In fact, we have

$$\mathbf{K}^{(n)}[f(X_t)] = \mathbf{P}^{(n)}[L_n[f(X_t)]] = \sum_{k=1}^{n} \mathbf{P}^{(n)}\left[\frac{1}{n}f(X_t(\omega_k))\right]$$

$$= \int f(x)\frac{1}{n}\sum_{k=1}^{n}\mathbf{P}^{(n)}[X_t(\omega_k) \in dx].$$

Hence, the distribution $\mu_n(t, dx)$ of the stochastic process $\{X_t, \mathbf{K}^{(n)}\}$ is

$$\mu_n(t, dx) = \mathbf{K}^{(n)}[X_t \in dx] = \frac{1}{n}\sum_{k=1}^{n}\mathbf{P}^{(n)}[X_t(\omega_k) \in dx];$$

the right-hand side represents the average of the distributions of the n-particles $X_t(\omega_1), \ldots, X_t(\omega_n)$.

We would like to know, assuming that the average random motion $\{X_t, \mathbf{K}^{(n)}\}$ of a particle in the cloud converges to a stochastic process $\{X_t, \mathbf{Q}\}$ as $n \to \infty$, what kind of stochastic process we get as the limit.

It is obvious that the ultimate stochastic process $\{X_t, \mathbf{Q}\}$ depends on the first Markov process

$$\left\{(X_t(\omega_1), \ldots, X_t(\omega_n)), (\omega_1, \ldots, \omega_n) \in \Omega^n, \mathbf{P}^{(n)}\right\}.$$

As an example, let $\{X_t, \Omega, \mathbf{P}\}$ be the Brownian motion with an initial distribution $\mu_a(x)\,dx = \mathbf{P}[X_a \in dx]$ and take the n-fold product

$$\{(X_t(\omega_1), \ldots, X_t(\omega_n)), (\omega_1, \ldots, \omega_n) \in \Omega^n, \mathbf{P}^n\}$$

of $\{X_t, \Omega, \mathbf{P}\}$ as the first Markov process. Applying the law of large numbers to

$$L_n(\overline{\omega}) = \sum_{k=1}^{n}\frac{1}{n}\delta_{\omega_k}, \quad \overline{\omega} = (\omega_1, \ldots, \omega_n) \in \Omega^n,$$

we see that $\{X_t, \mathbf{K}^{(n)}\}$ converges to the Brownian motion $\{X_t, \Omega, \mathbf{P}\}$ as $n \to \infty$. In other words, the ultimate stochastic process is the original Brownian motion $\{X_t, \Omega, \mathbf{P}\}$. Therefore, the terminal distribution $\mu_b(x)\,dx$ of the ultimate stochastic process is

$$\mu_b(x)\,dx = \mathbf{P}[X_b \in dx].$$

This is automatically determined by the initial distribution $\mu_a(x)\,dx = \mathbf{P}\,[X_a \in dx]$ of the first given Brownian motion $\{X_t, \Omega, \mathbf{P}\}$. So the density of the terminal distribution $\mu_b(x)$ cannot be chosen arbitrarily.

Then, are there situations where we can chose arbitrarily the density $\mu_b(x)$ of the terminal distribution of the ultimate stochastic process? In particular, when we take the random motion of one particle in the cloud, is it possible that the ultimate stochastic process will coincide with the one determined by the equations of motion (1.4.1)? This is the problem that Schrödinger studied in his paper published in 1931. Recall that when we describe random motions determined by the equations of motion we can choose a terminal distribution (exit function) arbitrarily.

In his paper of 1931 Schrödinger claimed that

"if we take the law of large numbers into account, the probability that when $n \to \infty$ the distribution converges to an arbitrarily chosen distribution will decrease exponentially. But occurrence of this is not completely impossible."

This is an idea connected with the large deviations of empirical distributions. At that time the principle of large deviation was not known, so this was a pioneering idea.

Let us present the large deviation principle for stochastic processes.

Let $\{X_t; \mathbf{P}_{(s,x)}, (s,x) \in [a,b] \times \mathbf{R}^d\}$ be the Markov process defined by the equation

$$\frac{\partial u}{\partial t} + \frac{1}{2}\sigma^2 \triangle u + \sigma^2 \mathbf{b}(t,x) \cdot \nabla u = 0,$$

and let

$$K_s^t = \exp\left(\int_s^t \bar{c}(r, X_r)\,dr\right)$$

be the associated Kac functional, where

$$\bar{c}(t,x) = c(t,x) + \frac{1}{2}\sigma^2\left(\nabla \cdot \mathbf{b}(t,x) + \mathbf{b}(t,x)^2\right).$$

Then the function

$$\phi(t,x) = \mathbf{P}_{(s,x)}\left[K_t^b f(b, X_b)\right]$$

satisfies the partial differential equation

$$\frac{\partial \phi}{\partial t} + \frac{1}{2}\sigma^2 \triangle \phi + \sigma^2 \mathbf{b}(t,x) \cdot \nabla \phi + \bar{c}(t,x)\,\phi = 0,$$

which is the first of the pair of equations of motion (1.4.1). In other words, ϕ is an evolution function. This was shown in Section 4.2.

Next, we introduce the measure

$$\mathbf{P}_{(s,x)}^K[F] = \mathbf{P}_{(s,x)}\left[K_s^b F\right].$$

Using the function

$$\xi(s, x) = \mathbf{P}^K_{(s,x)}[1] = \mathbf{P}_{(s,x)}\left[K^b_s\right]$$

we define the renormalized Kac functional by

$$N^t_s = \frac{1}{\xi(s, X_s)} K^t_s \xi(t, X_t)$$

and then set

$$\overline{\mathbf{P}}_{(s,x)}[F] = \mathbf{P}_{(s,x)}\left[N^t_s F\right].$$

This yields the renormalized stochastic process

$$\{X_t; \overline{\mathbf{P}}_{(s,x)}, (s, x) \in [a, b] \times \mathbf{R}^d\}.$$

Given an initial distribution $\mu_a(x)\, dx$, we put

$$\overline{\mathbf{P}} = \int \mu_a(x)\, dx\, \overline{\mathbf{P}}_{(a,x)};$$

this defines a stochastic process $\{X_t, \overline{\mathbf{P}} \in \mathbf{M}_1(\Omega)\}$ that will be used to formulate the large deviation principle.

The probability measures $\mathbf{P} \in \mathbf{M}_1(\Omega)$ we are interested in are those with a specified initial distribution $\mu_a(x)\, dx$ and a specified terminal distribution $\mu_b(x)\, dx$ as their marginal distributions. We denote the set of all such probability measures by

$$\mathbf{A}_{a,b} = \{\mathbf{P} \in \mathbf{M}_1(\Omega) : \mathbf{P}[X_r \in dx] = \mu_r(x)\, dx, \ r = a, b\}. \tag{5.5.1}$$

However, to discuss the large deviation principle for L_n, the conditions on the marginal distributions defining the set $\mathbf{A}_{a,b}$ are a bit too restrictive. We relaxing them somewhat to the following two conditions:

(1) Let $\mathcal{P}_k\left(\mathbf{R}^d\right) = \{B_1, \ldots, B_{2^k}\}$, $k = 1, 2, \ldots$ be a measurable partition of the space $\mathbf{R}^d$ such that $\sigma\left(\mathcal{P}_k\left(\mathbf{R}^d\right)\right) \uparrow \sigma\left(\mathbf{R}^d\right)$. For all $B_i \in \mathcal{P}_k\left(\mathbf{R}^d\right)$, the measure $\mathbf{P} \in \mathbf{M}_1(\Omega)$ satisfies

$$|\mathbf{P}[X_r \in B_i] - \mu_r(B_i)| \le \frac{\varepsilon}{2^k}, \ r = a, b.$$

(2) Let $\{X_t, \overline{\mathbf{P}} \in \mathbf{M}_1(\Omega)\}$ be the renormalized stochastic process defined above. Then $\mathbf{P} \in \mathbf{M}_1(\Omega)$ satisfies

$$\mathbf{P}[X_r \in dx] \ll \overline{\mathbf{P}}[X_r \in dx], \ r = a, b,$$

on $\sigma\left(\mathcal{P}_k\left(\mathbf{R}^d\right)\right)$.

Now let $\varepsilon > 0$ and we define the set $\mathbf{A}\left(\varepsilon, k\right)$ by

$$\mathbf{A}\left(\varepsilon, k\right) = \left\{\mathbf{P} \in \mathbf{M}_1\left(\Omega\right) \text{ satisfies conditions (1) and (2)}\right\}. \tag{5.5.2}$$

This definition means that the marginal distributions of any probability measure $\mathbf{P} \in \mathbf{A}\left(\varepsilon, k\right)$ are almost equal to $\mu_r\left(x\right) dx$, $r = a, b$. More precisely,

Lemma 5.5.1. *Let $\mathbf{A}_{a,b}$ and $\mathbf{A}(\epsilon, k)$ be the subsets of $\mathbf{M}_1\left(\Omega\right)$ defined in (5.5.1) and in (5.5.2), respectively. Then:*

(i) *when $k \uparrow$, $\mathbf{A}\left(\varepsilon, k\right) \downarrow$;*

(ii) *when $\epsilon \downarrow$, $\mathbf{A}\left(\varepsilon, k\right) \uparrow$;*

(iii) *for any $\varepsilon > 0$, $\mathbf{A}_{a,b} = \bigcap_{k=1,2,\ldots} \mathbf{A}\left(\epsilon, k\right)$.*

The following large deviation principle for stochastic processes is due to Nagasawa (1990) and Aebi and Nagasawa (1992).

Theorem 5.5.1 (Large Deviation Principle for Stochastic Processes).

(i) *Let $\left\{X_t, \overline{\mathbf{P}}\right\}$ be the renormalized stochastic process defined above and let $\mathbf{A}_{a,b}$ be the set defined in (5.5.1). Then there exist the projection $\mathbf{Q}$ of $\overline{\mathbf{P}} \in \mathbf{M}_1\left(\Omega\right)$ on $\mathbf{A}_{a,b}$, such that*

$$\inf_{\mathbf{P} \in \mathbf{A}_{a,b}} \mathrm{H}\left(\mathbf{P}|\overline{\mathbf{P}}\right) = \mathrm{H}\left(\mathbf{Q}|\overline{\mathbf{P}}\right). \tag{5.5.3}$$

(ii) *Consider the n-fold product n-fold product be $\left\{\left(X_t^1, \ldots, X_t^n\right), \overline{\mathbf{P}}^n\right\}$ of the renormalized stochastic process be $\left\{X_t, \overline{\mathbf{P}}\right\}$ and the subsets $\mathbf{A}_{a,b}$ and $\mathbf{A}\left(\epsilon, \mathbf{k}\right)$ of $\mathbf{M}_1\left(\Omega\right)$ defined in (5.5.1) and (5.5.2), respectively. Then*

$$\lim_{k \to \infty} \lim_{n \to \infty} \frac{1}{n} \log \overline{\mathbf{P}}^n[\{\overline{\omega} : L_n(\overline{\omega}) \in \mathbf{A}(\varepsilon, \mathbf{k})\}] = -\inf_{\mathbf{P} \in \mathbf{A}_{a,b}} \mathrm{H}(\mathbf{P}|\overline{\mathbf{P}}). \tag{5.5.4}$$

Note that, by (5.5.3), the right-hand side in (5.5.4) is equal to $-\mathrm{H}\left(\mathbf{Q}|\overline{\mathbf{P}}\right)$.

The proof of Theorem 5.5.1 is an application of the Entropy Projection Theorem by Csiszár, but we omit it; the reader is referred to Chapter 8 of Nagasawa (1993).

Equation (5.5.4) above is a generalization of the so-called "Sanov's property". The meaning of equality (5.5.4) is that the probability

$$\overline{\mathbf{P}}^n\left[\{\overline{\omega} : L_n(\overline{\omega}) \in \mathbf{A}(\varepsilon, k)\}\right]$$

that $L_n\left(\overline{\omega}\right) = \sum_{k=1}^n \frac{1}{n}\delta_{\omega_k}$ belongs to the set $\mathbf{A}\left(\varepsilon, k\right)$, that is, the probability that the initial and terminal distributions are near the respective prescribed measures $\mu_a\left(x\right) dx$ and $\mu_b\left(x\right) dx$ decreases exponentially. The rate of decrease is proportional to the distance of $\overline{\mathbf{P}}$ to the set $\mathbf{A}_{a,b}$, measured by the relative entropy. That is,

$$\overline{\mathbf{P}}^n\left[L_n \in \mathbf{A}\left(\epsilon, k\right)\right] \approx \exp\left(-n \inf_{\mathbf{P} \in \mathbf{A}_{a,b}} \mathrm{H}\left(\mathbf{P}|\overline{\mathbf{P}}\right)\right) = \exp\left(-n\,\mathrm{H}\left(\mathbf{Q}|\overline{\mathbf{P}}\right)\right).$$

Theorem 5.5.1 confirms the correctness of the pioneering view on the large deviation principle suggested by Schrödinger in his 1931 paper.

5.6 Kac's Phenomenon of Propagation of Chaos

The Schrödinger equation (for a complex evolution function) describes not only the motion of a single particle, but also that of a group of particles (cloud). That is, we can regard it as a Boltzmann equation, as will be shown below using the large deviation principle of empirical distributions (cloud) discussed in the preceding section.

Before that, let us describe the so-called phenomenon of "propagation of chaos". Kac (1959) considered a simple mathematical model for the Boltzmann equation governing the distribution of a rarefied gas. McKean (1966, 67) called this model "Kac's caricature" and discussed the "propagation of chaos" for n mutually interacting Markov processes.

Consider a multi-particle system consisting of n interacting particles (Markov processes) in one dimension (or, more generally, in higher dimension). Let the interaction between two particles be described by a function $b(x-y)$, and assume that a particle at the position x_j is under the average influence of the other particles at the positions y_i, i.e.,

$$\frac{1}{n-1} \sum_{i \neq j} b(x_j - y_i).$$

Then the random motion $X_j(t)$ of the j-th particle is described by the process

$$X_j(t) = X_j(0) + B_j(t) + \int_0^t \frac{1}{n-1} \sum_{i \neq j} b(X_j(s) - X_i(s)) ds.$$

Since our system depends on the number of particles n, we will denote it by

$$\left(X_1^{(n)}(t), X_2^{(n)}(t), \ldots, X_n^{(n)}(t) \right).$$

Now let the number of particles n go to ∞. Observe $m \leq n$ particles in it, with m is arbitrary, but kept fixed. Then

$$\left(X_1^{(n)}(t), X_2^{(n)}(t), \ldots, X_m^{(n)}(t) \right)$$

converges to a system consisting of m mutually independent particles

$$\left(X_1(t), X_2(t), \ldots, X_m(t) \right).$$

This is referred to as **"Kac's propagation of chaos"**.

The distribution density $\mu(t)$ of the process $X(t) = X_j(t)$, $j = 1, 2, \ldots, m$, satisfies the "Boltzmann equation"

$$\frac{\partial \mu}{\partial t} = \frac{1}{2} \frac{\partial^2 \mu}{\partial x^2} - \frac{\partial}{\partial x} (b[x, \mu]\mu), \tag{5.6.1}$$

where

$$b[x, \mu] = \int b(x - y)\mu(t, y)dy.$$

Equation (5.6.1) is called the McKean-Vlasov equation. It says that if we observe m fixed particles $(X_1^{(n)}(t), X_2^{(n)}(t), \ldots, X_m^{(n)}(t))$ and let $n \to \infty$, then the influence of the interaction between the m particles gradually becomes blurred and eventually one ends with a collection $(X_1(t), X_2(t), \ldots, X_m(t))$ of independent random motions, and the distribution density $\mu(t)$ of one particle and the distribution density $\mu(t)$ of the population of particles become the same. We call this **"propagation of chaos"**.

Next, let us consider "**inverse problem**" of the "propagation of chaos". Assume that individual particles in a group of independent particles execute random motions $\{X(t), \mathbf{Q}\}$ determined by the equations of motion (Schrödinger equation). In this case, does there exist an interacting n-particle system

$$\left(X_1^{(n)}(t), X_2^{(n)}(t), \ldots, X_n^{(n)}(t)\right)$$

that converge to $\{X(t), \mathbf{Q}\}$?

There are some aspects of this inverse problem one should be aware of. For the distribution density of the random motion $\{X(t), \mathbf{Q}\}$ determined by the equations of motion (Schrödinger equation) separation occurs in general as seen in Chapter 1 and Chapter 2. How do individual particles in the population realize such separated distributions?

Let us consider an example. In Section 1.7 we examined the stationary motion in one dimension under Hooke's force.

For simplicity, let $\kappa = 1$, $\sigma = 1$. Then we are dealing with the eigenvalue problem

$$-\frac{1}{2}\frac{\partial^2 \varphi}{\partial x^2} + \frac{1}{2}x^2\varphi = \lambda\varphi.$$

The smallest eigenvalue is $\lambda_0 = \frac{1}{2}$, with associated eigenfunction

$$\varphi_0(x) = \beta \exp(-x^2).$$

In this case the separation of the distribution does not occur, and it is not difficult to treat the inverse problem of the propagation of chaos. We will not discuss this here.

Now consider the eigenvalue $\lambda_1 = \frac{3}{2}$ and the associated eigenfunction

$$\varphi_1(x) = \beta x \exp\left(-\frac{1}{2}x^2\right).$$

The distribution density $\mu_1(x) = (\varphi_1(x))^2$ is separated at $x = 0$.

To solve the inverse problem in this case we must choose a system of n interacting particles $\left(X_1^{(n)}(t), X_2^{(n)}(t), \ldots, X_n^{(n)}(t)\right)$ so that the stochastic process obtained through the "propagation of chaos" coincides with the random motion $\{X(t), \mathbf{Q}\}$ determined by the eigenfunction $\varphi_1(x)$. Is this possible?

In the case under consideration the evolution drift of the random motion $\{X(t), \mathbf{Q}\}$ is

$$a(x) = \frac{d}{dx}(\log \varphi_1(x)).$$

Hence, by Theorem 4.3.2, the distribution density $\mu_1 = (\varphi_1)^2$ satisfies the equation

$$\frac{\partial \mu}{\partial t} = \frac{1}{2}\frac{\partial^2 \mu}{\partial x^2} - \frac{\partial}{\partial x}(a(x)\mu). \tag{5.6.2}$$

Therefore, equation (5.6.2) and the "Boltzmann equation", i.e., the McKean-Vlasov equation (5.6.1) must coincide.

Accordingly, $a(x) = b[x, \mu]$, i.e., we must have

$$\frac{d}{dx}(\log \varphi_1(x)) = \int b(x - y)\mu(t, y)dy, \tag{5.6.3}$$

where $b(x - y)$ is the function describing the interaction between pairs of particles of the Markov process $\left(X_1^{(n)}(t), X_2^{(n)}(t), \ldots, X_n^{(n)}(t)\right)$, which we used in our discussion of the propagation of chaos.

To find the interaction function $b(x - y)$, we must solve the equation (5.6.3). However, the eigenfunction $\varphi_1(x) = \beta x \exp\left(-\frac{1}{2}x^2\right)$ vanishes at the origin. As shown in Chapter 2, the random motion cannot approach the zero point of the distribution density $\mu_1 = (\varphi_1)^2$.

Therefore, in order to solve the equation (5.6.3), it is necessary to consider a model in which the multi-particle system is color-coded and separated into two groups of particles, because there is no solution unless one does so. If we divide the system into two groups, then the solution of equation (5.6.3) exists and

$$b(x) = \beta\left(\frac{3}{x^4} + \frac{2}{x^2} + b_0(x)\right),$$

with

$$b_0(x) = O\left(\frac{1}{x^2}\right), \quad x \downarrow 0,$$

this is a result due to Föllmer and Nagasawa; see Nagasawa and Tanaka (1985). This two-particle interaction $b(x - y)$ has a very strong singularity, which makes the problem difficult.

Nagasawa and Tanaka (1985) proved that the "propagation of chaos" holds in this example. (See Chapter VII of Nagasawa (1993).) The proof is very difficult and tedious due to the singularity of the interaction $b(x - y)$ and the color coding (grouping) of the multi-particle system. It would be very challenging to extend the method used therein to higher dimensions.

In order to discuss the general case, we have to resort to some other method. One such method is an application of Theorem 5.5.1 on the large deviation principle, that is, Theorem 5.6.1 below.

To this end, we first introduce the renormalized stochastic process $\{X_t, \overline{\mathbf{P}}\}$ used in Theorem 5.5.1 and its n-fold product $\{\mathbf{X}_t = (X_t^1, \ldots, X_t^n), \overline{\mathbf{P}}^n\}$. Further, we define a probability measure $\mathbf{P}_{(n,k)}[B]$ by

$$\mathbf{P}_{(n,k)}[B] = \frac{\overline{\mathbf{P}}^n[B \cap \{\overline{\omega} : L_n(\overline{\omega}) \in \mathbf{A}(\varepsilon, k)\}]}{\overline{\mathbf{P}}^n[\{\overline{\omega} : L_n(\overline{\omega}) \in \mathbf{A}(\varepsilon, k)\}]}, \tag{5.6.4}$$

where $\mathbf{A}(\varepsilon, k)$ is the set defined in (5.5.2) and one assumes that $\overline{\mathbf{P}}^n[\{\overline{\omega} : L_n(\overline{\omega}) \in \mathbf{A}(\varepsilon, k)\}] > 0$. Here $L_n(\overline{\omega})$ is "cloud of particle paths" (empirical distribution). Thus, the denominator is the probability that cloud $L_n(\overline{\omega})$ satisfies the conditions (1) and (2) that define the set $\mathbf{A}(\varepsilon, k)$ at $t = a, b$.

We will call the random motion $\{\mathbf{X}_t = (X_t^1, \ldots, X_t^n), \mathbf{P}_{(n,k)}\}$ "**conditional stochastic process**".

How does the conditional motion $\{\mathbf{X}_t = (X_t^1, \ldots, X_t^n), \mathbf{P}_{(n,k)}\}$ when we let k and n go to infinity.

Theorem 5.6.1 (Nagasawa (1990), Aebi and Nagasawa (1992)).

(i) *The conditional random motion $\{\mathbf{X}_t = (X_t^1, \ldots, X_t^n), \mathbf{P}_{(n,k)}\}$ is asymptotically independent with respect to the relative entropy.*

   Specifically, if $(\mathbf{P}_{(n,k)})_{(m)}$ denotes the merginal distribution of the probability measure $\mathbf{P}_{(n,k)} \in \mathbf{M}_1(\Omega^n)$ restricted to the m-fold product $\Omega_{mr+1} \times \cdots \times \Omega_{m(r+1)}$ of Ω, then

$$\lim_{k \to \infty} \lim_{n \to \infty} \mathrm{H}((\mathbf{P}_{(n,k)})_{(m)} | (\mathbf{Q}_{\epsilon,k})^m) = 0, \tag{5.6.5}$$

and

$$\lim_{k \to \infty} \mathrm{H}(\mathbf{Q} | \mathbf{Q}_{\epsilon,k}) = 0. \tag{5.6.6}$$

In more detail, the measure $\mathbf{Q}$ appearing in (5.6.6) is the projection of the renormalized stochastic processes $\overline{\mathbf{P}}$ on the set $\mathbf{A}_{a,b}$ defined in (5.5.1). Hence,

$\{X_t, \mathbf{Q}\}$ *is the random motion determined by the equations of motion* (1.4.1).

Moreover, since the relative entropy has the property that

$$\|\mathbf{Q} - \mathbf{P}\|^2 \le 2\mathrm{H}(\mathbf{Q}|\mathbf{P}),$$

we have:

(ii) *Let* $\left(bf P_{(n,k)}\right)_{(m)}$ *be the marginal distribution of the probability measure* $\mathbf{P}_{(n,k)} \in \mathbf{M}_1(\Omega^n)$ *restricted on* $\Omega_{mr+1} \times \ldots \times \Omega_{m(r+1)}$. *Then*

$$\lim_{k \to \infty} \lim_{n \to \infty} \left\|\left(\mathbf{P}_{(n,k)}\right)_{(m)} - \mathbf{Q}^m\right\| = 0 \qquad (5.6.7)$$

and

$$\lim_{k \to \infty} \lim_{n \to \infty} \left\|\mathbf{P}_{(n,k)}\left[L_n\right] - \mathbf{Q}\right\| = 0, \qquad (5.6.8)$$

where

$$L_n(\overline{\omega}) = \sum_{k=1}^{n} \frac{1}{n} \delta_{\omega_k}, \quad \overline{\omega} = (\omega_1, \ldots, \omega_n) \in \Omega^n$$

is "the cloud of n particle paths" (empirical distribution). Since $\mathbf{Q}$ *is the projection of the renormalized stochastic process* $\overline{\mathbf{P}}$ *on the set* $\mathbf{A}_{a,b}$, *the process* $\{X_t, \mathbf{Q}\}$ *is the random motion determined by the equations of motion* (1.4.1).

Equations (5.6.5) and (5.6.6) express the "propagation of chaos" in terms of convergence in relative entropy. Equation (5.6.7) and (5.6.8) express the "propagation of chaos" in terms of convergence in total variation of measures.

The proof is omitted; see Chapter VIII of Nagasawa (1993).

The conditional probability $\left(\mathbf{P}_{(n,k)}\right)_{(m)}$ converges to the set of independent random motions $\{X_t, \mathbf{Q}\}$ determined by the equations of motion (1.4.1).

For the "propagation of chaos" assertion (ii) of Theorem 5.6.1 is easier to understand. If we look at m-particles in $\left\{(X_t^1, \ldots, X_t^n), \mathbf{P}_{(n,k)}\right\}$, they converge to a set of m independent particles. Therefore, the motion $\mathbf{P}_{(n,k)}[L_n]$ of a single particle in the cloud converges to $\mathbf{Q}$, which is expressed by equation (5.6.8). This explains the "propagation of chaos" via the application of the large deviation principle.

Thus, according to Theorem 5.6.1, when the number of particles increases, the set of particles defined by equation (5.6.4) converge to a group of independent particles consisting of the process $\{X_t, \mathbf{Q}\}$ determined by the equations of motion (1.4.1). Hence, the cloud of random motions $\{X_t, \mathbf{Q}\}$ (empirical distribution) coincides with the distribution of the random motion $\{X_t, \mathbf{Q}\}$ of a single particle.

Thus, the distribution determined by a Schrödinger function $\psi_t(x, y, z)$ **describes a set of infinitely many particles, and** $|\psi_t(x, y, z)|^2$ **is the distribution density of the set of particles.** That is, we have

Proposition 5.6.1. *In the limit when the number* n *of particles goes to infinity, the cloud consists of independently moving infinitely many particles and the random motion* $\{X_t, \mathbf{Q}\}$ *of each particle in the cloud is the stochastic process determined by the equations of motion. Hence, the complex evolution equation (Schrödinger equation)*

$$i\frac{\partial \psi}{\partial t} + \frac{1}{2}\sigma^2(\nabla + i\mathbf{b}(t, x))^2\psi - V(t, x)\psi = 0 \qquad (5.6.9)$$

can be regarded as an equation that describes the motion of infinitely many particles.

Thus, we have shown that $|\psi_t(x, y, z)|^2$ is the "distribution density of an infinite number of particles". On the other hand, in the case of Born's statistical interpretation $|\psi_t(x, y, z)|^2$ is the "distribution density of a single particle". That is, we found that **both of the two points of view presented above are valid**.

Based on the above consideration, it became clear that the Schrödinger equation provides both the "description of the motion of one particle" and the "description of a group of particles." The latter is probably what Schrödinger called "clouds."

That is, the distribution density $\mu(t, x) = \psi(t, x)\bar{\psi}(t, x)$ determined by the solution $\psi(t, x)$ of equation (5.6.9) gives the distribution of a large number of populations of particles. Therefore, the Schrödinger equation (5.6.9) can be regarded as a kind of Boltzmann equation.

This confirmed the pioneering idea that Schrödinger had predicted in the 1931 paper.

Appendix 1. In the discussion up to the third chapter, the distribution defined by the Schrödinger function $\psi_t(x, y, z)$ is the distribution of the random motion $\{X_t, \mathbf{Q}\}$ of "a single particle". In this chapter, the Schrödinger function $\psi_t(x, y, z)$ describes the distribution of a population of infinitely many (multiple) particles, and $|\psi_t(x, y, z)|^2$ can be considered to be "the distribution density of a population of particles."

Let us note that the "description of population of particles" using the Schrödinger equation discussed in this section is a general theory. Thus, when discussing the motion of one electron by means of the Schrödinger equation, we are dealing with microscopic phenomena. On the other hand, when discussing the distribution of a population of particles by means of the Schrödinger equation, we are dealing with macroscopic phenomena. This means that the application of the Schrödinger equation is not limited to microscopic phenomena. This fact is important for applications of the theory of random motion. For example, the collective motion of electrons in the superconducting phenomenon discussed in Section 2.11 is one such application.

As another example, Nagasawa (1993) considered a one-dimensional model for mesons, in which a large number of gluons which perform zigzag motions are intertwined with each other. The gluon's group motion is generated by the entanglement of gluons and is described by the Schrödinger equation

$$i\frac{\partial \psi}{\partial t} + \frac{1}{2}\sigma^2\frac{\partial^2 \psi}{\partial x^2} - \kappa|x|\psi = 0, \tag{5.6.10}$$

and a solution ϕ of the eigenvalue problem

$$\frac{1}{2}\sigma^2\frac{\partial^2 \phi}{\partial x^2} + (\lambda - \kappa|x|)\phi = 0, \quad k > 0 \tag{5.6.11}$$

defines the distribution density $\mu = \phi^2$ of the motion of a gluon group, which is regarded as a meson corresponding to the eigenvalue λ. Calculating the energy of the gluon population (i.e. the eigenvalues of the stationary problem (5.6.11) above) and adding the mass of two quarks, almost all mass spectra of mesons were reproduced. We explained this in Section 2.10.

Furthermore, the usefulness of the Schrödinger equation is not limited to physics. It can be applied to various populations in which many particles interact while executing random zigzag motions; see the discussion in Nagasawa (1980).

Nagasawa (1981) treated septation of Escherichia coli, where E. coli was described by a one-dimensional model in which a large number of molecules perform intertwined random zigzag motions. This was discussed in Section 2.10. (See Nagasawa (2003).)

Appendix 2. Precautionary notes on a group of electrons. We explained above that the Schrödinger equation describes a group of particles (electrons), and a group (cloud) of paths of motion of particles (electrons) as well. In this way the concept of "electron wave" can be used effectively to describe the motion of such a group. (This is the same as handling collections of photons as "electromagnetic waves.") Then the Schrödinger function $\psi(t, x) = e^{R(t,x)+iS(t,x)}$ is the wave function of the electron wave, and $|\psi(t, x)|^2$ is the distribution function of the electrons, that is, the distribution density of the "electron wave".

It should be noted here that concepts referring to a population can not be reduced to concepts referring to the motion of a single particle. This is quite obvious in the case of electromagnetic fields: the concept of electromagnetic waves cannot be reduced to the motion of a single photon. However, with in the case of electrons, the concept of electronic waves, which works effectively for a group of electrons, is often reduced to the movement of a single electron. Typical is Bohr's argument that "electrons have a particle nature and wave nature." This is an incorrect claim, as Schrödinger pointed out earlier.

We must definitely distinguish between the motion of a single electron and that of a group of electrons. If we state that

"when handling a single electron, the motion of electrons can be discussed as a Markov process with the mechanics of random motion; when dealing with the motion of a group of electrons, the idea of electron wave may be effective for the group of electrons,"

this leads to no misunderstandings. In other words, the concept of "electron wave" is appropriate for a large group of electrons, and not a for a single electron.

The so-called "field theory" is available to handle groups of particles. Since individual particles perform random motions, one faces the challenge of reconstructing "field theory" as a theory of a group of particles. In such attempts, it is necessary to avoid assuming a priori that classical special relativity theory is valid. This is because, as shown in Section 2.6, random motion does not obey the theory of relativity in extremely small space regions and for extremely short time intervals. This topic is discussed in Chapters 6 to 10 of Nagasawa (2000).

When we deal with groups of particles, there arises the problem of extinction and proliferation of particles in a group. We will investigate this problem in the next chapter.

Chapter 6

Extinction and Creation

In the theory of Markov processes discussed in Chapter 4 we dealt with the motion of a single particle. In physical phenomena, and more generally in natural phenomena, particles may disappear suddenly and may multiply. In this chapter we will discuss problems of particle extinction (annihilation) and multiplication. First of all, we will deal with particle extinction in the theory of Markov processes. We will define the lifetime of a particle and let the motion terminate at the lifetime. Next, given a Markov process with finite lifetime, we will discuss the converse process of gluing one path to another to extend the life span. We will show that this "piecing-together method" can be applied to the multiplication of particles: one lets new particles start when the life of the observed particle is exhausted. The stochastic processes defined in this way are called branching Markov processes. We note that the extinction and multiplication of particles were traditionally dealt with in quantum field theory.

6.1   Extinction of Particles

Let the state space be the d-dimensional Euclidean space $\mathbf{R}^d$ and let $\{\Omega, X_t, t \geq 0; \mathrm{P}_x, x \in \mathbf{R}^d\}$ be a time-homogeneous Markov process as discussed in Chapter 4[1]. To make the story concrete, we assume that the process $\{\Omega, X_t, t \geq 0; \mathrm{P}_x, x \in \mathbf{R}^d\}$ is determined by the partial differential equation

$$\frac{\partial u}{\partial t} = \frac{1}{2}\sigma^2 \triangle u + \mathbf{b}(x) \cdot \nabla u. \tag{6.1.1}$$

(See Theorem 4.3.2 for this specific Markov process.) In terms of the fundamental solution (transition probability density) $p(t, x, y)$ of the equation (6.1.1) we have

$$u(t, x) = P_t f(x) = \int p(t, x, y) f(y) dy,$$

----

[1]Note that here we use the font P instead of $\mathbf{P}$.

© The Author(s), under exclusive license to Springer Nature Switzerland AG 2021
M. Nagasawa, *Markov Processes and Quantum Theory*,
Monographs in Mathematics 109, https://doi.org/10.1007/978-3-030-62688-4_6

where $\{P_t, t \geq 0\}$ denotes the associated semigroup of operators. Then, in terms of the Markov process $\{X_t, \mathrm{P}_x\}$,

$$u(t, x) = \mathrm{P}_x[f(X_t)],$$

where the right-hand side is the expectation with respect to the probability measure P_x.

Now, corresponding to equation (6.1.1), we consider an equation with an additional term $c(x)u$:

$$\frac{\partial u}{\partial t} = \frac{1}{2}\sigma^2 \triangle u + \mathbf{b}(x) \cdot \nabla u - c(x)u. \tag{6.1.2}$$

What does the presence of the function $c(x) \geq 0$ mean in this context? To explore this, suppose first that c is a constant. Then $u = e^{-ct}P_t f(x) = e^{-ct}\mathrm{P}_x[f(X_t)]$ is a solution of equation (6.1.2). Does this mean that the particle disappears somewhere and "is lost proportionally to e^{-ct}?" This idea is reasonable, and in fact the stochastic process described by equation (6.1.2) is a stochastic process with finite lifetime (it vanishes at a certain time).

If one looks at equation (6.1.2), it is natural to recall the twin equations of motion (1.4.1), because equation (6.1.2) is indeed the same as one of the equations (1.4.1). However, this connection between the twin equations of motion (1.4.1) and the equation (6.1.2) is irrelevant, because if we take only one of the twin equations (1.4.1), they loose their meaning. In the equations of motion (1.4.1), the function $c(x)$ was the potential function of external forces. It is clear from the discussion in Chapter 1 that the potential function of external force $c(x)$ does not model the "disappearance of particles." In this chapter we are not treating the "mechanics of random motion" and the function $c(x)$ has completely different meanings.

In physics, equation (6.1.2) has bee known for a long time. It is believed that it describes the phenomenon of evaporation of particles due to the applied high-temperature heat $c(x) \geq 0$. In order to model this phenomenon mathematically, it is necessary to define the lifetime of a particle, so that the motion of the particle will terminate at that point of time. However, the notion of the lifetime cannot be introduced arbitrarily if one wants to preserve the Markov property (i.e, that the motion does not depend on the past history). In other words, when we introduce the lifetime ζ in a Markov process $\{X_t, 0 \leq t < \infty\}$, the resulting new process $\{X_t, 0 \leq t < \zeta\}$ must be again a Markov process. This is a theoretical requirement for the lifetime ζ to be well defined.

We will present a method that allows us to construct, given the Markov process described by equation (6.1.1) (whose lifetime is infinite), a new Markov process with finite lifetime (i.e., which vanishes at a certain time).

First, we extend the path space Ω by introducing a new parameter r as

$$\Omega^0 = \Omega \times [0, \infty),$$

and we define a probability measure P_x^0 on Ω^0 by

$$P_x^0 = P_x \times p,$$

where $p(dr)$ is the probability measure on the half-line $[0, \infty)$ given by

$$p[(t, \infty)] = \exp(-t), \tag{6.1.3}$$

that is, the exponential distribution with rate equal to 1.

Further, using the given function $c(x)$, we put

$$c_t(\omega) = \int_0^t c(X_s(\omega))ds, \tag{6.1.4}$$

and define a random variable $\zeta(\omega, r)$ on Ω^0 as the first time $c_t(\omega)$ exceeds the value of the added parameter r, i.e.,

$$\zeta(\omega, r) = \inf\{t : c_t(\omega) > r\}. \tag{6.1.5}$$

We also introduce an extra point Δ (one can call it heaven or grave, according to taste!) that does not belong to $\mathbf{R}^d$.

Now we define a new stochastic process Y_t by the rule

$$Y_t(\omega, r) = X_t(\omega), \quad \text{if } t < \zeta(\omega, r), \tag{6.1.6}$$

$$Y_t(\omega, r) = \Delta, \quad \text{if } t \geq \zeta(\omega, r). \tag{6.1.7}$$

Thus, the random variable $\zeta(\omega, r)$ is the particle's lifetime (often called "killing" or "annihilation" time). The state space of the stochastic process Y_t is $\mathbf{R}^d \cup \{\Delta\}$.

Definition 6.1.1. The process $\{\Omega^0, P_x^0, x \in \mathbf{R}^d \cup \{\Delta\}; Y_t, \zeta\}$ with given $c(x) \geq 0$ is called a **stochastic process with killing**. We denote it simply by $\{\Omega^0, P_x^0, Y_t, \zeta\}$.

Remark. The recipe described above is easy to understand if you think about it as follows.

The first component ω of the pair $(\omega, r) \in \Omega^0 = \Omega \times [0, \infty]$ is the travel path chosen, while the second component r represents the total amount of travel expenses available. We can then interpret

$$c_t(\omega) = \int_0^t c(X_s(\omega))ds$$

as the total travel expenses up to time t for the route ω. Then, according to the definition (6.1.5), $\zeta(\omega, r)$ is the time when the total travel expenses exceed the specified allowed limit r. So $Y_t(\omega, r)$ moves along the chosen path $X_t(\omega)$ if $t < \zeta(\omega, r)$, a fact expressed by equation (6.1.6). However, if $t \geq \zeta(\omega, r)$, the wallet is already empty, so you cannot continue traveling: the trip ends at Δ, and you remain stuck there, i.e., (6.1.7) holds.

So, as already mentioned, the state space of the stochastic process with finite lifetime (with killing) is $\mathbf{R}^d \cup \{\Delta\}$.

Remark. Multiplication of particles will be treated in Section 6.3. The discussion of Markov processes with finite lifetime (with killing) is in fact preparation for that. The main subject is the phenomenon in which when one particle disappears at a point, a large number of new particles are generated there. To approach this subject, we first need to define the concept of a Markov process with finite lifetime (with killing).

Theorem 6.1.1. *Let* $\{\Omega, X_t, 0 \leq t < \infty, \mathrm{P}_x\}$ *be the Markov process defined by the equation* (6.1.1). *Then:*

(i) *The stochastic process* $\{\Omega^0, P^0_x, Y_t, \zeta\}$ *defined by the equations* (6.1.6) *and* (6.1.7) *is a Markov process with the transition probability*

$$P^0(t, x, \Gamma) = \mathrm{P}_x[\mathbf{1}_\Gamma(X_t) \exp(-c_t)], \quad x \in \mathbf{R}^d, \quad \Gamma \subset \mathbf{R}^d, \tag{6.1.8}$$

$$P^0(t, x, \{\Delta\}) = 1 - \mathrm{P}_x[\exp(-c_t)], \quad x \in \mathbf{R}^d, \tag{6.1.9}$$

$$P^0(t, \Delta, \{\Delta\}) = 1, \tag{6.1.10}$$

and

$$P^0_x[f(Y_t)] = \mathrm{P}_x[f(X_t) \exp(-c_t)] \tag{6.1.11}$$

for functions $f(x)$ *on* $\mathbf{R}^d$ *that satisfy* $f(\Delta) = 0$.

(ii) *The function* $u(t, x) = P^0_x[f(Y_t)]$ *satisfies the equation*

$$\frac{\partial u}{\partial t} = \frac{1}{2}\sigma^2 \triangle u + \mathbf{b}(x) \cdot \nabla u - c(x)u. \tag{6.1.12}$$

Proof. For functions $f(x)$ defined on $\mathbf{R}^d$ we have

$$P^0_x[f(Y_t)] = P^0_x[f(Y_t) : t < \zeta]$$

$$= \int_{\Omega \times [0,\infty)} (\mathrm{P}_x \times p)[d\omega dr]\, f(X_t(\omega))\mathbf{1}_{\{t < \zeta(\omega, r)\}}$$

$$= \int_\Omega \mathrm{P}_x[d\omega]\, f(X_t(\omega)) \int_{[0,\infty)} p[dr]\mathbf{1}_{(c_t(\omega),\infty)}(r),$$

which in view of (6.1.3) is equal to

$$\int_\Omega \mathrm{P}_x[d\omega] f(X_t(\omega)) \exp(-c_t(\omega)) = \mathrm{P}_x[f(X_t) \exp(-c_t)],$$

i.e., (6.1.11) holds. Setting $f = \mathbf{1}_\Gamma$, we obtain equality (6.1.8), and $P^0_x[\mathbf{1}_{\mathbf{R}^d}(Y_t)] = \mathrm{P}_x[\exp(-c_t)]$, and hence equality (6.1.11). Since

$$P^0(t, x, \{\Delta\}) = 1 - P^0_x[\mathbf{1}_{\mathbf{R}^d}(Y_t)] = 1 - \mathrm{P}_x[\exp(-c_t)],$$

equality (6.1.9) also holds. Equality (6.1.10) is an obvious consequence of (6.1.7). Finally, assertion (ii) is Kac's formula (see Theorem 6.1.2 below).   □

The Markov process $\{\Omega^0, P_x^0, x \in \mathbf{R}^d \cup \{\Delta\}; Y_t, \zeta\}$ with finite lifetime is called the **subprocess** of the Markov process $\{\Omega, X_t, t \geq 0; P_x, x \in \mathbf{R}^d\}$ **up to the lifetime** ζ. In this context, we have shown Kac's formula in Theorem 4.2.2. Namely, let $\{\Omega, X_t, 0 \leq t < \infty; P_x, x \in \mathbf{R}^d\}$ be the Markov process determined by equation (6.1.1), and set

$$c_t = \int_0^t c(X_r)dr.$$

Then if $f(x)$ is a bounded smooth function, the function $u(t, x)$ defined by the Kac formula,

$$u(t, x) = P_x[\exp(-c_t)f(X_t)], \tag{6.1.13}$$

satisfies the partial differential equation

$$\frac{\partial u}{\partial t} = \frac{1}{2}\sigma^2 \triangle u + \mathbf{b}(x) \cdot \nabla u - c(x)u.$$

There is another way to make the lifetime finite. For example, let $\{\Omega, X_t, 0 \leq t < \infty; P_x, x \in \mathbf{R}^d\}$ be the Markov process defined by equation (6.1.1) and let D be a domain of the Euclidean space $\mathbf{R}^d$ with boundary ∂D. We define the **first hitting time** (by the process X_t) of the boundary ∂D by

$$\tau_{\partial D}(\omega) = \inf\{t : X_t(\omega) \in \partial D\}, \tag{6.1.14}$$

and then define a new stochastic process X_t^0 by

$$X_t^0(\omega) = X_t(\omega), \quad \text{if } t < \tau_{\partial D}(\omega), \tag{6.1.15}$$

$$X_t^0(\omega) = \Delta, \quad \text{if } t \geq \tau_{\partial D}(\omega). \tag{6.1.16}$$

Thus, the stochastic process X_t^0 moves at the first hitting time $\tau_{\partial D}$ to the point Δ which does not belong to $D \cup \partial D$. Then the stochastic process $\{P_x, x \in D \cup \{\Delta\}; X_t^0, \tau_{\partial D}\}$ is a **"Markov process with finite lifetime $\tau_{\partial D}$."**

Remark. In general, ζ satisfies the "lifetime property"

$$\text{if } t < \zeta(\omega), \quad \text{then } \zeta(\omega) = t + \zeta(\theta_t\omega). \tag{6.1.17}$$

The lifetime ζ defined in equation (6.1.5) enjoys this property, and since "the first time $\tau_{\partial D}(\omega)$ when X_t hits the boundary ∂D", defined in (6.1.14) satisfies

$$\text{if } t < \tau_{\partial D}(\omega), \text{ then } \tau_{\partial D}(\omega) = t + \tau_{\partial D}(\theta_t\omega),$$

it also has the lifetime property.

Now let $f(x)$ be a bounded smooth function on a domain D such that $f(\Delta) = 0$, and set

$$u(t, x) = P_x[f(X_t^0)], \quad x \in D.$$

Then
$$u(t, x) = \mathrm{P}_x[f(X_t); t < \tau_{\partial D}(\omega)], \quad x \in D,$$
and $u(t, x)$ satisfies the equation
$$\frac{\partial u}{\partial t} = \frac{1}{2}\sigma^2 \triangle u + \mathbf{b}(x) \cdot \nabla u, \quad x \in D,$$
and the boundary condition
$$u(t, x) = 0, \quad x \in \partial D.$$

We call the stochastic process $\{\mathrm{P}_x, x \in D \cup \{\Delta\}; X_t^0, \tau_{\partial D}\}$ a **Markov process with absorbing boundary**.

After this preparatory discussion of Markov processes with finite lifetime (which vanish at some point) we will examine multiplication of particles in Section 6.3, i.e., the phenomenon that when one particle disappears, a large number of particles are generated, that is, multiplication occurs. The Markov process with absorbing boundary introduced above is suitable for modeling the phenomenon that when a particle collides with the boundary (wall) of a region, a large number of new particles are emitted from that boundary point (wall).

Since the lifetime is determined by a given function $c(x)$, the case of the lifetime ζ defined in (6.1.5) treats phenomena different from the boundary (wall) case.

6.2 Piecing-Together Markov Processes

In the preceding section we constructed a Markov process with extinction (killing) from a Markov process without extinction (killing). In this section, we go in the opposite direction and present a method for constructing a Markov process without extinction from Markov processes with extinction (i.e., with a finite lifetime) by piecing them together (Nagasawa (1977)). This is addressed not only mathematically for theoretical consistency, but also because the "piecing-together method" can be applied to study multiplication of particles, a topic that will be discussed in the next section.

Let us first define the appropriate notion of a Markov process with a finite lifetime. We let R denote the space in which particles move; it can be the Euclidean space $\mathbf{R}^d$, but also any space endowed with a distance or topology, which allows one to talk about limits. We will denote a path of the motion by $w(t)$. The path $w(t)$ is not necessarily continuous: jumps are allowed. However, it is assumed that $w(t)$ is right-continuous and admits left limits. The lifetime of w is denoted by $\zeta(w)$. It satisfies
$$\zeta(w) = \begin{cases} t + \zeta(\theta_t w), & \text{if } t < \zeta(w), \\ \Delta, & \text{if } t \geq \zeta(w). \end{cases}$$

Since we assumed that left limits exist,

$$\lim_{t \uparrow \zeta(w)} w(t) = w(\zeta(w)-) = w_{\zeta-}$$

exists.

Thus, the path space W of a Markov process with finite lifetime is

$$W = \{w : w(t) \text{ is right-continuous, admits left limits,}$$
$$\text{the lifetime } \zeta(w) < \infty \text{ exists, and } w(t) = \Delta \text{ if } t > \zeta(w)\}. \qquad (6.2.1)$$

We call the time-homogeneous Markov process

$$\{W, \mathcal{F}, X_t, \zeta, P_a, a \in R\}$$

with the path space W a **"Markov process with finite lifetime"**; here $\mathcal{F}$ is a σ-algebra on W. (To emphasize that R is not necessarily a Euclidean space, we denote its current point by a instead of x.)

Now as a special case suppose that the path $w(t)$ is continuous for $t < \zeta(w)$ and is killed at

$$\lim_{t \uparrow \zeta(w)} w(t) = w_{\zeta-}.$$

Choose another path $w_2 \in W$ which starts at $w_{\zeta-}$, glue w and w_2, and define

$$X_t(w, w_2) = \begin{cases} X_t(w), & \text{if } t < \zeta(w), \\ X_{t-\zeta(w)}(w_2), & \text{if } t \geq \zeta(w). \end{cases}$$

Iterating this procedure, we produce a Markov process without extinction

$$X_t(w, w_2, w_3, w_4, \ldots).$$

This is a result due to Volkonskiĭ (1960).

In the next section we will discuss particle multiplication, and we would like to apply Volkonskiĭ's method to it. For that purpose, the method, which connects the next path continuously from the killing point $w_{\zeta-}$, needs to be modified slightly. Our particle jumps from the killing point $w_{\zeta-}$ to another point and joins another path starting at that point. In order for the stochastic process thus obtained to enjoy the Markov property, jump conditions need to be imposed.

Definition 6.2.1. A kernel $N(w, db)$ is called a **"piecing-together law"**, if it is a measurable function of $w \in W$ and for each fixed w it is a probability measure on R with respect to db such that

$$N(\theta_t w, db) = N(w, db), \quad \text{for } t < \zeta(w), \qquad (6.2.2)$$

where θ_t is the shift operator, which acts on the path w by $\theta_t w(s) = w(t + s)$.

Condition (6.2.2) means that "$N(w, db)$ does not depend on the events before the lifetime $\zeta(w)$." It ensures that piecing together does not compromise the Markov property.

Let us give an example of piecing-together law. Let $p(a, db)$ be a measurable function on R as a function of a and a probability measure with respect to db. Then the kernel

$$N(w, db) = p(X_{\zeta-}(w), db)$$

is piecing-together law. If we set $p(a, db) = \delta_a(db)$, we recover Volkonskiĭ's case, which pieces together paths in a continuous manner.

Our aim is to construct a pieced-together Markov process. First of all, as a preparation, given the piecing-together law $N(w, db)$ we define a 1-step transition probability on the measurable space $\{W, \mathcal{F}\}$ by the formula

$$Q(w, dv) = \int_R N(w, db) \mathrm{P}_b[dv],$$

and consider the direct-product measurable space $\{W^\infty, \mathcal{F}^\infty\}$, with $W^\infty = W \times W \times W \times$ and $\mathcal{F}^\infty = \mathcal{F} \times \mathcal{F} \times \mathcal{F} \times \cdots$.

Theorem 6.2.1.

(i) *There exists a probability measure Π_w on the space $\{W^\infty, \mathcal{F}^\infty\}$ such that, for any bounded measurable function $f(v_1, v_2, \ldots, v_n)$ on $\{W^n, \mathcal{F}^n\}$,*

$$\Pi_w[f] = \int\int \cdots \int Q(w, dv_1) Q(v_1, dv_2) \cdots Q(v_{n-1}, dv_n) f(v_1, v_2, \ldots, v_n).$$
(6.2.3)

(ii) *Define a stochastic process Y_n by*

$$Y_n(\mathbf{v}) = v_n, \quad \mathbf{v} = (v_1, v_2, \ldots, v_n, \ldots) \in W^\infty.$$

Then $\{W^\infty, \mathcal{F}^\infty, Y_n, \Pi_w\}$ is a Markov chain on $\{W, \mathcal{F}\}$ with the one-step transition probability $Q(w, dv)$.

Proof. The existence of the probability measure Π_w on an n-fold direct-product space $\{W^n, \mathcal{F}^n\}$ such that (6.2.3) holds is readily established. However, extending the measure Π_w to the infinite product space $\{W^\infty, \mathcal{F}^\infty\}$ is not a trivial task. That this is possible is asserted by a theorem of Ionescu-Tulcea, for which we refer to p. 162 of Neveu (1965). Item (ii) is clearly a consequence of the equality (6.2.3) satisfied by the probability measure Π_w. □

Remark. Ionescu-Tulcea's theorem can be applied to prove the existence of a Markov chain on a state space. The latter may be any measurable space. When we discussed the existence (construction) of Markov processes in Section 1.2, we resorted to Kolmogorov's theorem to prove it; there the space in which the particle moves was the Euclidean space. The space $\{W, \mathcal{F}\}$ in Theorem 6.2.1 is not a Euclidean space, so Kolmogorov's theorem is not applicable. This is purely a matter of mathematics, but it is a point that cannot be dismissed in the discussion in this section.

Theorem 6.2.1 should actually be regarded as a preliminary result for the basic Theorem 6.2.2 (piecing together theorem).

Let us change notation and write the measurable product space $\{W^\infty, \mathcal{F}^\infty\}$ as $\{\mathscr{W}, \mathscr{F}\}$. Now we define a probability measure $\mathbf{P}_a$ on $\{\mathscr{W}, \mathscr{F}\}$ by the formula

$$\mathbf{P}_a[f] = \int \mathbf{P}_a[dw]\, \Pi_w[f], \quad a \in R,$$

for any bounded measurable function $f(v_1, v_2, \ldots, v_n)$ on $\{W^n, \mathcal{F}^n\}$, where Π_w is the probability measure on $\{\mathscr{W}, \mathscr{F}\}$ defined by formula (6.2.3).

First of all, for an element $\mathbf{w} = (w_1, w_2, \ldots,) \in \mathscr{W}$ we define the renewal times $r_k(\mathbf{w})$ by

$$r_1(\mathbf{w}) = \zeta(w_1), \ldots, r_k(\mathbf{w}) = \sum_{i=1}^{k} \zeta(w_k), \ldots, r_\infty(\mathbf{w}) = \lim_{k \to \infty} r_k(\mathbf{w}), \qquad (6.2.4)$$

and define a new stochastic process $\mathscr{X}_t(\mathbf{w})$ by

$$\mathscr{X}_t(\mathbf{w}) = \begin{cases} X_t(w_1), & \text{for } t < r_1(\mathbf{w}), \\ X_{t-r_1}(w_2), & \text{for } r_1(\mathbf{w}) \le t < r_2(\mathbf{w}), \\ \cdots\cdots\cdots \\ X_{t-r_{k-1}}(w_k), & \text{for } r_{k-1}(\mathbf{w}) \le t < r_k(\mathbf{w}), \\ \cdots\cdots\cdots \\ \Delta, & \text{for } t \ge r_\infty(\mathbf{w}). \end{cases}$$

Moreover, we introduce the shift operator θ_t acting on elements $\mathbf{w} = (w_1, w_2, \ldots)$ by

$$\theta_t \mathbf{w} = (\theta_{t-r_{k-1}} w_k, w_{k+1}, \ldots), \quad \text{for } r_{k-1}(\mathbf{w}) \le t < r_k(\mathbf{w}).$$

Then we can state the following theorem on piecing together Markov processes.

Theorem 6.2.2 (Piecing-Together Theorem). *Let* $\{W, \mathcal{F}, X_t, \zeta, \mathrm{P}_a, a \in R\}$ *be a Markov process with finite lifetime, and let* $N(w, db)$ *be a piecing-together law (see Definition 6.2.1). Then the stochastic process* $\{\mathscr{W}, \mathscr{F}, \mathscr{X}_t(\mathbf{w}), \theta_t, r_\infty, \mathbf{P}_a, a \in R\}$ *defined above by piecing together is a Markov process with the following properties:*

(i) *The process* $\{\mathscr{X}_t(\mathbf{w}), t < r_1, \mathbf{P}_a, a \in R\}$ *is equivalent to the Markov process* $\{W, \mathcal{F}, X_t, \zeta, \mathrm{P}_a, a \in R\}$ *with finite lifetime, that is,*

$$\mathbf{P}_a[f(\mathscr{X}_t) : t < r_1] = \mathrm{P}_a[f(X_t) : t < \zeta],$$

for any bounded measurable function f on the space R.

(ii) *For any bounded measurable function g on W and any bounded measurable function f on R,*

$$\mathbf{P}_a[g(w_1)f(\mathscr{X}_{r_1})] = \mathbf{P}_a\left[g(w_1)\int_R N(w_1, db)f(b)\right].$$

Proof. Theorem 6.2.2 readily follows from the construction of the Markov process by the piecing-together procedure. For details, we refer to Nagasawa (1977). □

In discussing the multiplication of particles in subsequent sections, the piecing-together Theorem 6.2.2 guarantees that the desired Markov process exists. In order to apply it, the most important point is to define an appropriate piecing-together law (joining rule) $N(w, db)$.

Now let us derive the basic equation that characterizes the Markov processes obtained by piecing together.

Theorem 6.2.3. *Let*

$$P_t f(a) = \mathbf{P}_a[f(X_t) : t < \zeta], \tag{6.2.5}$$

be the operator semigroup of the Markov process $\{W, \mathcal{F}, X_t, \zeta, \mathrm{P}_a, a \in R\}$ with finite lifetime and let

$$\mathbf{P}_t f(a) = \mathbf{P}_a[f(\mathscr{X}_t)] \tag{6.2.6}$$

be the operator semigroup of the stochastic process obtained by piecing together. Set

$$\phi(a, dr, db) = {}_a[r_1(w_1) \in dr, \mathscr{X}_{r_1} \in db]. \tag{6.2.7}$$

Let $f(a)$ is a bounded measurable function on the space R. Then the function $u(t, a) = \mathbf{P}_t f(a)$ satisfies the equation

$$u(t, a) = P_t f(a) + \int_0^t \int_R \phi(a, dr, db) u(t - r, b). \tag{6.2.8}$$

If $f(a) \geq 0$, then $u(t, a) = \mathbf{P}_t f(a)$ is the minimal solution of equation (6.2.8).

Proof. First of all, we decompose $u(t, a)$ as

$$u(t, a) = \mathbf{P}_a[f(\mathscr{X}_t)] = \mathbf{P}_a[f(\mathscr{X}_t) : t < r_1] + \mathbf{P}_a[f(\mathscr{X}_t) : t \geq r_1],$$

where the first term on the right-hand side is $P_t f(a)$, by assertion (i) of Theorem 6.2.2. Looking at how to construct a stochastic process by the piecing-together procedure, the second term on the right-hand side is

$$\mathbf{P}a[f(\mathscr{X}_t) : t \geq r_1] = \mathbf{P}_a[\mathbf{P}_{X_{r_1}}[f(\mathscr{X}_{t-r_1})] : t \geq r_1]$$

$$= \mathbf{P}_a[\mathbf{P}_{t-r_1} f(X_{r_1}) : t \geq r_1]$$

$$= \int_0^t \int_R \phi(a, dr, db)\mathbf{P}_{t-r} f(b)$$

$$= \int_0^t \int_R \phi(a, dr, db) u(t - r, b).$$

Therefore, (6.2.8) holds.

Next, set

$$P_t^{(0)} f(a) = P_t f(a)$$

and

$$P_t^{(k)} f(a) = \mathbf{P}_a[f(\mathscr{X}_t) : r_k \leq t < r_{k+1}], \quad k \geq 1.$$

Then

$$\mathbf{P}_t f(a) = \sum_{k=0}^{\infty} P_t^{(k)} f(a). \tag{6.2.9}$$

Since

$$P_t^{(k)} f(a) = \int_0^t \int_R \phi(a, dr, db) P_{t-r}^{(k-1)} f(b), \tag{6.2.10}$$

we see that $u(t, a) = \mathbf{P}_t f(a) = \sum_{k=0}^{\infty} P_t^{(k)} f(a)$ is a solution of equation (6.2.8) constructed by successive approximations. Therefore, if $f(a) \geq 0$, then $u(t, a)$ is the minimal solution of equation (6.2.8). $\qquad \square$

Remark. Equation (6.2.8) is completely determined by specifying $P_t f$ and $\phi(a, dr, db)$. In Theorem 6.2.3, we have shown that its solution is obtained by using the pieced-together Markov process as $\mathbf{P}_t f(a) = \mathbf{P}_a[f(\mathscr{X}_t)]$; in the proof of the theorem, the solution of equation (6.2.8) was obtained by successive approximations via formulas (6.2.9) and (6.2.10). This can be thought of as an analytic method to obtain the semigroup of operators $\mathbf{P}_t$ of the pieced-together Markov process.

Indeed, before the piecing-together Theorem 6.2.2 was known, the semigroup of operators $\mathbf{P}_t$ was usually constructed by means of the indicated successive approximation method (see Moyal (1962)), and then to obtain the corresponding stochastic process with operator semigroup $\mathbf{P}_t$, one applied the general theory. In our piecing-together method, the Markov process is obtained immediately by Theorem 6.2.2. This is a big advantage.

6.3 Branching Markov Processes

Consider the following situation: a particle performs a random motion and then disappears at a certain point and, say, two new particles are generated at that point. Alternatively, one can imagine that a particle splits into two particles. Let us assume that each particle moves randomly and independently, and the division is not influenced by other particles. In order to treat such a random motion mathematically, it is necessary to first define the state space in which particles move.

Definition 6.3.1. A measurable space S endowed with a product (multiplication) $a \cdot b$ defined for any two elements $a, b \in S$ is called a "**multiplicative space**" if the following conditions are satisfied:

(i) The multiplication mapping $S \times S \to S$, $(a, b) \mapsto a \cdot b$, is measurable.

(ii) There exists an element (point) $\partial \in S$ such that $\partial \cdot a = a \cdot \partial = a$ for any $a \in S$.

The point $\partial \in S$ is interpreted as "absence of particles", and should not be confused with the heaven Δ considered above. In fact, the point ∂ belongs to the state space S, whereas Δ does not.

Examples of multiplicative spaces encountered in applications:

(i) Let $\mathbf{N} = \{0, 1, 2, \ldots\}$, endowed with the product $a \cdot b = a + b$ for $a, b \in \mathbf{N}$, and with $\partial = 0$. Then $\mathbf{N}$ is a multiplicative space. In applications, $n \in \mathbf{N}$ represents the number of particles. This space is used when one is just interested in the increase or decrease of the number of particles.

(ii) Let R^n be the direct product of n copies of the measurable space R, and let $R^0 = \{\partial\}$, where the point ∂ does not belong to R. The product of $a = (x_1, x_2, \ldots, x_n) \in R^n$ and $b = (y_1, y_2, \ldots, y_m) \in R^m$ is defined as

$$a \cdot b = (x_1, x_2, \ldots, x_n, y_1, y_2, \ldots, y_m) \in R^{n+m}.$$

Then

$$S = \bigcup_{n=0}^{\infty} R^n$$

is a multiplicative space: indeed, $\partial \cdot a = a \cdot \partial = a$, for any $a \in S$. This is the most typical multiplicative space; $a \cdot b$ represents the group consisting of the particles in the groups a and b.

(iii) Let $\mathbf{M}$ be the space of finite measures on a measurable space, endowed with the product $a \cdot b = a + b \in \mathbf{M}$ for $a, b \in \mathbf{M}$, and with $\partial = 0 \in \mathbf{M}$. Then $\mathbf{M}$ is a multiplicative space. Let $\alpha_i > 0$ be constants such that $\sum_{i=1}^{\infty} \alpha_i < \infty$. Then the measure $\mu = \sum_{i=1}^{\infty} \alpha_i \delta_{x_i}$ belongs to $\mathbf{M}$. Hence, the space $\mathbf{M}$ is suitable for handling a group of infinitely many particles.

(iv) Let $S = \bigcup_{n=0}^{\infty} R^n$ be the space in example (ii). Let $a = (x_1, x_2, \ldots, x_n) \in R^n$ represent the positions of n indistinguishable particles, that is, we do not distinguish sorted coordinates. We denote this space by $\widetilde{R}^n$ and set $\widetilde{S} = \bigcup_{n=0}^{\infty} \widetilde{R}^n$. Then $\widetilde{S}$ is a multiplicative space. If we identify each point $a = (x_1, x_2, \ldots, x_n) \in \widetilde{R}^n$ with the point measure $\mu = \sum_{i=1}^{n} \delta_{x_i}$, then $\widetilde{S}$ can be embedded in the space $\mathbf{M}^{\mathrm{f}}$ of finite point measures.

Definition 6.3.2. If μ and ν are finite measures on a multiplicative space S, their **convolution** $\mu * \nu$ is defined by

$$\int \mu * \nu(db)\, g(b) = \int \mu(db_1)\nu(db_2)\, g(b_1 \cdot b_2), \qquad (6.3.1)$$

for any bounded measurable function $g \geq 0$.

Definition 6.3.3. If a transition probability $\mathbf{P}_t(a, db)$ on a multiplicative space S satisfies the conditions

$$\mathbf{P}_t(\partial, \{\partial\}) = 1$$

and

$$\mathbf{P}_t(a \cdot b, \cdot) = \mathbf{P}_t(a, \cdot) * \mathbf{P}_t(b, \cdot), \quad \forall a, b \in S, \tag{6.3.2}$$

then we say that $\mathbf{P}_t(a, db)$ has the **"branching property"**. A Markov process on the multiplicative space S with such a transition probability $\mathbf{P}_t(a, db)$ is called a **"branching Markov process"**.

The branching Markov process introduced in Definition 6.3.3 is a **"multiplicative Markov process"**.

Remark. The structure of formula (6.3.2) does not account for the "increase (or decrease)" of the number of particles. In example (ii) of the multiplicative space $S = \bigcup_{n=0}^{\infty} R^n$, equation (6.3.2) expresses the fact that a group a and a group b move and proliferate independently.

In fact, the multiplicative space $S = \bigcup_{n=0}^{\infty} R^n$ keeps track of the "increase (or decrease)" of the number of particles: if a group of particles $a = (x_1, x_2, \ldots, x_n) \in R^n$ jumps to another space R^m, it means that "the number of particles increases or decreases from n to m".

Definition 6.3.4. If a function f on a multiplicative space S satisfies the condition

$$f(a \cdot b) = f(a)f(b), \quad \forall a, b \in S, \tag{6.3.3}$$

then f is said to be **"multiplicative."**

A typical multiplicative function is given in Definition 6.3.5 below, and it is used to characterize the branching property in Theorem 6.3.2.

Theorem 6.3.1. *Suppose a transition probability $\mathbf{P}_t(a, db)$ has the branching property, that is, (6.3.2) holds. If $f(a)$ is a multiplicative bounded measurable function on the multiplicative space S, then the function $\mathbf{P}_t f(a)$ is also multiplicative:*

$$\mathbf{P}_t f(a \cdot b) = \mathbf{P}_t f(a) \mathbf{P}_t f(b), \quad \mathbf{P}_t f(\partial) = 1. \tag{6.3.4}$$

Proof. Let $f(a)$ be a multiplicative function. Then $f(\partial) = 1$. Since $\mathbf{P}_t(\partial, \{\partial\}) = 1$, we have $\mathbf{P}_t f(\partial) = 1$. In view of (6.3.2) and (6.3.3),

$$\mathbf{P}_t f(a \cdot b) = \int \int \mathbf{P}_t(a, dc_1) \mathbf{P}_t(b, dc_2) f(c_1) f(c_2),$$

that is, $\mathbf{P}_t f(a \cdot b) = \mathbf{P}_t f(a) \mathbf{P}_t f(b)$, as claimed. $\square$

In applications of branching processes, one often deals with groups of finitely many particles. In that case, one uses $S = \bigcup_{n=0}^{\infty} R^n$ (or $\tilde{S} = \bigcup_{n=0}^{\infty} \tilde{R}^n$) as a multiplicative space. Therefore, from now on we consider a branching Markov process $\{W, X_t, \mathbf{P}_a, a \in S\}$.

Definition 6.3.5. If f is a function on the space R, define the function $\widehat{f}(a)$ on the multiplicative space $S = \bigcup_{n=0}^{\infty} R^n$ (or $\widetilde{S} = \bigcup_{n=0}^{\infty} \widetilde{R}^n$) by

$$\widehat{f}(x_1, x_2, \ldots, x_n) = f(x_1)f(x_2)\cdots f(x_n) \quad \text{and} \quad \widehat{f}(\partial) = 1, \qquad (6.3.5)$$

for $a = (x_1, x_2, \ldots, x_n)$ and $a = \partial$, respectively.

Clearly, $\widehat{f}(a)$ is a multiplicative function on S (or $\widetilde{S}$). We let $B_1(R)$ denote the set of all measurable functions f on R such that $|f| \leq 1$ and set $\widehat{B}_1(S) = \{\widehat{f} : f \in B_1(R)\}$. Then Theorem 6.3.1 implies

Theorem 6.3.2. *Suppose a transition probability $\mathbf{P}_t(a, db)$ on the multiplicative space S has the branching property (i.e., (6.3.2) holds). Then $\mathbf{P}_t$ is an operator defined on the set $\widehat{B}_1(S)$ such that*

$$\mathbf{P}_t \widehat{f}(a) = \widehat{\mathbf{P}_t f}(a), \quad a \in S. \qquad (6.3.6)$$

That is, if $a = (x_1, \ldots, x_n)$, then the left-hand (resp., right-hand) hand side of equation (6.3.6) is $\mathbf{P}_t \widehat{f}(x_1, \ldots, x_n)$ (resp., $\prod_{k=1}^{n} \mathbf{P}_t \widehat{f}(x_k)$), and so (6.3.6) says that

$$\mathbf{P}_t \widehat{f}(x_1, \ldots, x_n) = \prod_{k=1}^{n} \mathbf{P}_t \widehat{f}(x_k).$$

Remark. Equation (6.3.6) was adopted as the "branching property" in the work of Skorokhod (1964) and Ikeda, Nagasawa and Watanabe (1967, 1968/69). As a matter of fact, equations (6.3.6) and (6.3.2) are equivalent.

Further, given a finite measure μ on the multiplicative space S, we set

$$\mathbf{P}_t(\mu, db) = \int \mu(da)\mathbf{P}_t(a, db).$$

Then

$$\mathbf{P}_t(\mu * \nu, \cdot) = \mathbf{P}_t(\mu, \cdot) * \mathbf{P}_t(\nu, \cdot). \qquad (6.3.7)$$

In other words, the branching (multiplicative) transition probability preserves the convolution product of measures.

The mapping

$$f(x) \mapsto \mathbf{P}_t \widehat{f}(x), \quad f \in B_1(R),$$

is a nonlinear operator on $B_1(R)$. We set

$$u_t(x) = \mathbf{P}_t \widehat{f}(x), \quad x \in R,$$

and look for a (nonlinear) equation that $u_t(x)$ satisfies. Then we will show that this equation characterizes the branching Markov processes.

First of all, if $\{W, X_t, \mathbf{P}_a, a \in S\}$ is a branching Markov process, we define

$$r_1(w) = \inf\{t : X_0(w) \in R, \ X_t(w) \in R^n, n \neq 1\}.$$

This is the time when a particle branches for the first time.

The stochastic process $\{W, X_t, t < r_1, \mathbf{P}_x, x \in R\}$ is a Markov process with finite lifetime r_1. We assume that it coincides with the Markov process with finite lifetime $\{\Omega^0, \mathrm{P}_x^0, x \in \mathbf{R}^d \cup \{\Delta\}; Y_t, \zeta\}$ defined in Section 6.1. Then

$$\mathbf{P}_x[f(X_t) : t < r_1] = \mathrm{P}_x^0[f(Y_t) : t < \zeta].$$

Set

$$u(t, x) = \mathbf{P}_x[f(X_t) : t < r_1] = \int_{\mathbf{R}^d} P_t^0(x, dy) f(y),$$

where $P_t^0(x, dy)$ is the transition probability of the Markov process Y_t. Then, by Theorem 6.1.1, the function $u(t, x)$ satisfies the partial differential equation

$$\frac{\partial u}{\partial t} = \frac{1}{2} \sigma^2 \triangle u + \mathbf{b}(x) \cdot \nabla u - c(x) u.$$

Next, we decompose $\mathbf{P}_x[\widehat{f}(X_t)]$ as

$$\mathbf{P}_x[\widehat{f}(X_t)] = \mathbf{P}_x[\widehat{f}(X_t) : t < r_1] + \mathbf{P}_x[\widehat{f}(X_t) : t \geq r_1]. \tag{6.3.8}$$

Since the first term on the right-hand side refers to history before multiplication or extinction occurs, we have

$$\mathbf{P}_x[f(X_t) : t < r_1] = \mathrm{P}_x^0[f(Y_t) : t < \zeta] = P_t^0 f(x). \tag{6.3.9}$$

By the Markov property at time r_1, the second term is

$$\mathbf{P}_x[\widehat{f}(X_t) : t \geq r_1] = \mathbf{P}_x[\mathbf{P}_{X_{r_1}}[\widehat{f}(X_{t-s})]|_{s=r_1} : t \geq r_1].$$

By the branching property (6.3.6),

$$\mathbf{P}_a[\widehat{f}(X_{t-s})] = \widehat{u_{t-s}}(a),$$

whence

$$\mathbf{P}_{X_{r_1}}[\widehat{f}(X_{t-s})]|_{s=r_1} = \widehat{u_{t-r_1}}(X_{r_1}).$$

Therefore, the second term on the right-hand side of (6.3.8) is

$$\mathbf{P}_x[\widehat{f}(X_t) : t \geq r_1] = \mathbf{P}_x[\widehat{u_{t-r_1}}(X_{r_1}) : t \geq r_1]$$

$$= \int_0^t \int_{\mathbf{R}^d} \mathbf{P}_x[r_1 \in ds, \ X_{r_1-} \in dy] \mathbf{P}_y[\widehat{u_{t-s}}(X_{r_1})].$$

Now note that

$$\mathbf{P}_y[\widehat{u}(X_{r_1})] = \sum_{n \neq 1}^{\infty} \mathbf{P}_y[X_{r_1} \in R^n] \int_{R^n} \mathbf{P}_y[X_{r_1} \in da | X_{r_1} \in R^n] \widehat{u}(a),$$

denote the probability that one particle branches into n particles by

$$q_n(y) = \mathbf{P}_y[X_{r_1} \in R^n],$$

and write the conditional distribution of n particles after branching as

$$\pi_n(y, da) = \mathbf{P}_y[X_{r_1} \in da | X_{r_1} \in R^n].$$

Set

$$F(y, u) = q_0(y) + \sum_{n=2}^{\infty} q_n(y) \int_{R^n} \pi_n(y, da) \widehat{u}(a). \tag{6.3.10}$$

Then

$$\mathbf{P}_y[\widehat{u}(X_{r_1})] = F(y, u).$$

Further, if we denote

$$K(x, ds, dy) = \mathbf{P}_x[r_1 \in ds, \ X_{r_1-} \in dy],$$

then the second term on the right-hand side of (6.3.8) takes on the form

$$\mathbf{P}_x[\widehat{f}(X_t) : t \geq r_1] = \int_0^t \int_{\mathbf{R}^d} K(x, ds, dy) F(y, u_{t-s}).$$

Thus, we have shown that if we set

$$u(t, x) = \mathbf{P}_t \widehat{f}(x), \quad x \in \mathbf{R}^d, \quad \text{for } f \in B_1(R),$$

where $\mathbf{P}_t$ is the operator semigroup of the branching Markov process $\{W, X_t, \mathbf{P}_a, a \in S\}$, then, by (6.3.8), $u(t, x)$ satisfies the equation

$$u(t, x) = P_t^0 f(x) + \int_0^t \int_{\mathbf{R}^d} K(x, ds, dy) F(y, u_{t-s}), \tag{6.3.11}$$

where $P_t^0 f(x) = \mathbf{P}_x[f(X_t) : t < r_1]$ and $F(y, u)$ is given by formula (6.3.10).

The next lemma provides a concrete expression for the kernel $K(x, ds, dy) = \mathbf{P}_x[r_1 \in ds, \ X_{r_1-} \in dy]$ which appears in equation (6.3.11).

Lemma 6.3.1. *Let* $P_t^0(x, dy)$ *be the transition probability of the Markov process* $\{\Omega^0, \mathbf{P}_x^0, x \in \mathbf{R}^d \cup \{\Delta\}; Y_t, \zeta\}$ *with finite lifetime defined in Section 6.1. Then*

$$K(x, ds, dy) = P_s^0(x, dy) c(y) ds.$$

Proof. Since we assumed that the two Markov processes $\{W, X_t, t < r_1, \mathbf{P}_x, x \in R\}$ and $\{\Omega^0, \mathrm{P}_x^0, x \in \mathbf{R}^d \cup \{\Delta\}; Y_t, \zeta\}$ are equivalent, we have

$$\mathbf{P}_x[r_1 \in ds, \ X_{r_1-} \in dy] = \mathrm{P}_x^0[\zeta \in ds, \ Y_{\zeta-} \in dy].$$

Therefore, for the Markov process with finite lifetime ζ, it is enough to show that

$$\mathrm{P}_x^0[\zeta \in ds, \ Y_{\zeta-} \in dy] = P_s^0(x, dy)c(y)ds.$$

But

$$\mathrm{P}_x^0[\zeta \in ds, \ Y_{\zeta-} \in dy] = \mathrm{P}_x^0[\zeta \le s + ds, \ Y_{\zeta-} \in dy] - \mathrm{P}_x^0[\zeta \le s, \ Y_{\zeta-} \in dy].$$

The first and second terms on the right-hand side are

$$\mathrm{P}_x^0[\zeta \le s + ds, \ Y_{\zeta-} \in dy] = \mathrm{P}_x^0[Y_{\zeta-} \in dy] - \mathrm{P}_x^0[s + ds < \zeta, \ Y_{\zeta-} \in dy],$$

and

$$\mathrm{P}_x^0[\zeta \le s, \ Y_{\zeta-} \in dy] = \mathrm{P}_x^0[Y_{\zeta-} \in dy] - \mathrm{P}_x^0[s < \zeta, \ Y_{\zeta-} \in dy],$$

respectively. Subtracting, we obtain the formula

$$\mathrm{P}_x^0[\zeta \in ds, \ Y_{\zeta-} \in dy] = -\mathrm{P}_x^0[s + ds < \zeta, \ Y_{\zeta-} \in dy] + \mathrm{P}_x^0[s < \zeta, \ Y_{\zeta-} \in dy].$$

Recall that the Markov process with finite lifetime $\{Y_t, \mathrm{P}_x^0\}$ is obtained by transforming the Markov process $\{X_t, \mathrm{P}_x\}$ determined by equation (6.1.1) by means of Kac's functional $K_t = \exp\left(-\int_s^t c(X_r)dr\right)$. Consequently,

$$\mathrm{P}_x^0[s + ds < \zeta, \ Y_{\zeta-} \in dy] = \mathrm{P}_x[K_{s+ds} : X_s \in dy]$$

and

$$\mathrm{P}_x^0[s < \zeta, \ Y_{\zeta-} \in dy] = \mathrm{P}_x[K_s : X_s \in dy].$$

We conclude that

$$\begin{aligned}
\mathrm{P}_x^0[\zeta \in ds, \ Y_{\zeta-} \in dy] &= -\mathrm{P}_x[K_{s+ds} : X_s \in dy] + \mathrm{P}_x[K_s : X_s \in dy] \\
&= -\mathrm{P}_x[dK_s : X_s \in dy] = \mathrm{P}_x[K_s c(X_s)ds : X_s \in dy] \\
&= \mathrm{P}_x^0[c(Y_s)ds : Y_s \in dy] = P_s^0(x, dy)c(y)ds.
\end{aligned}$$

The proof is complete. $\qquad\square$

Thus, we arrived at the following result.

Theorem 6.3.3 (Basic Theorem of Branching Markov Processes).
Let $\{W, X_t, \mathbf{P}_a, a \in S\}$ be a *branching Markov process on the multiplicative space* $S = \bigcup_{n=0}^{\infty} R^n$ (or $\widetilde{S} = \bigcup_{n=0}^{\infty} \widetilde{R}^n$), *with operator semigroup* $\mathbf{P}_t$. *Define*

$$u(t, x) = \mathbf{P}_t \widehat{f}(x), \quad x \in \mathbf{R}^d,$$

for $f \in B_1(R)$. Then the function $u(t, x)$ satisfies the "basic equation of branching Markov processes"

$$u(t, x) = P_t^0 f(x) + \int_0^t P_s^0 (cF(\cdot, u(t - s)))(x) ds, \qquad (6.3.12)$$

where

$$P_t^0 f(x) = \mathbf{P}_x[f(X_t) : t < r_1]$$

and $F(y, u)$ is given by formula (6.3.10).

If $u(t, x) = P_t \widehat{f}(x)$ is smooth, then it satisfies the equation

$$\frac{\partial u}{\partial t} = \frac{1}{2} \sigma^2 \triangle u + \mathbf{b}(x) \cdot \nabla u - c(x)u + c(x)F(x, u). \qquad (6.3.13)$$

In particular, if the branched particles start from where the original particle disappeared, then $u(t, x)$ satisfies the equation

$$u(t, x) = P_t^0 f(x) + \int_0^t P_s^0 \left(c\left\{ q_0 + \sum_{n=2}^{\infty} q_n u(t - s)^n \right\} \right)(x) ds. \qquad (6.3.14)$$

Therefore, when $u(t, x)$ is smooth, it is a solution of the equation

$$\frac{\partial u}{\partial t} = \frac{1}{2} \sigma^2 \triangle u + \mathbf{b}(x) \cdot \nabla u - c(x)u + c(x)\left\{ q_0(x) \sum_{n=2}^{\infty} q_n(x)u^n \right\}. \qquad (6.3.15)$$

Proof. To complete the proof of Theorem 6.3.3, it remains to show that equation (6.3.13) follows from equation (6.3.12). Indeed, (6.3.12) implies that

$$u(t + h, x) = P_{t+h}^0 f(x) + \int_0^{t+h} P_s^0 \big(cF(\cdot\, ; u(t + h - s)) \big)(x) ds. \qquad (6.3.16)$$

On the other hand, applying P_h^0 to both sides of equation (6.3.12) we obtain

$$\begin{aligned}
P_h^0 u(t, x) &= P_h^0 P_t^0 f(x) + \int_0^t P_h^0 P_s^0 \big(cF(\,\cdot\, ; u(t - s)) \big)(x) ds \\
&= P_{t+h}^0 f(x) + \int_h^{t+h} P_s^0 \big(cF(\,\cdot\, ; u(t + h - s)) \big)(x) ds.
\end{aligned} \qquad (6.3.17)$$

Combining (6.3.16) and (6.3.17), we see that

$$u(t + h, x) = P_h^0 u(t, x) + \int_0^h P_s^0 \big(cF(\cdot\, ; u(t + h - s)) \big)(x) ds.$$

Therefore,

$$\frac{u(t+h, x) - u(t, x)}{h} = \frac{P_h^0 u(t, x) - u(t, x)}{h} + \frac{1}{h} \int_0^h P_s^0 \big(cF(\cdot\, ; u(t + h - s)) \big)(x) ds.$$

Let $h \downarrow 0$. Then, if $u(t,x)$ is smooth, we have

$$\frac{\partial u}{\partial t} = \frac{1}{2}\sigma^2 \triangle u + \mathbf{b}(x) \cdot \nabla u - c(x)u + c(x)F(x;u),$$

i.e., (6.3.13) holds. □

For example, if we assume that $\mathbf{b}(x) = 0$ and a particle splits into two particles, then $u(t,x) = \mathbf{P}_t \widehat{f}(x)$ satisfies

$$\frac{\partial u}{\partial t} = \frac{1}{2}\sigma^2 \triangle u - c(x)u + c(x)u^2.$$

This equation is often encountered in applications of branching processes. The terms $-c(x)u$ and $+c(x)u^2$ on the right-hand side indicate that one particle disappears and two particles are created, respectively, so the sum $-c(x)u + c(x)u^2$ encodes the fact that one particle splits into two.

The nonlinear integral equation (6.3.12) and the semilinear partial differential equation (6.3.15) are basic equations that characterize branching Markov processes; they play an important role in applications. Let us give a typical example.

Let $\{W, X_t, \mathbf{P}_a, a \in S\}$ be a branching Markov process. When a particle starts at a point x, we denote the "probability that no particle exists at time t" by $u(t,x)$. The limit $e(x) = \lim_{t\to\infty} u(t,x)$ is called the "extinction probability."

Incidentally, since the function $u(t,x)$ is expressed as

$$u(t,x) = \mathbf{P}_t \widehat{0}(x) = \mathbf{P}_x[X_t = \partial] = \mathbf{P}_x[\text{no particle exists at time } t],$$

it is a solution of equation (6.3.15). What is crucial here is the fact that $u(t,x)$ is represented as $\mathbf{P}_t \widehat{0}(x)$.

The British statistician Galton and his compatriot mathematician Watson were the first to recognize the importance of this fact. In the late 19th century they were interested in the observation that families of British aristocrats were going extinct and proposed a model (Galton-Watson process) to explain this trend. In their model the multiplicative state space is $\mathbf{N} = \{0,1,2,\ldots\}$ and equation (6.3.15) becomes

$$\frac{\partial u}{\partial t} = -c(x)u + c(x)\Big\{q_0(x) + \sum_{n=2}^{\infty} q_n(x)u^n\Big\},$$

where $c(x) > 0$. Letting $t \to \infty$, we deduce that the extinction probability $e(x) = \lim_{t\to\infty} u(t,x)$ satisfies

$$-e + q_0 + \sum_{n=2}^{\infty} q_n(x)e^n = 0. \tag{6.3.18}$$

Since $q_0 + \sum_{n=2}^{\infty} q_n = 1$, equation (6.3.18) implies that "the extinction probability $e = 1$." In other words, Galton and Watson showed that it was natural that in the long run aristocratic families die out. This was the beginning of research on branching processes.

After the Galton-Watson process was introduced, various branching processes were studied. At that time a branching process was defined separately for each individual model and the general mathematical structure of such processes was not necessarily clear.

In this section, following Nagasawa (1972, 1977), we presented the general mathematical structure of branching Markov processes. For investigations of various branching processes, see, e.g., Bartlett (1955), Harris (1963), Sevastyanov (1958).

6.4   Construction of Branching Markov Processes

Let us prove that the branching Markov processes discussed in the preceding section do indeed exist. To this end we use the piecing-together theorem.

We first construct a Markov process that describes a population of particles with finite lifetime. Then we define a piecing-together law that governs the increase and decrease of the number of particles, and apply the piecing-together Theorem 6.2.2. Finally, we will show that the Markov process obtained in this manner is a branching process.

Given a Markov process $\{\Omega, X_t, \zeta, P_x, x \in R\}$ with finite lifetime, consider its n-fold direct product $\{\Omega^n, X_t, \zeta, P_a, a = (x_1, x_2, \ldots, x_n) \in R^n\}$. Let Δ be an extra point.

That is, for $w = (\omega_1, \omega_2, \ldots, \omega_n) \in \Omega^n$ we define the lifetime of w by

$$\zeta(w) = \min\{\zeta(\omega_k) : k = 1, 2, \cdots, n\},$$

and $X_t(w)$ by

$$X_t(w) = \begin{cases} (X_t(\omega_1), X_t(\omega_2), \ldots, X_t(\omega_n)), & \text{if } t < \zeta(w), \\ \Delta, & \text{if } t \geq \zeta(w). \end{cases}$$

Next, we introduce a probability measure P_a, $a = (x_1, x_2, \ldots, x_n) \in R^n$, by

$$P_a = P_{x_1} \times P_{x_1} \times \cdots \times P_{x_n}.$$

Here, the collective motion of particles is represented by w, and the path of one particle is written as $\omega \in \Omega$. Although using both w and ω might be confusing, this is to avoid confusion of later symbols.

Consider the Markov process

$$\{W, X_t, \zeta, P_a, a \in S\},$$

on the multiplicative state space $S = \bigcup_{n=0}^{\infty} R^n$ $(R^0 = \{\partial\})$, where $W = \bigcup_{n=1}^{\infty} \Omega^n$. This process describes the random motion of a population of particles before branching or extinction occurs. Particles belonging to a group are moving independently of one another.

Let $q_m(x) \geq 0$ be the probability that a particle will disappear at a point x and m particles will be generated at x. Then $\sum_{m=0}^{\infty} q_m(x) = 1$. However, $q_1(x) = 0$. The m particles generated at x are distributed in $db \in R^m$ according to the law $\pi_m(x, db)$. Moreover, let $\pi_0(x, db)$ be a point measure at ∂. Define

$$\pi(x, db) = q_0(x)\delta_\partial(db) + \sum_{m=2}^{\infty} q_m(x)\pi_m(x, db). \tag{6.4.1}$$

This definition shows that $\pi(x, db)$ is the distribution of particles just after the original particle disappeared at x, or branching has occurred.

Using the already constructed Markov process $\{W, X_t, \zeta, \mathrm{P}_a, a \in S\}$ on the multiplicative state space $S = \bigcup_{n=0}^{\infty} R^n$ (with $R^0 = \{\partial\}$) and the law $\pi(x, db)$ governing the increase and decrease of the number of particles given by formula (6.4.1), we define the piecing-together law $N(\mathbf{w}, db_1 db_2 \cdots db_n)$ as follows:
For $\mathbf{w} = (w_1, w_2, \ldots, w_n) \in W^n$, $n = 1, 2, \ldots$, set

$$N(\mathbf{w}, db_1 db_2 \cdots db_n) = \sum_{i=1}^{n} 1_{\{\zeta(\mathbf{w}) = \zeta(w_i)\}} \pi(X_{\zeta(w_i)-}(w_i), db_i)$$

$$\times \prod_{j \neq i} \delta_{\{X_{\zeta(\mathbf{w})-}(w_j)\}}(db_j). \tag{6.4.2}$$

What the piecing-together law $N(\mathbf{w}, db_1 db_2 \ldots db_n)$ says is that when n particles move independently, the i-th particle with shortest lifetime disappears with probability $q_0(X_{\zeta-}(w_i))$, or splits with probability $q_m(X_{\zeta-}(w_i))$ into m particles that are distributed in the interval db according to the law $\pi_m(X_{\zeta-}(w_i))), db)$. All the other particles (i.e., except particle i), continue to move as they were.

Further, applying the piecing-together Theorem 6.2.2, we obtain the pieced-together stochastic process

$$\{\mathscr{W}, \mathscr{X}_t(\mathbf{w}), \theta_t, r_\infty, \mathbf{P}_a, a \in S\}$$

on the multiplicative state space $S = \bigcup_{n=0}^{\infty} R^n$, with the associated sequence of branching times

$$r_1(\mathbf{w}) = \zeta(w_1), \ldots, r_k(\mathbf{w}) = \sum_{i=1}^{k} \zeta(w_k), \ldots, r_\infty(\mathbf{w}) = \lim_{k \to \infty} r_k(\mathbf{w}).$$

In fact, the stochastic process thus obtained is a **"branching Markov process"**. Let us verify that it indeed has the **"branching property"**:

Theorem 6.4.1. *The Markov process on the space S constructed by piecing together has the branching property. That is, the corresponding transition probability* $\mathbf{P}_t(a, db)$ *satisfies*

$$\mathbf{P}_t(a \cdot b, \cdot) = \mathbf{P}_t(a, \cdot) * \mathbf{P}_t(b, \cdot). \tag{6.4.3}$$

Proof. Set

$$P_t^k(a, db) = \mathbf{P}_a[\mathscr{X}_t \in db : \ r_k \le t < r_{k+1}].$$

Then

$$P_t^k(a \cdot b, \cdot) = \sum_{i=0}^{k} P_t^i(a, \cdot) * P_t^{k-i}(b, \cdot). \tag{6.4.4}$$

The left-hand side of relation (6.4.4) shows that k times splittings occur in two groups, a and b. This means that the i splittings occur in the group a and the $k - i$ splittings occur in the group b. We then take their sum and obtain the right-hand side.

(For details, see pp. 278–280, Lemma 12.2 of Nagasawa (1993).)

If we now use equation (6.4.4), we obtain

$$\mathbf{P}_t(a \cdot b, \cdot) = \sum_{k=0}^{\infty} P_t^k(a \cdot b, \cdot) = \sum_{k=0}^{\infty} \sum_{i=0}^{k} P_t^i(a, \cdot) * P_t^{k-i}(b, \cdot)$$

$$= \sum_{k=0}^{\infty} P_t^k(a, \cdot) * \sum_{k=0}^{\infty} P_t^k(b, \cdot) = \mathbf{P}_t(a, \cdot) * \mathbf{P}_t(b, \cdot).$$

The proof is complete.   □

6.5   Markov Processes with Age

So far we only observed the position of particles that undergo a random motion. In section endow particles with an "age" and assume that the age of a particle depends on its path. Hence, if we denote the age as k_t, it is a random variable. Depending on the chosen path, the observed particles can age faster or slower.

We will explain later on why we introduce age. Actually, in addition to introducing age, it is essential to attach to it a "weight."

First of all, let us prepare the underlying Markov process. In this section we write the Markov process determined by the equation

$$\frac{\partial u}{\partial t} = \frac{1}{2}\sigma^2 \triangle u + \mathbf{b}(x) \cdot \nabla u \tag{6.5.1}$$

as

$$\{\Omega, x_t, \mathbf{P}_x, x \in \mathbf{R}^d\}.$$

Let $p(t, x, y)$ denote the fundamental solution of equation (6.5.1). Then the corresponding transition probability $P(t, x, \Gamma)$ is given by

$$P(t, x, \Gamma) = \int_\Gamma p(t, x, y) dy.$$

Next, given a function $c(x) \geq 0$, the formula

$$c_t(\omega) = \int_0^t c(x_s(\omega)) ds \qquad (6.5.2)$$

determines the Markov process with finite lifetime

$$\{\Omega, x_t, \zeta, P_x, x \in \mathbf{R}^d \cup \{\Delta\}\}.$$

Its transition probability $P^0(t, x, \Gamma) = P_x[\mathbf{1}_\Gamma(x_t) : t < \zeta]$ is given by formulas (6.1.8), (6.1.9) and (6.1.10) of Theorem 6.1.1. For any function $f(x)$ on $\mathbf{R}^d$,

$$u(t, x) = P_x[f(x_t) : t < \zeta], \quad x \in \mathbf{R}^d,$$

satisfies the partial differential equation

$$\frac{\partial u}{\partial t} = \frac{1}{2}\sigma^2 \triangle u + \mathbf{b}(x) \cdot \nabla u - c(x)u. \qquad (6.5.3)$$

The Markov process with finite lifetime $\{\Omega, x_t, \zeta, P_x, x \in \mathbf{R}^d \cup \{\Delta\}\}$ is called a "subprocess", because $\{x_t, 0 \leq t < \zeta\}$ is a part of the underlying Markov process $\{x_t, 0 \leq t < \infty\}$.

To represent the "age" parameter we use the set $K = \{0, 1, 2, \ldots\}$, and then put

$$R = \mathbf{R}^d \times K \quad \text{and} \quad W = \Omega \times K.$$

A point $(x, k) \in R$ specifies that the position and age of a particle are x and k, respectively. In the same way, $w(t) = (\omega(t), k) \in W$ specifies that the age of a particle moving along the path $\omega(t)$ is k. Accordingly, we define a stochastic process $X_t(w)$, $w = (\omega, k) \in W$, with age k by

$$X_t((\omega, k)) = (x_t(\omega), k), \quad t < \zeta(w) = \zeta(\omega), \qquad (6.5.4)$$

and define a probability measure on $W = \Omega \times K$ by

$$P_{(x,k)}[dw] = P_{(x,k)}[d\omega \, d\ell] = P_x[d\omega] \times \delta_k(d\ell), \qquad (6.5.5)$$

where $\delta_k(d\ell)$ denotes the point measure at $k \in K$. That is, if the starting point is $(x, k) \in R$, the particle moves as $\{x_t(\omega), t < \zeta(\omega), P_x\}$, but the age k does not change.

Thus, $\{W, X_t, \zeta, P_{(x,k)}\}$ is a Markov process with finite lifetime on $R = \mathbf{R}^d \times K$.

Further, let
$$\pi((x,k), dy\, d\ell) = \delta_x(dy)\delta_{k+1}(d\ell) \tag{6.5.6}$$
(increase age by one), and for $w = (\omega, k) \in W$ set
$$N(w, dy\, d\ell) = \pi(X_{\zeta-}(w), dy\, d\ell). \tag{6.5.7}$$

Then $N(w, dy\, d\ell)$ is a piecing-together law, because it satisfies equation (6.2.2) of Definition 6.2.1.

Therefore, by Theorem 6.2.2, there exists a pieced-together Markov process
$$\{\mathcal{W}, \mathcal{X}_t = (x_t, k_t), \mathbf{P}_{(x,k)}, (x,k) \in R\},$$

which we call **"Markov process with age"**.

Earlier, for every element $\mathbf{w} = (w_1, w_2, \dots) \in \mathcal{W}$ we defined a series of renewal (piecing-together) times by equation (6.2.4). In the case of the Markov process with age,
$$r_1(\mathbf{w}) < r_2(\mathbf{w}) < \cdots < r_k(\mathbf{w}) < \cdots \tag{6.5.8}$$

is the sequence of times at which the age increases by one. This is obvious from formula (6.5.6).

Now for any bounded function $f(x)$ on $\mathbf{R}^d$ and any $\lambda > 0$ we define a function $f \cdot \lambda$ on $R = \mathbf{R}^d \times K$ by
$$(f \cdot \lambda)(x, k) = f(x)\lambda^k, \quad (x,k) \in R. \tag{6.5.9}$$

Then
$$(f \cdot \lambda)(X_t) = f(x_t)\lambda^{k_t},$$

so $\lambda > 0$ can be regarded as the "weight" of the age k_t. The idea of weighting the age is due to Sirao (1968). The following lemma clarifies the importance of weighting the age k_t.

Lemma 6.5.1. *For every bounded function $f(x)$ on $\mathbf{R^d}$ and every $\lambda > 0$,*
$$\mathbf{P}_{(x,k)}[(f \cdot \lambda)(X_t) : r_n \le t < r_{n+1}] = \mathrm{P}_x\left[f(x_t)e^{-c_t}\frac{(\lambda c_t)^n}{n!}\right]\lambda^k, \tag{6.5.10}$$

where r_n is the sequence of times at which the age increases by one (times of piecing together), and on the right-hand side $\{x_t, \mathrm{P}_x\}$ is the underlying Markov process. The function c_t is defined by formula (6.5.2). Moreover,
$$\mathbf{P}_{(x,0)}[(f \cdot \lambda)(X_t)] = \mathrm{P}_x[f(x_t)e^{(\lambda-1)c_t}]. \tag{6.5.11}$$

If $\lambda = 2$, the right-hand side becomes $\mathrm{P}_x[f(x_t)e^{c_t}]$.

Proof. We proceed by induction. First, looking at the construction plan, we see that

$$\mathbf{P}_{(x,k)}[(f \cdot \lambda)(X_t) : t < r_1] = \mathbf{P}_x[f(x_t) : t < \zeta\,]\lambda^k = \mathbf{P}_x[f(x_t)e^{-c_t}]\lambda^k.$$

Next, applying the Markov property at time r_1, we have

$$\mathbf{P}_{(x,k)}[(f \cdot \lambda)(X_t) : r_n \leq t < r_{n+1}]$$
$$= \mathbf{P}_{(x,k)}[\mathbf{P}_{(x_{r_1},k_{r_1})}[(f \cdot \lambda)(X_{t-s}) : r_{n-1} \leq t - s < r_n]|_{s=r_1} : r_1 \leq t],$$

which, since the age was k and then increased by one at r_1, so $k_{r_1} = k+1$, is equal to

$$\mathbf{P}_x\left[\int_0^t e^{-c_s} dc_s\, \mathbf{P}_{(x_s,0)}[f \cdot \lambda(X_{t-s}) : r_{n-1} \leq t - s < r_n]\Big|_{s=r_1}\right]\lambda^{k+1};$$

here we used the fact that, for any bounded measurable function $g(x)$,

$$\mathbf{P}_{(x,k)}[g(x_{r_1}) : r_1 \leq t] = \mathbf{P}_x\left[\int_0^t g(x_s)e^{-c_s} dc_s\right], \tag{6.5.12}$$

which we will prove shortly.

Now, assume that (6.5.10) holds for $n-1$. Then the last expression above (6.5.12) is equal to

$$\mathbf{P}_x\left[\int_0^t e^{-c_s} dc_s\, \mathbf{P}_{x_s}\left[f(x_{t-s})e^{-c_{t-s}}\frac{(\lambda c_{t-s})^{n-1}}{(n-1)!}\right]\right]\lambda^{k+1}$$
$$= \mathbf{P}_x\left[\int_0^t e^{-c_s} dc_s\, f(x_t)e^{-c_{t-s}(\theta_s\omega)}\frac{(\lambda c_{t-s}(\theta_s\omega))^{n-1}}{(n-1)!}\right]\lambda^{k+1}$$
$$= \mathbf{P}_x\left[e^{-c_t}f(x_t)\int_0^t \lambda\, dc_s \frac{(\lambda(c_t - c_s))^{n-1}}{(n-1)!}\right]\lambda^k$$
$$= \mathbf{P}_x\left[f(x_t)e^{-c_t}\frac{(\lambda c_t)^n}{n!}\right]\lambda^k,$$

where we used the equality $c_t(\omega) = c_s(\omega) + c_{t-s}(\theta_s\omega)$. Thus, (6.5.10) holds.

Finally, equality (6.5.11) is obtained by summing oven n both sides of equality (6.5.10).

It remains to verify that the equality (6.5.12) holds. Starting from its right-hand side, we have

$$\mathbf{P}_x\left[\int_0^t g(x_s)e^{-c_s} dc_s\right] = \mathbf{P}_x\left[\int_0^t g(x_s)d(-e^{-c_s})\right]$$
$$= \mathbf{P}_x\left[\lim_{h\downarrow 0}\sum_{ih<t} g(x_{ih})\left(e^{-c_{ih}} - e^{-c_{(i+1)h}}\right)\right]$$
$$= \lim_{h\downarrow 0}\sum_{ih<t}\mathbf{P}_{(x,k)}[g(x_{ih}) : ih < r_1 \leq (i+1)h]$$
$$= \mathbf{P}_{(x,k)}[g(x_{r_1}) : r_1 \leq t],$$

as needed.

The proof of the lemma is complete.   □

Remark. If in Lemma 6.5.1 we set $f(x) = 1$ and $\lambda = 1$, then (6.5.10) implies that

$$\mathbf{P}_{(x,0)}[r_n \leq t < r_{n+1}] = |rmP_x\left[e^{-c_t}\frac{(c_t)^n}{n!}\right].$$

Observe also that, since $\mathbf{P}_{(x,0)}[r_n \leq t < r_{n+1}] = \mathbf{P}_{(x,0)}[k_t = n]$, we can rewrite the above equality as

$$\mathbf{P}_{(x,0)}[k_t = n] = \mathbf{P}_x\left[e^{-c_t}\frac{(c_t)^n}{n!}\right]. \tag{6.5.13}$$

This shows that "k_t is a Poisson process with the time variable c_t." That is, $k_t = p(c_t)$, where $p(t)$ is the Poisson process with the probability law $\mathbf{P}_{(x,0)}$.

Now, let us assume that the age weight λ is equal to 2. For a bounded smooth function $f(x)$ on the space $\mathbf{R}^d$, we define

$$u(t,x) = \mathbf{P}_{(x,0)}[(f\cdot 2)(X_t)], \quad x \in \mathbf{R}^d,$$

with $(f\cdot\lambda)(x,k) = f(x)\lambda^k$ as in (6.5.9), and use equation (6.5.11) of Lemma 6.5.1. Then

$$u(t,x) = \mathbf{P}_x\left[f(x_t)e^{c_t}\right] = \mathbf{P}_x\left[f(x_t)e^{\int_0^t c(x_s(\omega))ds}\right].$$

Hence, by the Kac formula,

$$\frac{\partial u}{\partial t} = \frac{1}{2}\sigma^2\triangle u + \mathbf{b}(x)\cdot\nabla u + c(x)u. \tag{6.5.14}$$

Note that, in equation (6.5.14), the term $-c(x)u$ appearing in equation (6.1.2) has changed to $+c(x)u$. We have seen in Section 6.1 that equation (6.1.2) describes the annihilation (death) of particles. We arrived at equation (6.5.14) by weighting the age with $\lambda = 2$. If introducing the weight $\lambda = 2$ is interpreted as the statement that "the number of particles will be 2^{k_t}", we can also consider that equation (6.5.14) with the term $+c(x)u$ describes particle creation.

In this section, following Nagasawa (1968), we presented Markov processes with age.

6.6   Branching Markov Processes with Age

If in a branching process the motion of a single particle is governed by the Markov process with age constructed in the preceding section as, we call it "branching Markov process with age."

To obtain it, we can apply the method developed in Section 6.4 for constructing a branching Markov process to the Markov process with age

$$\{\mathbf{W}, X_t = (x_t, k_t), \mathbf{P}_{(x,k)}, (x,k) \in R = \mathbf{R}^d \times K\}.$$

To do so, we must deal with the presence of age and its weight. To simplify notation, we will write the path space with age as W; hence, $w \in W$ denotes a path with age.

Let us explain how to construct branching Markov processes with age and present some of their properties according to Nagasawa and Sirao (1967), Nagasawa (1968).

First of all, we use the function

$$c_t(\omega) = \int_0^t c(x_s(\omega))ds, \quad c(x) \geq 0,$$

to define a Markov process with age and finite lifetime $\zeta < \infty$,

$$\{W, X_t, \zeta, P_a, a \in R\}, \quad \text{where } R = \mathbf{R}^d \times K.$$

This is a Markov subprocess with lifespan $\zeta < \infty$, that is, a Markov process with age restricted to that extent (see Section 6.1 for subprocesses).

Let $\{W, X_t, \zeta, P_a, a \in R\}$ be a Markov process with age and finite lifetime and define its n-fold direct product $\{W^n, X_t, \zeta, P_{\mathbf{a}}, \mathbf{a} \in R^n\}$ as follows.

For $\mathbf{w} = (w_1, w_2, \ldots, w_n) \in \Omega^n$ we define the lifetime of $\mathbf{w}$ by

$$\zeta(\mathbf{w}) = \min\{\zeta(w_k) : k = 1, 2, \ldots, n\},$$

and put

$$X_t(\mathbf{w}) = (X_t(w_1), X_t(w_2), \ldots, X_t(w_n)), \quad \text{for } t < \zeta(\mathbf{w}),$$

$$X_t(\mathbf{w}) = \Delta \quad \text{(an additional point)}, \quad \text{for } t \geq \zeta(\mathbf{w}).$$

and define a probability measure $P_{\mathbf{a}}$, $\mathbf{a} = (a_1, a_2, \ldots, a_n) \in R^n$, by

$$P_{\mathbf{a}} = P_{a_1} \times P_{a_1} \times \cdots \times P_{a_n},$$

where $\mathbf{a} = (\mathbf{x}, \mathbf{k}) \in S$, and $S = \bigcup_{n=0}^{\infty} R^n$, with $R = \mathbf{R}^d \times K$ $(R^0 = \{\partial\})$.

Further, consider the Markov process with finite lifetime

$$\{\mathbf{W}, X_t, \zeta, P_{\mathbf{a}}, \mathbf{a} \in S\}$$

on the multiplicative state space S, where $\mathbf{W} = \bigcup_{n=1}^{\infty} W^n$. This Markov process describes the random motion of a population of particles before branching or extinction occurs. Particles belonging to the population are moving independently.

Now let $q_m(x) \geq 0$ denote the probability that a particle will disappear at the point x and m particles will be generated at x, distributed according to the law $\pi_m(x, d\mathbf{y})$ on $\mathbf{y} = (y_1, \ldots, y_m)$, where $\pi_0(x, dy)$ is a point measure at ∂. Furthermore, taking into account the age law, we put

$$\pi((x, \ell), db) = q_0(x)\delta_\partial(db) + \sum_{m=2}^{\infty} q_m(x)\pi_m(x, d\mathbf{y})\delta_{(\ell, 0, \ldots, 0)}(d\mathbf{k}), \tag{6.6.1}$$

where we write $b = (\mathbf{y}, \mathbf{k})$.

If a particle located at the point x and of age ℓ branches into a family of m members, one of these members is a parent of age ℓ and the remaining $m - 1$ particles are children of age 0. This is the meaning of the factor $\delta_{(\ell,0,\ldots,0)}(d\mathbf{k})$ in (6.6.1).

With the help of the branching law $\pi((x, \ell), db)$ of the Markov process with age given by formula (6.6.1), we define a piecing-together law $N(\mathbf{w}, db_1 db_2 \cdots db_n)$ by

$$N(\mathbf{w}, db_1 db_2 \cdots db_n) = \sum_{i=1}^{n} 1_{\{\zeta(\mathbf{w}) = \zeta(w_i)\}} \pi(X_{\zeta(w_i)-}(w_i), db_i) \times$$
$$\times \prod_{j \neq i} \delta_{\{X_{\zeta(\mathbf{w})-}(w_j)\}}(db_j), \tag{6.6.2}$$

where $\mathbf{w} = (w_1, w_2, \ldots, w_n) \in W^n$, $n = 1, 2, \ldots$.

The meaning of the piecing-together law is as follows. When n particles move independently, if the "i-th particle has the shortest lifetime, then it disappears with probability $q_0(X_{\zeta-}(w_i))$ or branches into m particles with probability $q_m(X_{\zeta-}(w_i))$, while the other particles will continue moving as they were."

Because the quantity $\zeta(\mathbf{w})$ appearing on the right-hand side of formula (6.6.2) is

$$\zeta(\mathbf{w}) = \min\{\zeta(w_i) : i = 1, 2, \ldots, n\},$$

$\{\zeta(\mathbf{w}) = \zeta(w_i)\}$ means that "the lifetime of the ith particle is the shortest, and at that time, because of (6.6.1), the ages of the particles are $(\ell, 0, \ldots, 0)$."

Thus, let us apply the piecing-together law $N(\mathbf{w}, db_1 db_2 \cdots db_n)$ defined by formula (6.6.2) to the Markov process $\{\mathbf{W}, X_t, \zeta, \mathbf{P_a}, \mathbf{a} \in S\}$ describing the random motion of the population of particles before extinction or multiplication occur, and use the piecing together Theorem 6.2.2. This yields the Markov process

$$\{\mathbf{W}, Y_t, \mathbf{P}_{(\mathbf{x}, \mathbf{k})}, (\mathbf{x}, \mathbf{k}) \in S\}, \quad \text{where } S = \bigcup_{n=0}^{\infty} R^n, \ R = \mathbf{R}^d \times K,$$

called **branching Markov process with age**.

Let us provide the characteristic property of branching Markov processes with age.

First of all, for any bounded function $f(x)$ on $\mathbf{R}^d$ and any age weight $\lambda > 0$ we define a function $\widehat{f \cdot \lambda}(\mathbf{x}, \mathbf{k})$ by

$$(\widehat{f \cdot \lambda})(\mathbf{x}, \mathbf{k}) = \widehat{f}(\mathbf{x}) \lambda^{|\mathbf{k}|}, \tag{6.6.3}$$

where $(\mathbf{x}, \mathbf{k}) = (x_1, \ldots, x_n, k_1, \ldots, k_n)$, $n = 1, 2, \ldots$, the exponent of λ is $|\mathbf{k}| = \sum_{j=1}^{n} k_j$, that is, the sum of ages, and

$$(\widehat{f \cdot \lambda})(\partial, k) = \lambda^k.$$

Theorem 6.6.1. *Let $\{\mathbf{W}, Y_t, \mathbf{P}_{(\mathbf{x},\mathbf{k})}, (\mathbf{x},\mathbf{k}) \in S\}$ be a branching Markov process with age and let*

$$P_t h(\mathbf{x}, \mathbf{k}) = \mathbf{P}_{(\mathbf{x},\mathbf{k})}[h(Y_t)] \tag{6.6.4}$$

be the associated operator semigroup. Then there holds "the branching property of the branching Markov process with age":

$$\mathbf{P}_t(\widehat{f \cdot \lambda})(\mathbf{x}, \mathbf{k}) = \widehat{u}_t(\mathbf{x}) \lambda^{|\mathbf{k}|}, \tag{6.6.5}$$

where

$$u_t(x) = \mathbf{P}_t(\widehat{f \cdot \lambda})(x, 0), \tag{6.6.6}$$

and $|\mathbf{k}|$ is the sum of ages. The right-hand side of equation (6.6.5) is $(\widehat{u_t \cdot \lambda})(\mathbf{x}, \mathbf{k})$.

Proof. Since the usual branching property (6.3.7) holds for the function $f \cdot \lambda$,

$$\mathbf{P}_t(\widehat{(f \cdot \lambda)})(\mathbf{x}, \mathbf{k}) = \prod_{j=1}^{n} \mathbf{P}(\widehat{f \cdot + \lambda})(x_j, k_j),$$

and, by Lemma 6.5.1,

$$\mathbf{P}_t(\widehat{f \cdot \lambda})(x_j, k_j) = \mathbf{P}_t(\widehat{f \cdot \lambda})(x_j, 0) \lambda^{k_j} = u_t(x_j) \lambda^{k_j},$$

i.e., (6.6.5) holds.   □

As evident from the proof, property (6.6.5) is an extension of the branching property (6.3.6) without age to the case of branching Markov processes with age. An important point here is that in equation (6.6.5) the dependence on the age weight λ is specified.

In the previous section, we have seen that equation (6.5.14) can be derived by setting the age weight λ equal to 2 in a Markov process with age. The aim was to add a term $+c(x)u$. Thanks to this, when we use the branched Markov process with age, the term of $-c(x)u$ disappears from the nonlinear equation (6.3.15). Let us show this.

To derive the basic equation that characterizes branching Markov processes with age, we can follow the discussion about the equation (6.3.8) in Section 6.3.

First of all we decompose $\mathbf{P}_{(x,0)}[(\widehat{f \cdot \lambda})(Y_t)]$ as

$$\mathbf{P}_{(x,0)}[(\widehat{f \cdot \lambda})(Y_t)] = \mathbf{P}_{(x,0)}[(\widehat{f \cdot \lambda})(Y_t) : t < r_1] + \mathbf{P}_{(x,0)}[(\widehat{f \cdot \lambda})(Y_t) : t \geq r_1]. \tag{6.6.7}$$

The first term on the right-hand side is

$$P_t^0(\widehat{f \cdot \lambda})(x, 0) = \mathbf{P}_{(x,0)}[(\widehat{f \cdot \lambda})(Y_t) : t < r_1]. \tag{6.6.8}$$

By the Markov property at the time r_1, the second term of the right-hand side of (6.6.7) is

$$\mathbf{P}_{(x,0)}[(\widehat{f \cdot \lambda})(Y_t) : t \geq r_1] = \mathbf{P}_{(x,0)}[\mathbf{P}_{Y_{r_1}}[(\widehat{f \cdot \lambda})(Y_{t-s})]|_{s=r_1} : t \geq r_1].$$

By the branching property (6.6.5),

$$\mathbf{P}_{(\mathbf{x},\mathbf{k})}[(\widehat{f\cdot\lambda})(Y_{t-s})] = \widehat{u_{t-s}}(\mathbf{x})\lambda^{|\mathbf{k}|},$$

so

$$\mathbf{P}_{(X_{r_1},k_{r_1})}[(\widehat{f\cdot\lambda})(Y_{t-s})]|_{s=r_1} = \widehat{u_{t-r_1}}(X_{r_1})\lambda^{|\mathbf{k}_{r_1}|},$$

where we wrote $Y_t = (X_t, k_t)$.

Therefore, the second term on the right-hand side of (6.6.7) is equal to

$$\mathbf{P}_{(x,0)}[(\widehat{f\cdot\lambda})(Y_t) : t \geq r_1] = \mathbf{P}_{(x,0)}[\widehat{u_{t-r_1}}(X_{r_1})\lambda^{|\mathbf{k}_{r_1}|} : t \geq r_1]$$

$$= \int_0^t \int_R \mathbf{P}_{(x,0)}[r_1 \in ds, \ Y_{r_1-} \in d(y,\ell)]\lambda^\ell \, \mathbf{P}_{(y,0)}[\widehat{u_{t-s}}(X_{r_1})],$$

where we used that $|\mathbf{k}_{r_1}| = \ell$ ($\mathbf{k}_{r_1} = (\ell, 0, \dots, 0)$.)

Moreover, if we denote

$$K((x,0), ds, d(y,\ell)) = \mathbf{P}_{(x,0)}[\, r_1 \in ds, Y_{r_1-} \in d(y,\ell)], \tag{6.6.9}$$

the second term on the right side of equation (6.6.7) is equal to

$$\int_0^t \int_R K((x,0), ds, d(y,\ell))\lambda^\ell \mathbf{P}_{(y,0)}[\widehat{u_{t-s}}(X_{r_1})]. \tag{6.6.10}$$

Next, note that $\mathbf{P}_{(y,0)}[\widehat{u}(X_{r_1})]$ can be expressed as

$$\sum_{\substack{n \neq 1}}^{\infty} \mathbf{P}_{(y,0)}[X_{r_1} \in (\mathbf{R}^d)^n] \int_{(\mathbf{R}^d)^n} \mathbf{P}_{(y,0)}[X_{r_1} \in d\mathbf{z}|X_{r_1} \in (\mathbf{R}^d)^n]\widehat{u}(\mathbf{z}),$$

Hence, upon introducing the probability that one particle splits into n particles by

$$q_n(y) = \mathbf{P}_{(y,0)}[X_{r_1} \in (\mathbf{R}^d)^n],$$

and the conditional distribution of n split particles by

$$\pi_n(y, d\mathbf{z}) = \mathbf{P}_{(y,0)}[X_{r_1} \in d\mathbf{z} \mid X_{r_1} \in (\mathbf{R}^d)^n],$$

and putting

$$F(y, u) = q_0(y) + \sum_{n=2}^{\infty} q_n(y) \int_{(\mathbf{R}^d)^n} \pi_n(y, d\mathbf{z})\widehat{u}(\mathbf{z}), \tag{6.6.11}$$

we have

$$\mathbf{P}_{(y,0)}[\widehat{u}(X_{r_1})] = F(y, u). \tag{6.6.12}$$

Therefore, the second term on the right-hand side of equation (6.6.7) is

$$\mathbf{P}_{(x,0)}[(\widehat{f\cdot\lambda})(Y_t) : t \geq r_1] = \int_0^t \int_R K((x,0), ds, d(y,\ell))\lambda^\ell F(y; u_{t-s}). \tag{6.6.13}$$

Thus, we proved the following result.

Theorem 6.6.2. *Let $f(x)$ be a function on the d-dimensional Euclidean space $\mathbf{R}^d$ such that $|f(x)| \leq 1$, and let $\lambda = 2$. Then the function*

$$u(t, x) = \mathbf{P}_{(x,0)}[(\widehat{f \cdot 2})(Y_t)], \quad x \in \mathbf{R^d}, \tag{6.6.14}$$

satisfies the basic equation of branching Markov processes with age:

$$u(t, x) = P_t^0(\widehat{f \cdot 2})(x, 0) + \int_0^t \int_R K((x, 0), ds, d(y, \ell)) 2^\ell F(y; u(t - s, \cdot)), \tag{6.6.15}$$

where $P_t^0(\widehat{f \cdot 2})(x, 0)$ is given by formula (6.6.8) and $F(y, u)$ is given by formula (6.6.11).

Note that since the integrand $\widehat{f \cdot 2}$ in equation (6.6.14) is not bounded, $u(t, x)$ is well defined only when the integral on the right-hand side of equation (6.6.14) is finite. So, when one applies Theorem 6.6.2, one has to check first whether $\mathbf{P}_{(x,0)}[(\widehat{f \cdot 2})(Y_t)]$ is finite or not.

Furthermore, the right-hand side of equation (6.6.15) can be expressed in terms of the "basic Markov process" $\{\Omega, x_t, P_x, x \in \mathbf{R}^d\}$ determined by the equation

$$\frac{\partial u}{\partial t} = \frac{1}{2}\sigma^2 \triangle u + \mathbf{b}(x) \cdot \nabla u.$$

Let us prove this as two lemmas dealing with the first and second terms on the right-hand side of equation (6.6.15).

Lemma 6.6.1. *Let $\lambda = 2$. Then*

$$P_t^0(\widehat{f \cdot 2})(x, 0) = P_x[f(x_t)], \tag{6.6.16}$$

where on the right side $\{\Omega, x_t, P_x, x \in \mathbf{R^d}\}$ is the basic Markov process.

Proof. The left-hand side of equation (6.6.16) is

$$P_t^0(\widehat{f \cdot 2})(x, 0) = \mathbf{P}_{(x,0)}[(\widehat{f \cdot 2})(Y_t) : t < r_1].$$

Since on the right-hand side of this equation $t < r_1$, branching does not occur. Therefore Y_t on the right-hand side coincides with the Markov process X_t with age. Consequently,

$$\mathbf{P}_{(x,0)}[(\widehat{f \cdot 2})(Y_t) : t < r_1] = \mathbf{P}_{(x,0)}[(\widehat{f \cdot 2})(X_t) : t < \zeta] = \mathbf{P}_{(x,0)}[(\widehat{f \cdot 2})(X_t)e^{-c_t}],$$

where we set $\lambda = 2$ and use formula (6.5.11). We see that the last expression is equal to

$$P_x[f(x_t)e^{-c_t}e^{c_t}] = P_x[f(x_t)],$$

as claimed. □

Lemma 6.6.2.

$$\int_R K((x,0), ds, d(y, \ell)) 2^\ell F(y) = P_x[c(x_s)F(x_s)]ds, \qquad (6.6.17)$$

where $\{\Omega, x_t, P_x, x \in \mathbf{R}^d\}$ on the right-hand side is the basic Markov process.

Proof. For the sake of convenience, denote

$$J = \int_R K((x,0), ds, d(y, \ell)) 2^\ell F(y).$$

By (6.6.9),

$$J = \mathbf{P}_{(x,0)}[(\widehat{f \cdot 2})(Y_t) : t < r_1]_{(x,0)}[(F \cdot 2)(Y_{r_1-});\ r_1 \in ds].$$

Since we consider times before branching occurs, the process $\{Y_t, t < r_1\}$ coincides with the Markov process with age $\{X_t, t < \zeta\}$. Hence

$$J = \mathbf{P}_{(x,0)}[(F \cdot 2)(X_s);\ \zeta \in ds] = \mathbf{P}_{(x,0)}[(F \cdot 2)(X_s)e^{-c_s}dc_s],$$

and since $dc_s = c(x_s)ds$, we see that

$$J = \mathbf{P}_{(x,0)}[c(F \cdot 2)(X_s)e^{-c_s}]ds,$$

where we put $\lambda = 2$ and used formula (6.5.11). Then e^{c_s} is added, and so

$$J = P_x[c(x_s)F(x_s)e^{-c_s}e^{c_s}]ds = P_x[c(x_s)F(x_s)]ds,$$

as claimed. $\square$

Theorem 6.6.3. *Let $f(x)$ be a function on the d-dimensional Euclidean space $\mathbf{R}^d$ such that $|f(x)| \le 1$ and let $\lambda = 2$. Set*

$$u(t, x) = {}_{(x,0)}[(\widehat{f \cdot 2})(Y_t)], \quad x \in \mathbf{R}^d. \qquad (6.6.18)$$

Then, when the function $u(t, x)$ is well defined, it satisfies the equation

$$u(t, x) = P_t f(x) + \int_0^t P_s(cF(\,\cdot\,; u(t-s)))(x)ds, \qquad (6.6.19)$$

where $P_t f(x) = P_x[f(x_s)]$ is the operator semigroup of the basic Markov process and $F(y, u)$ is given by formula (6.6.11).

Proof. It suffices to rewrite the right-hand side of formula (6.6.15) in Theorem 6.6.2 using Lemma 6.6.1 and Lemma 6.6.2. $\square$

In order to transform the integral equation (6.6.19) into a differential equation, analytic conditions must be imposed on $F(x; u)$, conditions that must be verified in each individual case. First, equation (6.6.19) implies that

$$u(t+h, x) = P_{t+h}f(x) + \int_0^{t+h} P_s(cF(\,\cdot\,; u(t+h-s)))(x)ds. \qquad (6.6.20)$$

Applying P_h to the equation (6.6.19) we have

$$\begin{aligned} P_h u(t, x) &= P_h P_t f(x) + \int_0^t P_h P_s(cF(\,\cdot\,; u(t-s)))(x)ds \\ &= P_{t+h}f(x) + \int_h^{t+h} P_s(cF(\,\cdot\,; u(t+h-s)))(x)ds. \end{aligned} \qquad (6.6.21)$$

Combining (6.6.20) and (6.6.21), we obtain

$$u(t+h, x) = P_h u(t, x) + \int_0^h P_s(cF(\,\cdot\,; u(t+h-s)))(x)ds.$$

Therefore,

$$\begin{aligned} \frac{u(t+h, x) - u(t, x)}{h} &= \frac{P_h u(t, x) - u(t, x)}{h} \\ &\quad + \frac{1}{h}\int_0^h P_s(cF(\,\cdot\,; u(t+h-s)))(x)ds. \end{aligned}$$

Let $h \downarrow 0$. Then, under a suitable condition on $F(x; u)$, we deduce that $u(t, x)$ satisfies the partial differential equation

$$\frac{\partial u}{\partial t} = \frac{1}{2}\sigma^2 \triangle u + \mathbf{b}(x) \cdot \nabla u + c(x)F(x; u). \qquad (6.6.22)$$

If $\pi_m(x, d\mathbf{y}) = \delta_{(x,\ldots,x)}(d\mathbf{y})$, then

$$F(x, u) = q_0(x) + \sum_{m=2}^{\infty} q_m(x)u^m,$$

and so equation (6.6.22) becomes

$$\frac{\partial u}{\partial t} = \frac{1}{2}\sigma^2 \triangle u + \mathbf{b}(x) \cdot \nabla u + c(x)\{q_0(x) + \sum_{m=2}^{\infty} q_m(x)u^m\}. \qquad (6.6.23)$$

A key observation is that equation (6.6.23) does not contain the term $-c(x)u$.

In the case of the normal branching Markov process discussed in Section 6.3, equation (6.3.15) contains the term $-c(x)u$, which in principle canot be removed.

This is a strong restriction (for the theory of semilinear parabolic differential equations). By introducing age and weight, this restriction is eliminated. Note that $q_m(x) \geq 0$ is a function which can be selected freely, as needed.

It is even preferable to remove the condition $q_m(x) \geq 0$. Actually, even when the function $q_m(x)$ takes negative values, it can be handled in the same way as introducing age (another structure with a "sign" must be added). For this, the reader is referred to Sirao (1968) and Nagasawa (1968).

In equation (6.6.23), the nonlinear term appears in the form of u^m. We have shown that we can treat this case by means of a branching Markov process (with age). So, is there a stochastic process that allows us to treat an equation with nonlinear terms such as $(\nabla u)^n$? The answer is affirmative: just like we added age, by introducing a suitable space to describe ∇u, we can extend the branching property further. For details, see Nagasawa (1972).

As already noted, $u_t(x)$ may blow-up in finite time. To study this situation, as an application of branching Markov processes with age Nagasawa and Sirao (1968) discussed the blow-up problem for the equation

$$\frac{\partial u_t(x)}{\partial t} = \frac{1}{2}\sigma^2 \triangle u_t(x) + \mathbf{b}(x) \cdot \nabla u_t(x) + c(x)(u_t(x))^\beta. \qquad (6.6.24)$$

Fujita (1966) treated this problem analytically.

For semilinear parabolic differential equations and their application to biological problems, one should consult the paper by Kolmogoroff, Petrovsky and Piscounoff (1937).

In this section, we focused on explaining the mathematical structure of branching Markov processes. However, we did not provide enough applications of branching processes. As mentioned earlier, the reader is referred to, e.g., Harris (1963) and Sevastyanov (1958).

Nevertheless, let us mention briefly one important application to particle annihilation and creation processes in cosmic ray cascades. Harris (1963) proposed a mathematical model that relies on a branching process. In that case, if the cross-section is infinite, that is, if each electron produces infinitely many photons in every interval of thickness (every finite time interval), one must use the multiplicative space $\mathbf{M}$ of finite measures introduced in Section 6.3. This is because the branching Markov process on the multiplicative space $\mathbf{M}$ is suitable for treating a population of infinitely many particles:

$$\mu = \sum_{i=1}^{\infty} x_i \delta_{x_i} \in \mathbf{M}, \quad \text{with} \quad \sum_{i=1}^{\infty} x_i < \infty.$$

Here, x_i stand for the the energy of a particle instead of the its position.

This problem was subsequently discussed by Jirina (1964), Lamperti (1967), Watanabe (1968), and Fujimagari and Motoo (1969, 1971).

Bibliography

List of documents related to the theme of this book.

Aebi, R. (1996): *Schrödinger Diffusion Processes*, Birkhäuser, Basel.

Aebi, R., Nagasawa, M. (1992): Large deviations and the propagation of chaos for Schrödinger processes, *Probab. Th. Rel. Fields* **94**, 53–68.

Aharonov, Y., Bohm, D. (1959): Significance of Electro-magnetic Potentials in Quantum Mechanics, *Phys. Rev.* **115**, 498.

Aspect, A., Dalibard, J., Roger, G. (1982): Experimental test of Bell's inequalities using time-varying analyzers, *Phys. Rev. Letters* **49**, 1804–1807.

Ballentine, L.E. (1988): *Foundations of quantum mechanics since the Bell inequality*, American Association of Physics Teachers, College Park, MD.

Bardeen, J., Cooper, N., Schrieffer, J.R. (1957): Theory of superconductivity, *Phys. Rev.* **108**, 1175–1204.

Bartlett, M.S. (1955): *An introduction to stochastic processes with special reference to methods and applications*, Cambridge University Press.

Bauer, H. (1981): *Probability Theory and Elements of measure Theory*, Academic Press, London, New York, Toronto, Sydney, San Francisco.

Bell, J.S. (1964): On the Einstein Podolsky Rosen Paradox, *Physics* **1**, 195–202.

Bharucha-Reid, A.T. (1960): *Elements of the theory of Markov processes and their applications*, Mcgraw-Hill.

Billingsley, P. (1968): *Convergence of probability Measures*, John Wiley and Sons Inc., New York.

Blumenthal, R.M., Getoor, R.K. (1968): *Markov Processes and Potential Theory*, Academic Press, New York, London.

Bochner, D. (1949): Diffusion equations and stochastic processes, *Proc. Nat. Acad. Sci. USA* **35**, 368–370.

© The Editor(s) (if applicable) and The Author(s), under exclusive license
to Springer Nature Switzerland AG 2021
M. Nagasawa, *Markov Processes and Quantum Theory*,
Monographs in Mathematics 109, https://doi.org/10.1007/978-3-030-62688-4

Bochner, D. (1960): *Harmonic analysis and theory of probability*, Univ. Calif. Press, Berkeley, Los Angeles, CA.

Bohm, D. (1952): A suggested interpretation of the quantum theory in terms of "hidden" variables I, *Phys. Rev.* **85**, 166–179.

Born, M. (1926): Zur Quantenmechanik der Stossvorgänge, *ZS. für Phys.* **37**, 863–867.

Born, M., Jordan, W. (1925): Zur Quantenmechanik, *ZS. für Phys.* **34**, 858–888.

Born, M., Heisenberg, W., Jordan, W. (1926): Zur Quantenmechanik II, *ZS. für Phys.* **35**, 557–615.

Cameron, R.H., Martin, W.T. (1944): Transformation of Wiener integrals under translations, *Ann. of Math.* **45**, 386–396.

Carlen, E. (2006): A look back and a look ahead. In *Diffusion, Quantum Theory, and Radically Elementary Mathematics*, ed. by Faris, W.G., Princeton Univ. Press, Mathematical Notes 47.

Cinlar Erhan (2010): *Probability and Stochastics*, Graduate Texts in Mathematics 261, Springer New York, Dordrecht, Heidelberg, London.

Clauser, J.F., Horne, M.A. (1974): Experimental consequences of objective local theories, *Phys. Rev.* **D10**, 526–535.

Condon, E.U. (1962): 60 years of quantum physics, *Physics Today* **15**, 37–49.

Cornish, S.L., Roberts, J.L., Cornell, E.A., Wieman, C.E., Donley, E.A., Claussen, N.R. (2001): Dynamics of collapsing and exploding Bose-Einstein condensates, *Nature* **412**, 295

Csiszar, I. (1975): I-divergence geometry of probability distribution and minimization problems, *The Annals of Probability* **3**, 146–158.

Csiszar, I. (1984): Sanov property, generalized I-projection and a conditional limit theorem, *The Annals of Probability* **12**, 768–793.

Dawson, D., Gorostiza, L., Wakolbinger, A. (1990): Schrödinger processes and large deviations, *J. Math. Phys.* **31**, 2385–2388.

de Broglie, L. (1923): Ondes et quanta, *Comptes rendus* **177**, 507–508.

de Broglie, L. (1924): *Recherche sur la theorie des quanta*, Thése, Université de Paris.

de Broglie, L. (1953): *Physicien et Penseur*, Albin Michel, Paris.

de Broglie, L. (1943): *La physique nouvelle et let quanta*, Paris.

Deuschel, J.-D., Stroock, D.W. (1989): *Large Deviations*, Academic Press, Boston, San Diego, New York, Berkeley, London, Sydney, Tokyo.

Dirac, P.A.M. (1930, 58): *The Principle of Quantum Mechanics*, 1st, 4th ed., Oxford, New York.

Dirac, P.A.M. (1972): Relativity and Quantum Mechanics, *Fields and Quanta* **3**, 139–164.

Doob, J.L. (1953): *Stochastic Processes*, John Wiley and Sons, New York.

Domenig, T., Nagasawa, M. (1994): A Skorokhod problem with singular drift and its application to the origin of Universes, *Proc. of Japan Acad. Science* **70**, Ser. A, No. 4, 88–93.

Douglas, R.G. (1964): On external measures and subspace density, *Michigan Math. J.* **11**, 243–246.

Dressel, F.G. (1946): The fundamental solution of parabolic equations 2, *Duke Math. J.* **13**, 61–77.

Dynkin, E.B. (1965): *Markov Processes*, Vol. I, II. Springer-Verlag, Heidelberg.

Eddington, A.S. (1928): Gifford lecture, Referred by Schrödinger (1932).

Einstein, A. (1905, a): Über einen die Erzeugung und Verwandlung des Lichtes betreffenden heuristischen Gesichtspunkt, *Annalen der Physik* **17**, 132–148.

Einstein, A. (1905, b): Über die von der molekularkinematischen Theorie der Wärme geforderte Bewegung von ruhenden Flüssigkieiten suspendierten Teilchen, *Annalen der Physik* **17**, 549–560.

Einstein, A. (1905, c): Zur Electrodynamik bewegter Körper, *Annalen der Physik* **17**, 891–921.

Einstein, A. (1906): Zur Theorie der Lichterzeugung und Lichtabsorption, *Annalen der Physik* **20**, 199–206.

Einstein, A. (1924): Quantentheorie des einatomigen idealen Gases, *Sitzungsberichte der preussischen Akad, der Wissenschaften Physikalisch-Matematische Klasse*, 261–267.

Einstein, A. (1925): Quantentheorie des einatomigen idealen Gases. Zweite Abhandlung, *Sitzungsberichte der preussischen Akad, der Wissenschaften Physikalisch-Matematische Klasse* **20**, 3-14.

Einstein, A., Podolsky, B., Rosen, N. (1935): Can quantum-mechanical description of physical reality be considered complete? *Phys. Rev.* **47**, 777–780.

Faris, W.G. (ed.) (2006): *Diffusion, Quantum Theory, and Radically Elementary Mathematics*, Mathematical Notes 47, Princeton University Press.

Feller, W. (1936): Zur Theorie der stochastichen Prozesse (Existenz und Eindeutigkeit), *Math. Ann.* **13** (3), 113–160.

Feller, W. (1954): Diffusion processes in one dimension, *Trans. Amer. Math. Soc.* **77**, 1–31.

Feller, W. (1968): *An Introduction to Probability Theory and Its Applications, Vol.* 1, 3rd ed., Wiley, New York.

Feller, W. (1971): *An Introduction to Probability Theory and Its Applications, Vol.* 2, 2nd ed., Wiley, New York.

Fényes, I. (1952): Eine wahrscheinlichkeitstheoritische Begründung und Interpretation der Quantenmechanik, *ZS. für Physik* **132**, 81–106. Feynman, R.P. (1948): Space-time approach to non-relativistic quantum mechanics, *Reviews of Modern Phys.* **22**, 367–387.

Feynman, R.P. (1964): *The Character of Physical Law*, Modern Library.

Fine, A. (1986, 88): *The Shaky Game, Einstein Realism and the Quantum Theory*, The University of Chicago Press, Chicago, IL.

Föllmer, H. (1986): Time reversal on Wiener spaces, in: *Bibos-Symp. Stochastic Processes in Math. Phys.*, Lecture Notes in Math. **1158**, Springer, Heidelberg, pp. 119–129.

Föllmer, H. (1988): *Random fields and diffusion processes*, Ecole d'été de Saint Flour XV–XVII (1985–87), Lecture Notes in Math. **1362**, Springer, Heidelberg.

Friedman, A. (1964): *Partial differential equations of parabolic type*, Prentice-Hall, Inc., Englewood Cliffs.

Fujimagari, T., Motoo, M. (1969): On cascade processes, *Proc. Int. Conf. on functional analysis and related topics*, 383–391.

Fujimagari, T., Motoo, M. (1971): Cascade semigroups and their characterization, *Kodai Math. Sem. Reports* **23**, 402–472.

Fujita, H. (1966): On the blowing up of solutions of Cauchy problem for $u_t = \triangle u + u^{1+\alpha}$, *J. Fac. Sci. Univ. Tokyo* **13**, 109–124.

Fukushima, M. (1980): *Dirichlet Forms and Markov Processes*, Kodansha LTD, Tokyo, North-Holland Publ. Co., Amsterdam, Oxford, New York.

Girsanov, I.V. (1960): On transforming a certain class of stochastic processes by absolutely continuous substitution of measures, *Theor. Probab. Appl.* **5**, 285–301.

Görlitz, A., Vogels, J.M, Leanhardt, A.E., Raman, C., Gustavson, T.L, Abo-Shaeer, J.R., Chikkatur, A.P., Gupta, S., Inouye, S, Rosenband, T., Ketterle, W. (2001): Realization of Bose-Einstein condensates in lower dimensions, *Phys. Rev. Letters* **87**, 130402-1/130402-4.

Gudder, S.P. (1970): On hidden-variable theories, *J. Math. Phys.* **11**, 431–436.

Guth, A.H. (1981): Inflationary universe: A possible solution to the horizon and flatness problems, *Phys. Rev. D* **23**, 347–356.

Harris, T.E. (1963): *The theory of branching processes*, Springer, New York.

Hasegawa, Y., Rauch, H. (2006): Quantum contextual phenomena observed in single-neutron interferometer experiments, in: *Quantum theory: Reconsideration of foundations*, ed. G. Adenier, A.Y.Khrennikov and Theo. M. Nieuwenhuizen, AIP conference proceedings 810.

Heisenberg, W. (1925): Über quanten-theoretische Umdeutung kinetischer und mechanischer Beziehungen, *ZS. für Phys.* **33**, 879–893.

Heisenberg, W. (1927): Über den anschaulichen Inhalt der quantentheoretischen Kinematik und Mechanik, *ZS. für Phys.* **43**, 172–198.

Ichinose, T. (1994): Some results on the relativistic Hamiltonians: selfadjointness and imaginary-time path integral, *Proc. Int. Conf. Univ. Alabama.*

Ichinose, T., Tamura, H. (1984): Propagation of Dirac particle. A path integral approach, *J. Math. Phys.* **25**, 1810–1819.

Ikeda, N., Nagasawa, M, Watanabe, S. (1968/69): Branching Markov processes, I, II, III, *J. Mat. Kyoto Univ.* **8**, 233–278, 365–410, **9**, 95–160.

Ikeda, N., Watanabe, S. (1981/89): *Stochastic Differential Equations and Diffusion Processes*, Kodansha/North-Holland, Tokyo, Amsterdam.

Ionescu Tulcea, C. (1949): Mesures dans les éspaces produits, *Atti Acad. Naz. Lincei Rend.* **7**, 208–211.

Itô, K. (1942): Differential equations determining a Markov process, *Journ. Pan-Japan Math. Coll.* **1077** (Zenkoku Sizyo Sugaku Danwakai-si).

Itô, K. (1951): *On stochastic differential equations*, Memoirs of the AMS 4, American Math. Soc.

Itô, K. (1961): *Lectures on Stochastic Processes*, Tata Institute, Bombay.

Itô, K. (1975): Stochastic differentials, *Applied Mathematics and Optimization* **1**, 374–381.

Itô, K., McKean, H.P. (1965): *Diffusion Processes and Their Sample Paths*, Springer, New York.

Ito, S. (1957): Fundamental solutions of parabolic differential equations and boundary value problems, *Japanese J. Math.* **27**, 55–102.

Jacob, N. (1996): *Pseudo-Differential Operators and Markov Processes*. Akademie-Verlag, Berlin.

Jirina, M. (1964): Branching processes with measure-valued states, 3rd Prague Conference, 333–357.

Jammer, M. (1974): *The Philosophy of Quantum Mechanics. The Interpretations of Quantum Mechanics in Historical Perspective*, John Wiley and Sons, Inc., New York.

Jeulin, Th., Yor, M. (1988/89): Filtration des points browniens et équations differentielles stochastiques linaires, *Sém. de Probab.* **24**, 227–265. Lect. Notes in Math. 1426, Springer-Verlag, Heidelberg.

Jin, D.S. et al (1996): Collective excitation of a Bose-Einstein condensate in a dilute gas, *Phys. Rev. letters* **77**, 420–423.

Jönsson, C. (1961): Eectroneninterferenzen an mehreren künstlich hergestellten Feinspalten, *ZS für Physik* **161** (4), 454–474.

Jönsson, C. (1974): Electron Diffraction at Multiple Slits, *American Journal of Physics* **42** (1), 4–11.

Kac, M. (1949): On distributions of certain Wiener functionals, *Trans. Amer. Math. Soc.* **65**, 1–13.

Kac, M. (1951): On some connections between probability theory and differential and integral equations, *Proc. Second Berkeley Symp. on Math. Stat. Probab.* 189–215, Univ. Calif. Press.

Kac, M. (1959): Probability and related topics in physical sciences, New York.

Kennard, E.H. (1927): Zur Quantenmechanik einfacher Bewegungstypen, *ZS. für Phys.* **44**, 326–352.

Kim, Y-H., Yu, R., Kulik, S.P., Shih, Y., Scully, M.O. (2000): Delayed "choice" quantum eraser, *Phy. Rev. letters* **84**, 1–5.

Kochen, S., Specker, E.P. (1967): The problem of hidden variables in quantum mechanics, *J. Math. and Mech.* **1** (7), 59–87.

Kolmogoroff, A. (1931): Über die Analytischen Methoden in Wahrscheinlichkeitsrechnung, *Math. Ann.* **104**, 415–458.

Kolmogoroff, A. (1933): *Grundbegriffe der Wahrscheinlichkeitsrechnung*, Ergebn. d. Math. **2**, Heft 3, Springer-Verlag, Heidelberg.

Kolmogoroff, A. (1936): Zur Theorie der Markoffschen Ketten, *Math. Ann.* **112**, 155–160.

Kolmogoroff, A. (1937): Zur Umkehrbarkeit der statistischen Naturgesetze, *Math. Ann.* **113**, 766–772.

Kolmogoroff, A. Petrovsky, I., Piscounoff, N. (1937): Etude de l'équation de la diffusion avec croissance de la quantité de matière et son application à un problème biologique, *Bulletin de Université d'é'tat a Moscou* **1**, Fasc. 6, 1–25.

Kolmogoroff, A., Dmitriev, N.A. (1947): Branching stochastic processes, *Doklady* **56**, 5–8.

Kunita, H., Watanabe, T. (1966): On certain reversed processes and their applications to potential theory and boundary theory, *J. Math. Mech.* **15**, 398–434.

Kunita, H., Watanabe, S. (1967): On square-integrable martingales, *Nagoya J. Math.* **30**, 209–245.

Lamb, W.E. (1969): An operational interpretation of nonrelativistic quantum mechanics, *Physics Today* **22** (4), 23–28.

Lamperti, J. (1967): Continuous state branching processes, *Bull. Amer. Math. Soc.* **73** (3), 382–386.

Lévy, P. (1948/65): *Processes Stochastiques et Mouvement Brownien*, Gauthier-Villars, Paris.

Lévy, P. (1951): Wiener's random function and other Laplacian random functions, *Proc. 2nd Berkeley Symp. on Math. Statist. and Probab.*, Univ. California Press, pp. 171–186.

Lindenstrauss, J. (1965): A Remark on extreme doubly stochastic measure, *American Math. Monthly* **72**, 379–382.

Lorentz, H.A. (1915): *The theory of Electrons*, 2nd ed., Dover Publ. Inc., New York, 1952.

Maruyama, G. (1954): On the transition probability functions of the Markov processes, *Nat. Sci. Rep. Ochanomizu Univ.* **5**, 10–20.

McKean, H.P. (1960): The Bessel Motion and a singular integral equation, *Mem. Coll. Sci. Univ. Kyoto, Ser. A, Math.* **33**, 317–322.

McKean, H.P. (1966): A class of Markov processes associated with non-linear parabolic equations, *Proc. Natl. Acad. Sci.* **56**, 1907–1911.

McKean, H.P. (1967): Propagation of chaos for a class of nonlinear parabolic equations, *Lecture Series in Differential Equations*, Catholic Univ., pp. 41–57.

McKean, H.P. (1969): *Stochastic Integrals*, Academic Press, New York.

Merli, P.G., Missiroli, G.F., and G. Pozzi (1976): On the statistical aspect of electron interference phenomena, *American Journal of Physics* **14** (2), 178–194.

Mewes, M.O. et al (1996): Collective excitation of a Bose-Einstein condensation in a magnetic trap, *Phys. Rev. letters* **77**, 988–991.

Meyer, P.A. (1966): *Probability and Potential*, Blaisdell Publ. Co., Waltham, Mass.

Moore, W. (1989): *Schrödinger, life and thought*, Cambridge Univ. Press.

Moyal, J.E. (1962): The general theory of stochastic population processes, *Acta Math.* **108**, 1–31.

Murakami, H. (1957): On non-linear partial differential equations of parabolic types, I, II, III, *Proc. Japan Acad.* **33**, 530–535, 616–621, 622–627.

Nagasawa, M. (1959): Isomorphisms between commutative Banach algebras with an application to rings of analytic functions, *Kodai Math. Sem. Rep.* **11**, 182–188.

Nagasawa, M. (1961): The adjoint process of a diffusion process with reflecting barrier, Kodai Math. Sem. Rep. **13**, 235–248.

Nagasawa, M. (1964): Time reversal of Markov processes, Nagoya Math. Jour. **24**, 177–204.

Nagasawa, M. (1968): Construction of branching Markov processes with age and sign, Kodai Math. Sem. Rep. **20**, 469–508.

Nagasawa, M. (1969): Markov processes with creation and annihilation, *Z. Wahrscheinl. Verw. Geb.* **14**, 49–60.

Nagasawa, M. (1972): Branching property of Markov processes, *Sém. de Probabilités Strasbourg* **VI**, Springer-Verlag, Heidelberg, pp. 177–197.

Nagasawa, M. (1973): Multiplicative excessive measures of branching processes, *Proc. Japan Acad.* **49**, 497–499.

Nagasawa, M. (1974, a): Multiplicative excessive measures and duality between equation of Boltzmann and of branching processes, *Sem. de prob.* **IV**, Springer-Verlag, Heidelberg, pp. 471–485.

Nagasawa, M. (1974, b): *Duality and time reversal of Markov processes*, Lecture Notes at Tokyo Institute of Technology.

Nagasawa, M. (1974, c): *The last occurrence times and time reversal of Markov processes*, Lecture Notes at Tokyo Institute of Technology.

Nagasawa, M. (1975): A probabilistic approach to non-linear Dirichlet problem, *Sém. de Probab.* **X**, Springer-Verlag, Heidelberg, pp. 184–193.

Nagasawa, M. (1975): Note on pasting of two Markov processes, *Sém. de Probab.* **X**, Springer-Verlag, Heidelberg.

Nagasawa, M. (1977): Basic Models of Branching Processes, *Bull. Int. Stat. Inst.* **XLVII** (2), 423–445. Proceedings of the 41st Session of the International Statistical Institute New Delhi.

Nagasawa, M. (1979): Segregation model and its application, Preprint of Sem. für Angewandte Math. Univ. of Zürich.

Nagasawa, M. (1979): Applications of diffusion models to problems in biology, Preprint of Sem. für Angewandte Math. Univ. of Zürich.

Nagasawa, M. (1980): Segregation of a population in an environment, *J. Math. Biol.* **9**, 213–235.

Nagasawa, M. (1981): An application of segregation model for septation of Escherichia Coli, *J. Theoret. Biol.* **90**, 445–455.

Nagasawa, M. (1982): Applications of diffusion models to problems in biology, *Univ. Bielfeld, Schwerpunkt Mth.* **38**, 338–347.

Nagasawa, M. (1985): *Macroscopic, intermediate, microscopic and mesons*, Lect. Notes Phys. 262, Springer, pp. 427–437.

Nagasawa, M. (1987): The propagation of chaos for diffusion processes with bad drift coefficients, in: *Conference on Stoch. differential systems, Eisenach, GDR*, 1986, Lect. Notes Control Inf. Sci. 96, pp. 77–81.

Nagasawa, M. (1988): A statistical model of segregation of a population, in: *Stochastic modelling in Biology*, ed. P. Tautu, World Scientific, Singapore.
Nagasawa, M. (1989, a): Transformations of diffusion and Schrödinger processes, *Probab. Th. Rel. Fields* **82**, 109–136.

Nagasawa, M. (1989, b): Stochastic variational principle of Schrödinger processes, in: *Seminar on stochastic processes*, ed. Cinlar, Chung, Getoor, Birkhäuser, Basel.

Nagasawa, M. (1990): Can the Schrödinger equation be a Boltzmann equation? Evanston 1989, in: *Diffusion processes and Related problems in Analysis*, ed. M. Pinsky, Birkhäuser, Basel, Boston, Berlin.

Nagasawa, M. (1991): The equivalence of diffusion and Schödinger equations: A solution to Schrödinger's conjecture, in: *Proceedings of Locarno Conference* 1991, ed. S. Albeverio, World Scientific, Singapore.

Nagasawa, M. (1992): A principle of superposition and interference of diffusion processes, in: *Third European Symposium on Analysis and Probability*, 1992 *at the Institute Henri Poincar*, Sém. de Probabilités **XXVII**, Lecture Notes in Math. 1557, Springer-Verlag, Heidelberg, 1993, pp. 1–14.

Nagasawa, M. (1992): Branching property of Markov processes, *Séminaire de probabilités (Strasbourg)* **6**, 177–197.

Nagasawa, M. (1993): *Schrödinger Equations and Diffusion Theory*, Monographs in Mathematics 86, Birkhäuser Verlag, Basel.

Nagasawa, M. (1995): *The locality of hidden-variable theories in quantum physics*, Lecture Notes at the University of Zürich.

Nagasawa, M. (1996): Quantum theory, theory of Brownian motions, and relativity theory, *Chaos, Solitons and Fractals* **7**, 631–643.

Nagasawa, M. (1997, a): Time reversal of Markov processes and relativistic quantum theory, *Chaos, Solitons and Fractals* **8**, 1711–1772. Erratum to the paper, *Chaos, Solitons and Fractals* **11** (2000), 2579.

Nagasawa, M. (1997): On the locality of hidden variable theories in quantum physics, *Chaos, Solitons and Fractals* **8**, 1773–1792.

Nagasawa, M. (2000): *Stochastic Processes in Quantum Physics*, Monographs in Mathematics 94, Birkhäuser Verlag, Basel, Boston, Berlin.

Nagasawa, M. (2002): On quantum particles, *Chaos, Solitons and Fractals* **13**, 1393–1405.

Nagasawa, M. (2002): A note on a remark by Landau regarding a charged particles in a magnetic field, *Chaos, Solitons and Fractals* **14**, 1065–1070.

Nagasawa, M. (2007): Dynamic theory of stochastic movement of systems, in: *Stochastic Economic Dynamics*, ed. by B.J. Jensen and T. Palokangas, Copenhagen Business School Press, pp. 133-164.

Nagasawa, M. (2009): A note on the expectation and deviation of physical quantities, *Chaos, Solitons and Fractals* **39**, 2311–2315. doi:10.1016/j.chaos. 207.06. 129

Nagasawa, M. (2012): On Heisenberg's inequality and Bell's inequality, *Kodai Math. J.* **35**, 33–51.

Nagasawa, M. (2012*): A mathematical theory for double-slit experiments of Walborn et al., *Kodai Math. J.* **35**, 589–612.

Nagasawa, M. (2014): A simple mathematical model for high temperature superconductivity, *Kodai Math. J.* **37** (2), 247–259.

Nagasawa, M., Aebi, R. (1992): Large deviations and the propagation of chaos for Schrödinger processes, *Probab. Th. Rel. Fields* **94**, 53–68.

Nagasawa, M., Barth, Th., Wakolbinger, A. (1981): Mathematisches Modellieren: Erregungszustände von Affen am Futterplatz order von Molekülen in einer Zelle, University of Zürich.

Nagasawa, M., Domenig, T. (1996): Diffusion processes on an open time interval and their time reversal, in: *Itô's Stochastic Calculus and Probability Theory*, ed. Ikeda, Watanabe, Fukushima and Kunita, Springer-Verlag, Heidelberg, pp. 261-280.

Nagasawa, M, Ikeda, N., Watanabe, S. (1968/69): Branching Markov processes, I, II, III, *J. Mat. Kyoto Univ.* **8**, 233–278, 365–410, **9**, 95–160.

Nagasawa, M., Maruyama, T. (1979): An application of time reversal of Markov processes to a problem of population genetics, *Advances in Appl. Probab.* **11**, 457–478.

Nagasawa, M., Sato, K. (1963): Some theorems on time change and killing of Markov processes, *Kodai Math. Sem. Rep.* **15**, 195–219.

Nagasawa, M., Schröder, K. (1997): A note on the locality of Gudder's hidden-variable theory, *Chaos, Solitons and Fractals* **8**, 1793–1805.

Nagasawa, M., Sirao, T. (1969): Probabilistic treatment of the blowing up of solutions for a non-linear integral equation, *Trans. American Math. Soc.* **139**, 301–310.

Nagasawa, M., Tanaka, H. (1985): A diffusion process in a singular meandrift-field, *Z. Wahrsch. verw. Gebiete* **68**, 247–269.

Nagasawa, M., Tanaka, H. (1986): Propagation of chaos for diffusing particles of two types with singular mean field interaction, *Probab. Th. Rel. Fields* **71**, 69–83.

Nagasawa, M., Tanaka, H. (1987, a): Diffusion with interactions and collisions between coloured particles and the propagation of chaos, *Probab. Th. Rel. Fields* **74**, 161–198

Nagasawa, M., Tanaka, H. (1987, b): A proof of the propagation of chaos for diffusion processes with drift coefficients not of average form, *Tokyo J. Math.* **10**, 403–418.

Nagasawa, M., Tanaka, H. (1999, a): Stochastic differential equations of pure-jumps in relativistic quantum theory, *Chaos, Solitons and Fractals* **10** (8), 1265–1280.

Nagasawa, M., Tanaka, H. (1999, b): The principle of variation for relativistic quantum particles, in: *Sém. de probab.* **XXXV**, pp. 1–27, Lect. Notes in Math. 1775, Springer-Verlag, Heidelberg.

Nagasawa, M., Tanaka, H. (1999, c): Time dependent subordination and Markov processes with jumps, in: *Sém. de probab.* **XXXIV**, pp. 257-288, Lect. Notes in Math. 1729, Springer-Verlag, Heidelberg.

Nagasawa, M., Tanaka, H. (1999, d): The principle of variation for relativistic quantum particles, in: *Sém. de probab.* **XXXV**, pp. 1-27, Lect. Notes in Math. 1775, Springer-Verlag, Heidelberg,

Nagasawa, M., Tanaka, H. (1999*): Concave majorants of Lévy processes, (unpublished), cf. Nagasawa (2000).

Nagasawa, M., Uchiyama, K. (1976): A remark on the non-linear Dirichlet problem of Branching Markov processes, in: *Proceedings of the third Japan-USSR symposium on probability theory*, Lect. Notes in Math. 550, Springer-Verlag, Heidelberg.

Nagasawa, M., Yasue, K. (1982): A statistical model of mesons, *Publ. de l'Inst. Rech. Math. Avan. (CNRS)* **33**, 1–48.

Nagayama, K. (1999): Complex observation in electron microscopy. I. Basic scheme to surpass the Scherzer limit, *Journ. Phys. Soc. Japan* **68**, 811–822.

Nelson, E. (1966): Derivation of Schrödinger equation from Newtonian mechanics, *Phys. Rev.* **150**, 1076–1085.

Neveu, J. (1965): *Mathematical Foundations of the Calculus of Probability*, Holden-Day, Inc., San Fransisco, London, Amsterdam.

Paley, R.E.A.C., Wiener, N. (1934): Fourier transformations in complex domain, *Amer. Math. Soc. Coll. Publ.*

Luis de la Pena, Ana Maria Cetto, Andre Valdes Hernandez (2015): *The emerging Quantum*, Springer.

Parthasarathy, K.R. (1967): *Probability measures on Metric Spaces*, Acad. Press, New York, London.

Pauling, L., Wilson, E.B. (1935): *Introduction to Quantum Mechanics With Applications to Chemistry*, McGraw-Hill Book Co. Inc., New York.

Pethick, C.J. Smith, H. (2002): *Bose-Einstein condensation in dilute gases*, Cambridge University Press.

Phillips, R.S. (1952): On the generation of semigroups of linear operators, *Pacific J. Math.* **2**, 343–369.

Planck, M. (1900): Zur Theorie des Gesetzes der Energieverteilung im Normalspectrum, *Verh. der D. Phys. Ges.* **2**, 237–245.

Planck, M. (1900): Über eine verbesserung der Wien'shen Spectralgleichung, *Verh. der D. Phys. Ges.* **2**, 202–204.

Planck, M. (1901): Über das Gesetz der Energieverteilung im Normalspectrum, *Annalen der Phys.* **4**, 553–563.

Revuz, D., Yor, M. (1991, 1999): *Continuous Martingales and Brownian Motions*, Springer-Verlag, Berlin, New York, London, Paris, Tokyo.

Robertson, H.R. (1929): The uncertainty principle, *Phys. Rev.* **34**, Letters to the editor, 163–164.

Rosa, R. (2012): The Merli-Missiroli-Pozzi Two-Slit Electron-interference Experiment, *Physics in Perspective* **14** (2), 178–194.

Sato, K. (1981): Cosmological baryon-number domain structure and the first order phase transition of a vacuum, *Physics Letters* **99** B, 66–70.

Sato, K-i. (1999): *Lévy Processes and Infinitely Divisible Distributions*, Cambridge University Press.

Scully, M.O., Drühl, K. (1982): Quantum eraser: A proposed photon correlation experiment concerning observation and "delayed choice" in quantum mechanics, *Phy. Rev. A* **24**, 2208–2213.

Schrödinger, E. (1926, I): Quantisierung als Eigenwertproblem (1. Mitteilung), *Ann. der Physik* **79**, 336–376.

Schrödinger, E. (1926): Über das Verhältnis der Heisenberg-Born-Jordanschen Quantenmechanik zu der meinen, *Ann. der Physik* **79**, 734–756.

Schrödinger, E. (1926, IV): Quantisierung als Eigenwertproblem (4. Mitteilung), *Ann. der Physik* **81**, 109–139.

Schrödinger, E. (1931): Über die Umkehrung der Naturgesetze, *Sitzungsberichte der preussischen Akad. der Wissenschaften Physikalisch-Mathematische Klasse*, 144–153.

Schrödinger, E. (1932): Sur la théorie relativiste de l'électron et l'interpretation de la mécanique quantique, *Ann. Inst. H. Poincaré* **2**, 269–310.

Selleri, F. (1983): Die Debatte um Quantentheorie, Frieder. Vieweg und Sohn Verlagsgesellschaft mbH, Braunschweig, Germany.

Sevastyanov, B.A. (1958): Branching stochastic processes for particles diffusing in a restricted domain with absorbing boundaries, *Theory Prob. Appl.* **3**, 121–136.

Sevastyanov, B.A. (1958): *Branching processes* (Russian), Nauka, Moscow.

Shandarin, S.F., Zeldovich, Ya.B. (1989): The large scale structure of the universe: Turbulence, intermittency, structures in self-gravitating medium, *Rev. of Modern Physics* **61**, 185–220.

Shiryayev, A.N. (1984): *Probability*, Springer-Verlag, New York, Berlin, Heidelberg, Tokyo.

Sirao, T. (1968): On signed branching Markov processes with age, *Nagoya Mat. J.* **32**, 155–225.

Skorokhod, A.V. (1964): Branching diffusion processes, *Theory Prob. Appl.* **9**, 492–497.

Stroock, D.W., Varadhan, S.R.S. (1969): *Multidimensional Diffusion Processes*, Springer-Verlag, Berlin, Heidelberg.

Tanaka, H. (1979): Stochastic differential equations with reflecting boundary condition in convex regions, *Hiroshima Math. J.* **9**, 163–177.

Tanaka, H. (1997): Limit theorems for Brownian motion with drift in a white nois environment, *Chaos, Solitons and Fractals* **8**, 1807–1816.

Thirring, W.E. (1987): Consequences of the Schrödinger equation for atomic and molecular physics, in: *Schrödinger, centenary celebration of polymath*, ed. Klimister, C.W., Cambridge University Press.

Tomonaga, S. (1946): On a relativistically invariant formulation of the quantum theory of wave fields, *Progress of Theoretical Physics* **1**, 27–39.

Tonomura, A. (1990): Electron Holography: A new view of the microscopic, *Physics Today* **43** (4), 22–29.

Tonomura, A. (1994): *Electron Holography*, Springer, Heidelberg.

Ueda, M., Saito, H. (2002): Mean-field analysis of collapsing and exploding Bose-Einstein condensates, *Physical Review A* **65**, 033624-1–033624-6.

Varadhan, S.R.S. (1984): *Large deviations and applications*, Soc. Ind. Appl. Math., Philadelphia.

Vilenkin, A. (1982): Creation of universes from nothing, *Physics Lett.* **117** B, 25–28.

Vilenkin, A. (1983): Birth of inflationary universes, *Phys. Rev. D* **27**, 2848–2858.

Volkonsky, V.A. (1960): Additive functionals of Markov processes, *Trudy Moscow Mat. Obsc.* **9**, 153–189.

Von Neumann, J. (1932): *Mathematische Grundlagen der Quantenmechanik*, Springer-Verlag, Heidelberg.

Walborn, S.P., Terra Cunha, M.O., Padua, M.O., Monken, C.H. (2002): Double-slit quantum eraser, *Physical Review A* **65**, 033818-1–033818-6.

Watanabe, S. (1968): A limit theorem of branching processes and continuous state branching processes, *J. Math. Kyoto Univ.* **8** (1), 141–167.

Wiener, N. (1923): Differential space, *J. Math. Phys.* **2**, 131–174.

Wiener, N. (1924): Un problème de probablités dénombrables, *Bull. Soc. Math.* **11**,

Yang, C.N. (1987): Square root of minus one, complex phases and Erwin Schrödinger, in: *Schrödinger, centenary celebration of polymath*, ed. Klimister, C.W., Cambridge University Press.

Yor, M. (1992): *Some aspects of Brownian motion, Part 1, Some special functionals*, Birkhäuser Verlag, Basel, Boston, Berlin.

Yosida, K. (1948): On the differentiability and representation of one-parameter semi-group of linear operators, *J. Math. Soc. Japan* **1**, 15–21.

Yosida, K. (1965): *Functional Analysis*, Springer-Verlag, Berlin, Heidelberg.

Index

Printed in the United States
by Baker & Taylor Publisher Services